T0262846

Multimedia Communications and Networking

Multimedia Communications and Networking

Mário Marques da Silva

CRC Press
Taylor & Francis Group
Boca Raton   London   New York

CRC Press is an imprint of the
Taylor & Francis Group, an **informa** business

CRC Press
Taylor & Francis Group
6000 Broken Sound Parkway NW, Suite 300
Boca Raton, FL 33487-2742

© 2012 by Taylor & Francis Group, LLC
CRC Press is an imprint of Taylor & Francis Group, an Informa business

No claim to original U.S. Government works

Printed in the United States of America on acid-free paper
Version Date: 2012907

International Standard Book Number: 978-1-4398-7484-4 (Hardback)

Library of Congress Cataloging-in-Publication Data

Silva, Mario Marques da.
    Multimedia communications and networking / Mario Marques da Silva.
    pages cm
    Includes bibliographical references and index.
    ISBN 978-1-4398-7484-4 (hardcover : alk. paper)
    1. Multimedia communications. I. Title.

TK5105.15.S58 2012
004.6--dc23   2012027083

Visit the Taylor & Francis Web site at
http://www.taylorandfrancis.com

and the CRC Press Web site at
http://www.crcpress.com

To my family and friends

Contents

Preface

This textbook aims to present a detailed and comprehensive description of the most important state-of-the-art fundamentals and system details in the fields of multimedia, networking, digital communications, and network security. It describes several subjects that are key aspects in the development and understanding of current and emergent services.

The objective of this textbook is to concentrate in a single book the author's view about current and emergent multimedia, digital communications, and networking services and technologies. Different bibliographic sources normally cover each one of these topics independently, without the required relationship. On the one hand, the proposed book allows the reader to reduce the time and cost required to learn and improve skills and knowledge in the covered fields, with a relationship between the covered topics. Moreover, this book presents a compilation of the latest developments in the area, which is the outcome of several years of research and participation in many international projects.

This textbook covers most of the relevant subjects with examples and end of chapter questions, properly designed to be used for BSc or MSc courses in computer science or electrical engineering. The approach used in this textbook aims to facilitate the learning of the covered subjects by students of several disciplines such as multimedia, networking, telecommunications, and network security. Moreover, this book may also be used by academic, institutional, or industrial professionals to support the planning, design, and development of multimedia, telecommunications, or networking systems.

About the Author

Mario Marques da Silva is a professor at Universidade Autónoma de Lisboa (CESITI) and at the Portuguese Naval Academy (CINAV). He is a researcher at the Portuguese Instituto de Telecomunicações. He received his BSc in electrical engineering in 1992 and MSc and PhD degrees in telecommunications/electrical engineering, respectively, in 1999 and 2005, both from the Universidade Técnica de Lisboa. Between 2005 and 2008, he was with NATO Air Command Control & Management Agency (NACMA) in Brussels (Belgium), where he managed the deployable communications of the new Air Command and Control System Program. He has been involved in several telecommunications projects, either as a researcher or as project leader, including the involvement in activities such as research, architecture, development, analysis, simulation and testing of networking, HF, V/UHF, satellite and cellular/UMTS communications systems. His research interests include networking (e.g., TCP/IP, network security, mobile ad-hoc networking) and mobile communications, including block transmission techniques (OFDM, SC-FDE), WCDMA, multiuser detection, interference cancellation, space–time coding, MIMO systems, smart and adaptive antennas, channel estimation, and software-defined radio. He is the author of the book *Transmission Techniques for Emergent Multicast and Broadcast Systems* (CRC Press, 2010) and of several dozen journal and conference papers. Mario Marques da Silva is a member of the Institute of Electrical and Electronics Engineers (IEEE) and a member of Armed Forces Communications and Electronics Association (AFCEA). He is also a reviewer of many international scientific journals and conferences.

AN INTRODUCTION TO MULTIMEDIA COMMUNICATIONS AND NETWORKING

1.1 Fundamentals of Communications

Communication systems are used to enable the exchange of data between two or more entities (humans or machines). As can be seen from Figure 1.1, data consists of a representation of an information source, whose transformation is performed by a source encoder. An example of a source encoder is a thermometer, which converts temperatures (information source) into voltages (data). A telephone can also be viewed as a source encoder, which converts the analog voice (information source) into a voltage (data) before being transmitted along the telephone network (transmission medium). In case the information source is analog and the transmission medium is digital, a CODEC (coder and decoder) is employed to perform the digitization. A VOCODER (voice coder) is a codec specific for voice, whose functionality consists of converting analog voice into digital at the transmitter side and the reciprocal at the receiver side.

The emitter of data consists of an entity responsible for the insertion of data into the communication system and for the conversion of data into signals. Note that signals are transmitted, not data. Signals consist of an adaptation* of data, such that their transmission is facilitated in accordance with the used transmission medium. Similarly, the receiver is responsible for converting the received signals into data.

The received signals correspond to the transmitted signals subject to attenuation and distortion and added with noise and interferences. These channel impairments originate that the received signal differs from that transmitted. In the case of analog signals, the resulting signal levels do not exactly translate the original information source. In the case of digital signals, the channel impairments originate corrupted bits. In both cases, the referred channel impairments originate a degradation of the signal-to-noise plus interference ratio (SNIR).† A common performance indicator of

---

* Signals can be, for example, a set of predefined voltages, which represent bits used in transmission.

† In linear units, the SNIR is mathematically given by $SNIR = \dfrac{S}{N+I}$, where S stands for the power of signal, N expresses the power of noise, and I the power of interferences. For the sake of simplicity, the SNIR is normally referred to as only SNR (signal-to-noise ratio), but where the interference is also taken into account (in this case N stands for the power of noise and interferences). Furthermore, both SNIR (or SNR) are normally expressed in logarithmic units as $SNIR_{dB} = 10\log_{10}\left(\dfrac{S}{N+I}\right)$.

Figure 1.1   A generic block diagram of a communication system.

digital communication systems is the bit error rate (BER). This corresponds to the number of corrupted bits divided by the total number of transmitted bits over a certain time period.

A common definition associated with information is knowledge. It consists of a person's ability to have access to the right information at the right time. The conversion between information and knowledge can be automatically performed using information systems, whereas the information can be captured by sensors and distributed using communication systems.

1.1.1 Analog and Digital Signals

Analog signals present a continuous amplitude variation over time. An example of an analog signal is the voice. Contrarily, digital signals present time discontinuity (e.g., voltages or light pulses). The bits* generated in a computer are examples of digital data. The text is another example of digital data. Examples of analog and digital signals are depicted in Figure 1.2.

Digital signals present several advantages (relating to analog). The following advantages can be listed:

- Error control is possible in digital signals: corrupted bits can be detected and/or corrected.
- Since they present only two discrete values, the consequences of channel impairments can be more easily detected and avoided (as compared to analog signals).
- Digital signals can be regenerated, almost eliminating the effects of channel impairments. Contrarily, the amplification process of analog signals results in the amplification of signals, noise, and interferences, keeping the SNR relationship unchanged.†
- The digital components are normally less expensive than the analog components.
- Digital signals facilitate cryptography and multiplexing.
- Digital signals can be used to transport different sources of information (voice, data, multimedia, etc.) in a transparent manner.

---

* With logic states 0 or 1.

† In fact, the amplification process results even in a degradation of the SNR, as it adds the amplifier's internal noise to the signal at its input. This subject is detailed in Chapter 3.

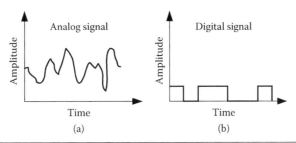

Figure 1.2 Example of (a) analog and (b) digital signals.

However, digital signals present an important disadvantage:

- For the same information source, the bandwidth required to accommodate a digital signal is typically higher than the analog counterpart.* This results in a higher level of attenuation and distortion.

1.1.2 Modulator and Demodulator

As can be seen from Figure 1.3, when the source (e.g., a computer) generates a digital stream of data and the transmission medium is analog, a modem (modulator and demodulator) is used to perform the required conversion. The modulator converts digital data into analog signals, whereas the demodulator (at the receiver) converts analog signals into digital data. An example of an analog transmission medium is radio transmission, whose signals consist of electromagnetic waves (presenting a continuous variation in time).

A modem (e.g., asynchronous digital subscriber line [ADSL] or cable modem) is responsible for modulating a carrier wave with bits, using a certain modulation scheme.† The reverse of this operation is performed at the receiver side. Moreover, a modem allows sending a signal modulated around a certain carrier frequency, which can be another reason for using such device.

In case the data is digital and the transmission medium is also digital, a modem is normally not employed, as the conversion between digital and analog does not need to be performed. In this case, a line encoder/decoder (sometimes also referred to as digital modem, nevertheless not accurately) is used. This device adapts the original digital data to the digital transmission medium‡ and adapts parameters such as levels, pulse duration, and so on. Note that in using such digital encoders, the signals are transmitted in the baseband.§

---

* As an example, analog voice is transmitted in a 3.4 kHz bandwidth, whereas the digital pulse code modulation requires a bandwidth of 32 kHz (64 kbps).

† Using amplitude, frequency, or phase shift keying. Advanced modems make use of a combination of these elementary modulation schemes.

‡ Using line codes such as return to zero, nonreturn to zero, Manchester, and so on (as detailed in Chapter 6).

§ Instead of carrier-modulated (bandpass), as performed by a modem.

Figure 1.3   A generic communication system incorporating a modem.

The output of a line encoder consists of a digital signal, as it comprises discrete voltages that encode the source logic states. Consequently, it can be stated that the line encoder is used when the transmission medium is digital. On the other hand, the output of a modulator consists of an analog signal, as it modulates a carrier, which is an analog signal.

In case of high data rate, the required bandwidth necessary to accommodate such signal is also high,* and the medium may originate high level of attenuation or distortion at limited frequency components of the signal. In such case, the use of a modem can be a good choice, which allows the modulation of the signal around a certain carrier frequency. The carrier frequency can be carefully selected such that the channel impairments in the frequencies around it (corresponding to the signal bandwidth) do not seriously degrade the SNR. The reader should refer to Chapter 6 for a detailed description of the modulation schemes used in modems and a description of digital encoding techniques.

1.1.3 Transmission Mediums

Transmission mediums can be classified as cable or wireless. Examples of cable transmission mediums include twisted pair cables, coaxial cables, multimode or single-mode optical fiber cables, and so on. In the past, local area networks (LAN) were made of coaxial cables. These cables were also used as a transmission medium for medium- and long-range analog communications. Although coaxial cables were replaced by twisted pair cables in LAN, the massification of cable television enabled their reuse.

As a result of the telephone cablings, twisted pairs are still the dominant transmission medium in houses and offices. These cablings are often reused for data. With the improvement of isolators and copper quality, as well as the development of shielding, the twisted pair became widely used for providing high-speed data communications, in addition to its initial use for analog telephony.

Currently, multimode optical fibers have been installed more and more in homes, which allows reaching throughputs of the order or several gigabits per second (Gbps). Moreover, single-mode optical fibers are the most used transmission medium in transport networks. A transport network consists of the backbone (core) network, used for transferring high amounts of data among different main nodes. These main nodes are then connected to secondary nodes, and then finally connected to customer nodes.

---

* According to the Nyquist theorem, as detailed in Chapter 3.

A radio or wireless communication system is composed of a transmitter and a receiver, using antennas to convert electrical signals into electromagnetic waves and vice versa. These electromagnetic waves are propagated over the air. Note that wireless transmission mediums can be either guided or unguided. In the former case, directional antennas are used at both the transmitter and the receiver sides, such that electromagnetic waves propagate directly from the transmitting antenna into the receiving antenna. The reader should refer to Chapter 4 for a detailed description of cable transmission mediums. Chapter 5 introduces wireless transmission mediums.

1.1.4 Synchronous and Asynchronous Communication Systems

Synchronous and asynchronous communications refer to the ability or inability to have information about the start and end of bit instants.* Using asynchronous communications, the receiver does not achieve perfect time synchronization with the transmitter, and the communication accepts some level of fluctuation. Consequently, start and stop bits are normally included in a frame† to periodically achieve bit synchronization of the receiver with the transmitter. Note that between the start and the stop bit, the receiver of an asynchronous communication suffers from a certain amount of time shift. The referred periodic synchronization using start and stop bits is normally included as part of the functionalities implemented by a modem when establishing a communication in asynchronous mode of operation. Normally, asynchronous communications do not accommodate high-speed data rates. They are normally used for random (not continuous) exchanges of data (at low rate).

On the other hand, synchronous communications consider a receiver that is bit-synchronized with the transmitter. This bit synchronization can be achieved using one of the following methods:

- By sending a clock signal multiplexed with the data or using a parallel dedicated circuit
- When the transmitted signal presents a high zero crossing rate, such that the receiver can extract the start and end of bit instants from the received signal

Synchronous communications are normally used in high-speed lines and for the transmission of large blocks of data. Examples of synchronous communication system are the synchronous digital hierarchy (SDH) networks, used for the transport of large amounts of data in a backbone.

1.1.5 Simplex and Duplex Communications

A simplex communication consists of a communication between two or more entities, where the signals only flow in a single direction. In this case, only one entity acts as a

---

* Nevertheless, frame synchronization is required in either case.
† A group of exchanged bits.

transmitter and the other(s) as a receiver. This can be seen in Figure 1.4. Note that the transmitter may be transmitting signals to more than one receiver.

When the signals flow in a single direction, but with alternation in time, the communication is half duplex. Although both entities act simultaneously as transmitter and as receiver (at different time instants), instantaneously, each host acts as either a transmitter or as a receiver. Half-duplex communication is depicted in Figure 1.5.

Finally, when the communication is simultaneous in both directions, it is in full-duplex mode. In this case, two or more entities act simultaneously as both transmitter and receiver. Full-duplex communication is depicted in Figure 1.6. Full-duplex communications normally require two parallel transmission mediums (e.g., two pairs of wires): one for transmission and another for reception.

1.1.6 Communications and Networks

Point-to-point communication establishes a direct connection (link) between two adjacent end stations, two adjacent network nodes (e.g., routers), or an end station and an adjacent node.

A network can be viewed as a cloud composed of several nodes and end stations, where each node is responsible for switching the data, such that an end-to-end connection is

Figure 1.4   Simplex communication.

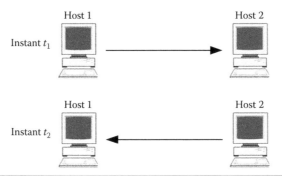

Figure 1.5   Half-duplex communication.

Figure 1.6   Full-duplex communication.

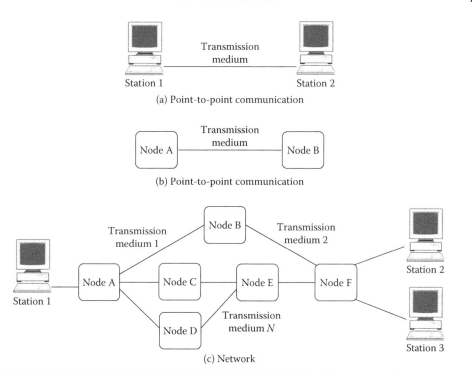

Figure 1.7 Examples of (a) and (b) point-to-point communication and a (c) network.

established between two end stations. Examples of point-to-point communications and of a network are depicted in Figure 1.7. An end-to-end network connection consists of a concatenation of several point-to-point links, where each of these links can be implemented using a different transmission medium (e.g., satellite, optical fiber, etc.).

A node of a network can be a router or a private automatic branch exchange (PABX). The former device switches packets (packet switching), while the latter is responsible for physically establishing permanent connections, such that a phone call between two end entities is possible (circuit switching). This subject is detailed in Section 2.

Depending on the amount of destination stations of data involved, a communication can be classified as unicast, multicast, or broadcast. Unicast stands for a communication whose destination is a single station. In case the destination of data is all the network stations, the communication is referred to as broadcast. Very often broadcast communications are established in a single direction (i.e., there is no feedback from the receiver into the transmitter). Finally, when the destination is more than a single station, but less then all network stations, the communication is referred to as multicast.

1.1.7 Switching Modes

1.1.7.1 Circuit Switching Circuit switching establishes a permanent physical path between the origin and the destination. This is the switching mode used in classic telephone networks. Only after startup is a synchronous exchange of data allowed. This end-to-end path (circuit) is permanently dedicated until the connection ends.

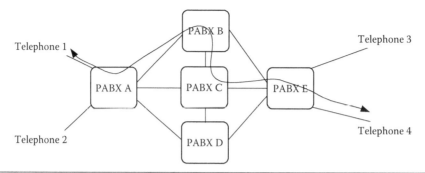

Figure 1.8   Example of a circuit-switching (telephone) network.

The time to establish the connection is high, but a delay is assured only due to the propagation speed of signals. This kind of switching is ideal for delay-sensitive communications, such as voice. If the connection cannot be established due to a lack of resources, it is said that the call was blocked, but once established, congestion does not occur. All the bandwidth available is assigned to a certain connection which, for long time periods, may not be used and, in other periods, may not be enough (e.g., if that connection sends variable data rates). For this reason, it is of high cost. In telephone networks, the switching is physically performed by operators using PABX. This consists of a switch whose functionality is typically achieved using space and/or time switching. The space switching consists of establishing a physical shunt between one input and one output. Since digital networks normally incorporate multiplexed data into different time slots* (each telephone connection is transported in a different time slot), there is the need to switch a certain time slot from one physical input into another time slot of another physical output. This is performed by the time- and space-switching functionality of a digital PABX. An example of a circuit-switching (telephony) network is depicted in Figure 1.8.

1.1.7.2 Packet Switching   With the introduction of data services, the notion of packet switching has arrived. Packet switching considers the segmentation of a message into parts, where each part is referred to as a packet (with fixed[†] or variable[‡] length). As can be seen from Figure 1.9, a digital message is composed of many bits, while a packet consists of a smaller amount of these bits.

Packets are forwarded and switched independently through the nodes of a network between the origin and the destination. Each packet transports enough information to allow its routing (end destination address included in a header).

While the nodes of a circuit-switching network establish a permanent shunt between one input and one output, and since packet switching considers a number of

---

* This is normally referred to as time division multiplexing.
[†] For example, asynchronous transfer mode (ATM).
[‡] For example, multiprotocol label switching or Internet protocol (IP).

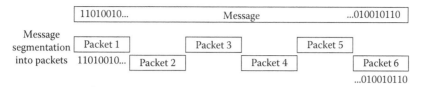

Figure 1.9 The segmentation of a message into packets.

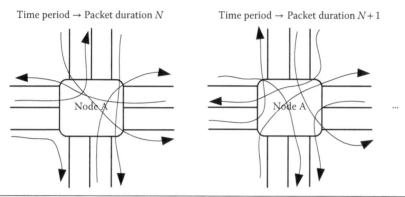

Figure 1.10 Switching of packets in different instants.

bits grouped into a packet, the nodes of a packet-switching network switch data only for the duration of a packet transmission. The following packet that uses the same input or output of a node may belong to a different end-to-end connection. This is depicted in Figure 1.10. Consequently, packet-switching networks make much better usage of the network resources (nodes) than circuit-switching networks. Note that a node of a packet-switching network is typically a router.

Each node of the network is able to store packets, in case it is not possible to send it due to temporary congestion. In this case, the time for message transmission is not guaranteed, but this value is kept within reasonable limits, especially if quality of service (QoS) is offered. Packet switching is of lower cost than circuit switching and is ideal for data transmission, since it allows a better management of the resources available (a statistical multiplexing is performed). Moreover, with packet switching, it is not needed to assign all the resources (i.e., bandwidth) available to a certain user who, for long periods, does not make use of it, being the network resources shared among several users, as a function of the resources available and of the users' need. The network resources are made available as a function of each user need and as a function of the instantaneous network traffic.

There are different packet-switching protocols, such as ATM, IP, Frame Relay, X.25, and so on. The IP version 4 (IPv4) does not introduce the concept of QoS, since it does not include priority rules to avoid delays or jitter (e.g., for voice). Moreover, it does not avoid loss of data for certain types of services (e.g., for pure data communication) and it does not allow the assignment of higher bandwidth to certain services, relating to other (e.g., multimedia vs. voice). On the other hand, ATM and IP version 6 have mechanisms to improve the QoS.

1.1.8 Connection Modes

Depending on the end-to-end service provided, the connection modes through networks can be of two types: connectionless and connection oriented. These modes can be used in any of the layers of a network architecture (e.g., the open system interconnection reference model [OSI-RM] or the transmission control protocol [TCP]/IP stack).

1.1.8.1 Connection Oriented   To provide a connection-oriented service, there is the need to previously establish a connection before data is exchanged and to terminate it after data exchange. The connection is established between entities, incorporating the negotiation of the QoS and cost parameters of the service being provided. The communication is bidirectional, and the data are delivered with reliability. Moreover, to prevent a faster transmitter to overload a slower receiver, flow control is used (to prevent overflow situations). Example of a connection-oriented service is the telephone network, where a connection is previously established before voice exchange. In the telephone network, taking, as a reference, two words transmitted one after the other, we do not experience an inversion of the correct sequence of these words (e.g., receiving the second word before the first one). The TCP of the TCP/IP stack is an example of a connection-oriented protocol.

A connection-oriented service is always confirmed,* as the transmitter has information about whether or not the data reached the receiver free of errors, correcting the situation in case of errors. This can be performed using positive confirmation, such as the positive acknowledgement with retransmission (PAR) procedure, or using negative confirmation, such as the negative acknowledgement (NAK).

In the PAR case, when the transmitter sends a block of data, it initiates a chronometer and expects for the correct reception of an acknowledgement (ACK) message from the receiver within a certain time frame. In case such ACK message is not received in time, the transmitter assumes that the message was received corrupted and performs the retransmission of the block of data. In case the ACK message is received, the transmitter proceeds with the remaining transmission of data. The advantage of this procedure is that the ACK message sent by the receiver to the transmitter allows two confirmations: (1) the data was properly received (error control), and (2) the receiver is ready to receive more data (flow control).

In the case of the NAK, the receiver only sends a message in case the data is received with errors; otherwise, the receiver does not send any feedback to the transmitter. The advantage is the lower amount of data exchanged. The disadvantage is that, in the PAR case, flow control is performed together with error control, whereas in the NAK situation, only error control is performed. The reader should refer to Chapter 2 for a detailed description of the service primitives used in connection-oriented services.

---

* On the other side, the connectionless service can be confirmed or nonconfirmed.

1.1.8.2 Connectionless The connectionless mode does not perform the previous establishment of the connection before data is exchanged. Therefore, data is sent directly without prior connection establishment. As the connection-oriented mode requires a handshaking between the transmitter and the receiver,* this introduces delays in signals. Consequently, for services that are delay-sensitive, the connectionless mode is normally used. The connectionless mode is also used in scenarios where the experienced error probability is reduced (e.g., the transmission of bits in an optical fiber).

Depending on whether the service is confirmed or not, data reliability may or may not be assured. Even if data reliability is not assured, such functionality can be provided by an upper layer of a multilayer network architecture. In such scenario, there is no need to execute the same functionalities twice.

The connectionless mode can provide two different types of services:

- Confirmed service
- Nonconfirmed service

In the case of the nonconfirmed service, the transmitter does not have any feedback about whether the data reached the receiver free of error. Contrarily, in the case of the confirmed service, although a connection establishment is not required before the data is exchanged (as in the case of the connection-oriented service), the transmitter has feedback from the receiver about whether the data reached the receiver free of errors. The reader should refer to the description of the confirmation methods used in confirmed services presented for the connection-oriented service, namely the PAR and NAK.

As an example, the IP telephony service is normally supported by the nonconfirmed connectionless user datagram protocol (UDP). However, in IP telephony, the reordering of packets is performed by the application layer.† Another example of a nonconfirmed connectionless mode is the IPv4, which does not provide reliability to the delivered datagrams and which does not require the previous establishment of the connection before data is sent. In case such reliability is required, the TCP is used as an upper layer (instead of the UDP). The serial line IP is an example of a data link layer protocol, which is nonconfirmed and connectionless.

The reader should refer to Chapter 2 for a detailed description of the service primitives used in confirmed and nonconfirmed connectionless services.

1.1.9 Network Coverage Areas

Packet-switching networks may also be classified as a function of the coverage area. Three important areas of coverage exist: local area networks (LANs), metropolitan area networks (MANs), and wide area networks (WANs).

---

* For example, implementing data repetitions to assure data reliability.
† These functions are carried out by the real-time protocol (RTP).

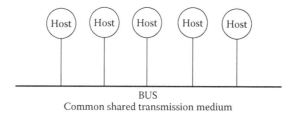

Figure 1.11   Bus topology.

A LAN consists of a network that covers a reduced area like a home, office, or small group of buildings (e.g., an airport) using high-speed data rates. A MAN consists of a backbone (transport network) used to interconnect different LANs within a coverage area of a city, campus, or something similar. Such a backbone is typically implemented using high-speed data rates. Finally, a WAN consists of a transport network (backbone) used to interconnect different LAN and MAN, whose area of coverage typically goes beyond 100 km. While the transmission medium used in LAN is normally the twisted pair, optical fiber, or wireless, the optical fiber is among the most used transmission medium in MAN and WAN.

1.1.10 Network Topologies

The network topology stands for the organization of the devices within a network. The topology concept is applicable to a LAN, a MAN, or a WAN. In the case of a LAN, such topology refers to the way hosts and servers are linked together, while in the MAN and WAN cases, this refers to the way nodes (routers) are linked together. For the sake of simplicity, this description refers to hosts and servers (in the case of LAN) and nodes (in the case of MAN and WAN) just as hosts.*

A bus stands for the topology where all hosts are connected to a common and shared transmission medium. The bus topology can be seen from Figure 1.11. In this case, the signals are transmitted to all hosts and, since the host's network interface cards (NIC) are permanently listening to the transmitted data, they detect whether they are the destination of the data. If the response is positive, the NIC passes the data to the host; otherwise, the data is discarded. This topology presents the advantage that, even if a host fails, the rest of the network keeps running without problems. The main disadvantage of this topology relies on the high overload of the whole network (including all network hosts), which results from the fact that all data is sent to all network hosts.

In the ring topology, the cabling is common to all the hosts, but the hosts are connected in serial. The ring topology is depicted in Figure 1.12. Each host acts as a repeater: each host retransmits in a termination the data received in the other termination. The main disadvantage of this topology is that, if a host fails, the rest of the network is placed out of order. This topology is normally used in SDH networks (MAN

---

* In fact, both computers and routers are hosts.

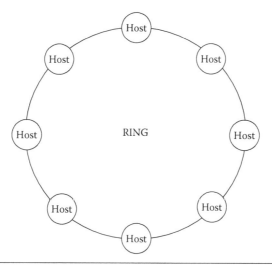

Figure 1.12   Ring topology.

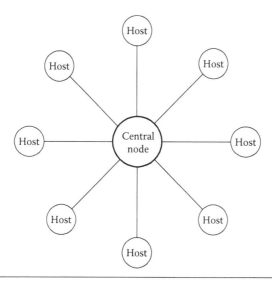

Figure 1.13   Star topology.

and WAN), where double rings are normally employed to improve redundancy. The token ring technology used in LAN is also based on the ring topology.

The star topology includes a central node connected to all other hosts. The central node repeats or switches the data from one host into one or more of the other hosts. Since all data flows through this node, this represents a single point of failure. The star topology is depicted in Figure 1.13.

The tree topology is a variation of the star topology. In fact, the tree topology consists of a star topology with several hierarchies. The tree topology is depicted in Figure 1.14. The central node is responsible for repeating or switching the data to the hosts within each hierarchy. In case the destination of the data received by a certain central node refers to another hierarchy, such central node forwards the data to the

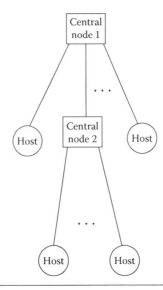

Figure 1.14   Tree topology.

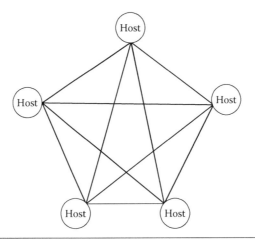

Figure 1.15   Mesh topology.

corresponding hierarchy central node, which is then responsible for forwarding the data to the destination host.

Finally, the mesh topology considers each host connected to all* or part† of the other hosts in the network. This is depicted in Figure 1.15. The advantage of such configuration relies on the existence of many alternative pathways for the data transmission. Even if one or more paths are interrupted or overloaded, the remaining redundancies represent alternative paths for the data transmission. The drawback of such topology is the high amount of cabling necessary to implement it.

It is worth noting that there are two different types of topologies: the physical topology and the logical topology. The physical topology refers to the real cabling distribution

---

* Complete mesh topology.
† Incomplete mesh topology.

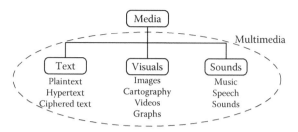

Figure 1.16 Basic types of media.

along the network, while the logical topology stands for the way the data is exchanged in the network. A physical star topology with a repeater (a hub)* as a central node presents a medium common and shared by all network hosts. In such case, the logical topology is the bus (common and shared medium). On the other hand, a physical star topology with a switch† as a central node corresponds, as well, to a logical star topology. Moreover, a physical star topology with a central node that rigidly switches the data to the adjacent host (left or right) corresponds to a logical ring topology.

1.1.11 Classification of Media and Traffic

Different media can be split into three groups [Khanvilkar 2005]:

- Text: plaintext, hypertext, ciphered text, and so on
- Visuals: images, cartography, videos, videoteleconference (VTC), graphs, and so on
- Sounds: music, speech, other sounds, and so on

While the text is inherently digital data (mostly represented using a string of 7-bit ASCII characters), the visuals and sounds are typically analog signals, which need to be digitized first to allow its transmission through a digital network, such as an IP-based network (e.g., the Internet or an Intranet). As can be seen from Figure 1.16, multimedia is simply the mixture of different types of media, such as speech, music, images, text, graphs, videos, and so on.

When media sources are exchanged through a network, it is generically referred to as traffic. As depicted in Figure 1.17, the traffic can be considered real time (RT) or nonreal time (NRT). While RT traffic is delay-sensitive, NRT media is not. An example of RT traffic is the telephony or the VTC, whereas a file transfer or web browsing can be viewed as NRT traffic.

RT traffic can also be classified as continuous or discrete. Continuous RT traffic consists of a stream of elementary messages with interdependency. An example of continuous RT traffic is the telephony, whereas the chat is an example of discrete RT traffic.

---

* A hub/repeater repeats in all other outputs the bits received in one input. In addition, it acts as regenerator.
† A switch only switches data to the output where the destination host is located. This is performed based on the destination's address.

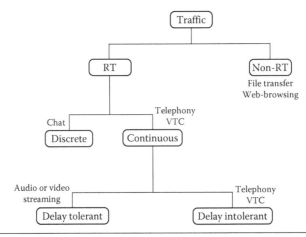

Figure 1.17   Classification of traffic.

Finally, RT continuous traffic can still be classified as delay-tolerant or delay-intolerant. RT continuous delay-tolerant traffic can accommodate a certain level of delay in signals, without sudden performance degradation. Such tolerance to delays results from the use of a buffer, which stores in memory the difference between the received data and the played data. In case the transfer of data is suddenly delayed, the buffer accommodates such delay, and the media presented to the user does not translate such delay introduced by the network. The video streaming is an example of a delay-tolerant media. Contrarily, the performance of delay-intolerant traffic degrades heavily when the data transfer is subject to delays (or variation of delays). An example of RT continuous and delay-intolerant media is the telephony or the VTC. The Internet telephony (IP telephony) or the VTC allows a typical maximum delay of 200 ms to achieve an acceptable performance.

1.2 Present and Future of Telecommunications

Current and emergent communication systems tend to be IP based and are meant to provide acceptable QoS in terms of speed, BER, end-to-end packet loss, jitter, and delays for different types of traffic.

Many technological achievements were made in the last few years in the area of communications, and others are planned for the future to allow for new and emergent services. However, while in the past new technologies pushed new services, nowadays the reality is the opposite: end users want services to be used on a day-by-day basis, whatever the technology that supports it. Users want to browse over the Internet, get e-mail access, use the chat, and establish a VTC, regardless of the technology used (e.g., fixed or mobile communications). Thus, services must be delivered following the concept of "anywhere" and at "anytime." Figure 1.18 presents the bandwidth requirements for different services.

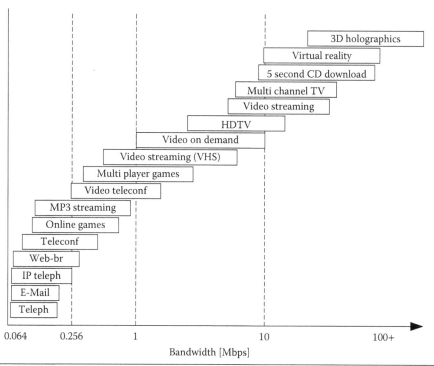

Figure 1.18 Bandwidth requirements of the different services.

1.2.1 The Convergence

The main objective of the telecommunications industry is to create conditions to make the convergence a reality [Raj 2010]. The convergence of telecommunications can be viewed in different ways. It can be viewed as the convergence of services, that is, the creation of a network able to support different types of service, such as voice, data (e-mail, web browsing, data base access, files transfer, etc.), and multimedia, in an almost transparent way to the user [Raj 2010].

The convergence can also be viewed as the complement between telecommunications, information systems, and multimedia, in a way to achieve a unique objective: make the information available to the user with reliability, speed, efficiency, and at a low price. According to the Gilder's law, the speed of telecommunications will increase three times every year over the next 20 years, and according to the Moore's law, the speed of microprocessors will duplicate every 18 months.

The convergence can also be viewed as the integration of different networks in a single one, in a transparent way to the user. The convergence can also be viewed as the convergence between fixed and mobile concepts [Raj 2010], as the mobile is covering indoor environments (e.g., femtocells* of long-term evolution), allowing data and television/multimedia services, traditionally provided by fixed services, whereas fixed

---

* A femtocell is a cellular base station for use at home or in offices, which creates an indoor cell in locations where cellular coverage is deficient, inexistent, or to provide high-speed data services. It typically interconnects with the service provider via broadband xDSL or cable modem.

telecommunications provide mobility with the cordless systems, whose example is the digital European cordless telephone (DECT) standard. There are terminals able to operate as cellular phones or as fixed network terminals. New televisions receive the TV broadcast and also allow browsing over the Internet.

The convergence is viewed by many people as the convergence of all the convergences, which will lead to a deeply different society, whose results can already be observed nowadays with the use of the following services:

- Telework
- Telemedicine
- Web-TV
- E-banking
- E-business
- Remote control over houses, cars, offices, machines, and so on
- VTC

Human lives, organizations, and companies will tend to increase their efficiency with the new communication means, the increase of the available information, and with multicontact.

With technological evolutions—increase of user data rates, improved spectral efficiency, better performances (lower BER), increase of network capacity, and decrease of latency (RT communications)—and with the massification of telecommunications as a result of lower prices (as a result of technologic evolution and the increase of competition), it is expected that virtual reality and three-dimensional (3D) holographic will be a reality in the near future.

1.2.2 *Collaborative Era of the Network Applications*

While the convergence approach was based on the ability to allow information sharing, the new approach consists of the use of the network as an enabler to allow sharing of knowledge. It consists of the ability to provide the right information to the right person at the right time. For this to be possible, a high level of interactivity is required to be made available to each Internet user. In parallel, business intelligence are important platforms that allow decision makers to receive the filtered information* required for the decision to be made at the correct moment.

We observe, nowadays, an explosion of ad-hoc applications, which allows any Internet user to inject nonstructured information (e.g., Wikipedia) into the Internet world, in parallel with an increase of peer-to-peer applications such as Torrent, eMule, and so on. Social networks, which are currently being used by millions of people, allow the exchange of unmanaged multimedia by groups of people just to share information or by groups interested in the same subject. Note that this multimedia exchange can

---

* For example, key performance indicators.

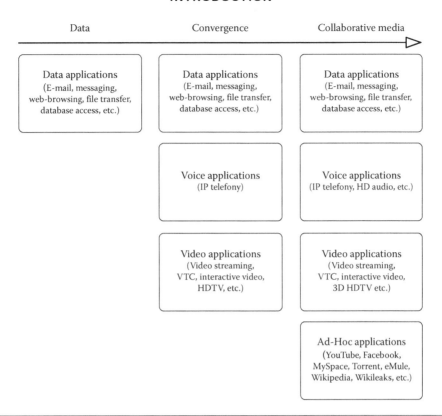

Figure 1.19 Evolution of network applications: from data to collaborative tool.

be text, audio, video, multiplayer games, and so on. This can only be possible with the ability of the IP to support all types of services in parallel with the provision of QoS by the network, that is, with the convergence as a support platform. This is the new paradigm of the modern society: the collaborative era. The collaborative era of the Internet can also be viewed as the transformation of man-to-man communication into man-to-machine and machine-to-machine communication, using several medias, and where the source or destination party can be a group instead of a single entity (person or equipment).

Figure 1.19 shows the evolution of the network usage. Initially, this was viewed merely for the data applications. Afterwards, as referred to in Section 1.2.1, convergence was an important issue to allow a better usage of the network. Increase in the level of Internet users' interactivity made the Internet world a space for deep collaboration between entities, but with a higher level of danger as well.

1.2.3 Transition Toward the Collaborative Era

To reach demands of the modern society, in terms of both convergence and collaborative services, several problems need to be solved from the scientific and industrial community. From Section 1.2.2, we may conclude that the convergence can be viewed as an important requirement to support the collaborative services.

Although we observe an enormous demand for convergence, we see that there are still problems that need to be solved. An example is the universal mobile telecommunication system (UMTS), which still treats voice and data in different ways, as data is IP based while voice is still circuit switching based. The long term evolution (LTE) is the cellular standard that deals with this issue and makes the all over IP a reality.

From the services viewpoint, the total digitalization of the several information sources and the use of efficient encoding and compressing data algorithms are very important. The information sources can be voice, fax, images, music, videoconference, e-mail, web browsing, positioning systems, high definition television, and pure data transmission (database access, file transfer, etc). Different services need different transmission rates, different margin of latencies and jitter, different performances, or even fixed or variable transmission rates. The several moving picture experts group (MPEG) protocols for voice or video, those already existent and those which are still in research and development phase, intend to perform an adaptation of the several information sources to the transmission media, allowing a reduction of the number of encoded bits to be transmitted.

Different services present different QoS requirements, namely the following:

- Voice communications are delay sensitive, but present low sensitivity to loss of data and require low data rate but are approximately constant.
- Iterative multimedia communications (e.g., web browsing) are sensitive to loss of data, requiring considerable data rate with a variable transmission rate, and are moderately delay-sensitive.
- Pure data communications (e.g., database access, file transfer) are highly sensitive to loss of data, requiring relatively variable data rate, without sensitivity to delay.

Jitter is defined as the delay variation through the network. Depending on the application, jitter can be a problem or jitter issues can be disregarded. For instance, data applications, which deliver their information to the user only if the data is completely received (reassembling of data), pay no attention to the jitter issues (e.g., file transfer). This is totally different if voice and video applications are considered; those applications degrade immediately if jitter occurs.

The transmission of data services (e.g., pure data communications, web browsing, etc.) through most of the reliable mediums (e.g., optical fiber, twisted pair, etc.) usually considers error detection algorithms jointly with automatic repeat request* (ARQ), instead of error correction (e.g., block coding or forward error correction). This happens because these services present very rigid requirements in terms of BER, whereas not very demanding in terms of delay sensitivity (in this case, stopping the

---

* ARQ works associated with error detection. The transmitter sends groups of bits (known as frames), which are subject to an encoding in the transmitter. The decoding process performed in the receiver allows this station to gain knowledge about whether there was an error in the propagation of the frame. In the case of error, the receiver requests a repetition of the frame from the transmitter.

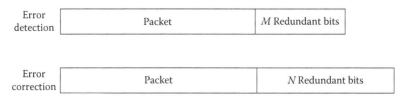

Figure 1.20 Types of error control: the error detection and error correction. Their differences in terms of the amount of redundant bits ($N > M$).

transmission and requesting for repetitions is not crucial). Note that the utilization of error correction requires more redundant bits per frame than error detection (amount of additional data beyond the pure information data). This can be seen in Figure 1.20.

Nevertheless, the transmission of data services through a nonreliable medium (e.g., wireless) is normally carried out using error correction, as the number of repetitions would be tremendous, creating much more overhead (and corresponding reduction of performance due to successive repetitions) than that necessary to encode the information data with error correction techniques. Similar principle is applied to services that are delay-sensitive (voice) where, to reduce latency, error correction is normally a better choice, instead of error detection.

These are the notions that introduce the QoS concept, which implies that each service will impose certain requirements. For the convergence to become a reality, the network should be able to take all these requirements into account.

Taking into account all the previously described factors, one that presents a great contribution to support the new collaborative services is the maximum transmission rate, as it is associated with the user data rate. The factors that limit the use of higher transmission rates are the several sources of interference and the noise. The effects of noise can be minimized through the use of regenerators, as well as advanced detection algorithms (e.g., matched filters). Interferences tend to increase with the increase in the used bandwidth (which corresponds to an increase of transmission rates), this being the main limitation of the use of higher data rates.

The challenge facing the today's telecommunications industry is how to continually improve the end-user experience and to offer appealing services through a delivery mechanism that offers improved speed and service attractiveness and interaction. To deliver the required services to the users with the minimum cost, the technology should allow better and better performances, higher throughputs, improved capacities, and higher spectral efficiencies.

What can be done to increase the throughput of a wireless communication system? One can choose a shorter symbol duration T_S. This, however, implies that a larger fraction of the frequency spectrum will be occupied, since the bandwidth required by a system is determined by the baud rate $1/T_S$. Wireless channels are normally characterized by multipath propagation caused by reflections, scattering, and diffraction in the environment. Shorter symbol duration might, therefore, cause an increased degree of intersymbol interference (ISI) and thus performance loss. As an alternative to shorter symbol duration, one may choose using a multicarrier approach, multiplexing data into multiple narrow sub-bands, as adopted by orthogonal frequency division multiplexing (OFDM) [Marques da Silva et al. 2010]. OFDM technique has been selected for LTE, as opposed to wideband code division multiple access, which is the air interface technique that has been selected by the European Telecommunications Standard Institute for UMTS. Thus, the problem of ISI can be mitigated. But still, the requirement for increased bandwidth remains, which is crucial in regard to the fact that frequency spectrum has become a valuable resource. This imposes the need to find schemes able to reach improved spectral efficiencies such as higher-order modulation schemes, the use of multiple antennas at transmitter and at receiver such as multiple input multiple output systems, more efficient error control, and so on [Marques da Silva et al. 2010].

End of Chapter Questions

1. What are the advantages of using digital communications relating to analog communications? What are the disadvantages?
2. What are the reasons that may imply the use of modem?
3. What is the difference between simplex, half-duplex, and full-duplex communication?
4. What is the physical topology used to implement a logical bus? In such case, what is the central node?
5. What is the difference between unicast, multicast, and broadcast communication?
6. What is the difference between an analog and a digital signal?
7. What is the difference between a LAN, MAN, and WAN?
8. What is the difference between connectionless and connection-oriented service?
9. What is the difference between a point-to-point communication and a network?
10. What is the difference between a circuit-switching and a packet-switching network? Give examples of networks based on these two switching types?
11. Which QoS requirements do you know?
12. What is the convergence of telecommunications?
13. What is the difference between physical topology and logical topology?
14. Which types of media do you know?
15. How can the different types of traffic be considered grouped as?
16. What is the collaborative era of the telecommunications?

2

NETWORK PROTOCOL ARCHITECTURES

2.1 Introduction to Protocol Architecture Concept

The problem of interconnecting terminals in a network is a complex task. The approach of trying to solve all problems without segmentation of functions in groups becomes an equation with a very difficult solution. Therefore, the traditional solution is to group functionalities into different layers and allocate each group to a different layer. This is called network protocol architecture, also commonly known as network architecture. This approach defines only "what" is to be done by each layer, but not "how" such functionalities are to be implemented by the layer, whose responsibility belongs to the protocol of the layer individually. This approach leaves room for a layer to improve (e.g., due to technological evolutions), without implications in the remaining layers, as long as the interface between a certain layer and its adjacent layers are kept as specified by the protocol architecture. In this sense, the protocol architecture defines the number of layers, what is to be done by each layer, and the interface between different layers. Note that a network architecture not based on layers would not allow changing the "how to do" without changing the architecture itself and without changing the remaining functions of the network architecture.

There are different network architectures. The International Standardization Organization created the widely known open system interconnection–reference model (OSI-RM), as depicted in Figure 2.1. Layers can also be identified by numbers, starting from the lower layer (physical layer–layer 1) and up to the upper layer (application layer–layer 7).

This seven-layer architecture model defines and describes a group of concepts applicable to communication between real systems composed of hardware, physical processes, application processes, and human users [Stallings 2010].

This architecture can be split into two groups: the four lower layers (from the physical up to the transport layer), which are responsible for assuring a reliable communication of data between terminal equipments; and the three upper layers (from the session up to the application layer), with a higher level of logical abstraction, interfacing with the user application [Monica 1998]. Note that the OSI-RM is only a reference model, and the systems implemented use more or less parts of this model. Therefore, we may view the transmission control protocol/Internet protocol (TCP/IP) stack, in use in the Internet world, as the most used real implementation closer to the OSI-RM.

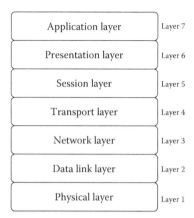

Figure 2.1   Layers of the open system interconnection reference model.

To better understand this network architecture concept, let us consider Peter in Boston who writes a letter to Christine in Bristol. The letter is written using a specific protocol, starting by "Dear Christine" and ending by "Warm regards" and signature. The letter is inserted into an envelope, to which the destination address (Christine's address) is written in a specific location. The envelope is transported and delivered to the post office in Boston, where a stamp is added in a specific location. This post office will send the letter through the post office network, which may use several means of transportations (van, bus, airplane, train, etc.) to allow the delivery of the letter at the post office in Bristol. In Bristol, the letter will be distributed between zones based on the destination addressed and delivered to the postman, who will post the letter in Christine's mailbox.

We may view the transportation of the letter as a process composed of several layers. An upper layer (application layer), which is the communication between Peter and Christine, and uses a specific protocol (the letter starts by "Dear Christine" and ends with "Warm regards" and a signature). This protocol specifies what and where in the letter it is to be written and only refers to the agents of this layer (Peter and Christine), not any of the intermediate agents (e.g., post office, plane, postman, etc.). This is control data, that is, overhead. Moreover, this communication follows a protocol, which consists of a set of procedures that are to be followed between the two entities.

Although the communication between them is supported by the lower layers (envelopes, post, airplane, etc.), one may say that there is a virtual circuit between Peter and Christine. In Figure 2.2, the application layer of the origin and destination is linked by a dashed line that represents a virtual circuit. As in the case of the circuit between Peter and Christine, there is no direct connection between them. The lower layer (presentation) is used as a service (which is also supported by another lower layer) to allow the data to arrive at the destination.

On the second stage, the letter was inserted into an envelope and the destination address was written on the proper location. This second stage also has its own protocol, which includes the added overhead (destination address) to be read by the post

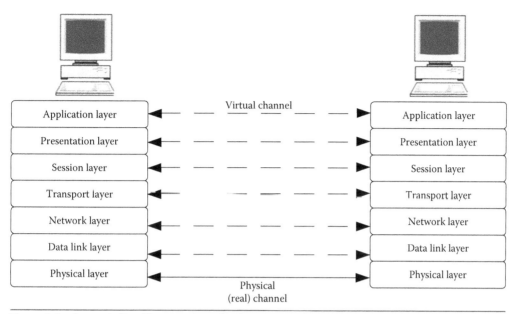

Figure 2.2 Communication of two computers using the open system interconnection reference model.

office network (lower layer), which is essential to allow the letter being forwarded to the destination. Once the letter reaches the post office in Boston, a stamp is added to the envelope on a specific location. This is a procedure that is recognized by worldwide post office protocol, and is essential to allow the letter to be forwarded from the United States/Boston to the destination address in the United Kingdom/Bristol. An employee at the post office in Boston will take the letter (together with many others) into the airport using a van as a mean of transportation. In Bristol, another employee will collect the letter from the airport and transport it to the post office by van.

The address is composed of two parts: the country/city and the street and number. The country/city pair is the information necessary to decide about the route to be used to forward the letter from the origin to the destination, through the worldwide post office network. This can be viewed as the network layer control data (overhead). A decision has to be made about whether to follow a direct flight from Boston to Bristol (in case there is one) or to send it through London. In this latter case, London acts as an intermediate node (router). Such a node receives the letter, reads the destination address, and decides about how to forward it to Bristol. This is the function of a router, which belongs to the network layer (layer 3). It receives the data from the physical layer (bits), it goes up the several layers up to the network layer, and removes the layer 1 and layer 2 control data. It reads the destination address (country/city) and decides about the best output interface to forward the data into the destination. Again, the router adds new layer 2 and layer 1 control data and sends it through the physical layer.

Returning to our example, the street name/number pair is the information necessary to decide about how to forward the letter within the destination city (Bristol). Note that this additional control data is to be processed by a lower layer. In this case, one may say that this control data is to be processed by the data link layer (switch), which is responsible for forwarding the letter within the city (point-to-point, between Bristol post office and Christine's house). A switch belongs to the data link layer and is responsible for forwarding data within a local area network (LAN), whereas a router is responsible for forwarding data between different LANs. The switch receives the data from the physical layer (bits), removes the layer 1 control data, and reads the layer 2 control data. Using the example, it reads the destination address (street name/number) and decides about the best output interface to forward the data to the destination. We may conclude that the country/city pair can be viewed as a LAN (layer 3 address) and the interconnection between different LANs (cities) is performed at layer 3 routing. Similarly, the street name/number can be viewed as the physical address of the terminal (layer 2 address), and the interconnection within the city (i.e., between streets and numbers) is performed at layer 2 switching.

From the example, we conclude that each layer of a network architecture performs different functionalities, each one presenting its own protocol, and a different overhead.

2.2 Open System Interconnection–Reference Model

Section 2.1 described that the network protocol architecture deals with functionalities performed by each layer and the type of interfaces between different layers. It was seen that a specific layer provides services to the upper layer and makes use of the services provided by the lower layer. The definition of how these functionalities are carried out by each layer is not specified by the protocol architecture. This is specified by the protocol adopted by each layer. The OSI-RM consists of an abstract network architecture model, being implemented in part by the different network protocol architectures. As will be described, the TCP/IP includes many of the concepts specified by the OSI-RM.

As can be seen from Figure 2.3, the message format of each layer is referred to as a protocol data unit (PDU), preceded by a letter corresponding to the layer. The message format of the application layer is application PDU (APDU). The message format of the presentation layer is the presentation PDU (TPDU). The message format of the session layer is the session PDU (SPDU). The message format of the transport layer is the transport PDU (TPDU).* The message format of the network layer is the network PDU (NPDU) also known as packet.† The message format of the data link layer is the link PDU (LPDU), also known as frame.‡ Finally, the message format of the physical layer is the bit. As can be seen from Figure 2.3, the nth layer service data unit (SDU), corresponds $n + 1$-PDU received from the upper layer, being encapsulated into the n-PDU. Moreover, the nth layer protocol control information (PCI) corresponds to

---

* In the TCP/IP, TPDU is called segment.
† Its main purpose is to allow it being forwarded throughout the entire network (i.e., between LANs).
‡ A frame is composed of a group of information bits to which control bits are added to allow performing error control and flow control.

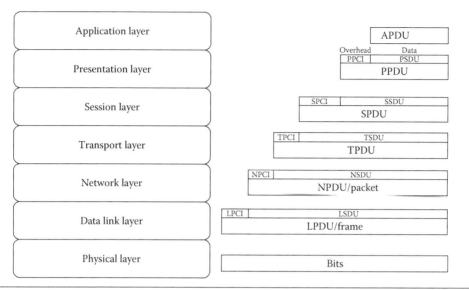

Figure 2.3 Protocol data units of different layers.

the overhead generated in the nth layer, which may include fields such as source and destination addresses, redundant bits for error detection or correction, acknowledgement numbers for error control and flow control, and so on.

2.2.1 The Seven Layer Open System Interconnection–Reference Model

Figure 2.4 depicts the several layers of the OSI-RM, generically describing the functions provided by each layer. A brief description of each layer is provided in this section.

2.2.1.1 Physical Layer This is layer 1 of the seven-layer OSI model of computer networking, also known as OSI-RM. This layer is responsible for the transmission of the data received from the upper layer (data link layer), in the form of bits, between adjacent nodes (point-to-point*). Looking at Figure 2.5, the link between adjacent nodes can be the link between station 1 and node A, or between node A and node B, and so on. It is responsible for the representation of bits to be transmitted through a transmission medium.† Such representation includes the type of digital encoding or modulation scheme to use (voltages, pulse duration, amplitude, frequency or phase modulation, etc.) [Marques da Silva et al. 2010]. This layer also specifies the type of interface between the equipment and the transmission medium, including the mechanical interfaces (e.g., RJ45 connector).

---

* In the sense of a network, point-to-point refers to the interconnection between two adjacent routers (nodes), between a host and an adjacent router or between two adjacent hosts.

† Such transmission medium can be a twisted pair, coaxial cable, fiber optic cable, satellite link, wireless, and so on.

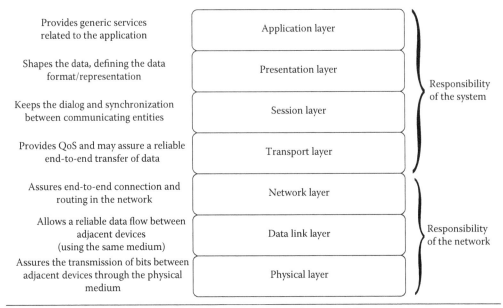

Provides generic services related to the application — Application layer

Shapes the data, defining the data format/representation — Presentation layer

Keeps the dialog and synchronization between communicating entities — Session layer

Provides QoS and may assure a reliable end-to-end transfer of data — Transport layer

} Responsibility of the system

Assures end-to-end connection and routing in the network — Network layer

Allows a reliable data flow between adjacent devices (using the same medium) — Data link layer

Assures the transmission of bits between adjacent devices through the physical medium — Physical layer

} Responsibility of the network

Figure 2.4 Generic description of the open system interconnection reference model layers.

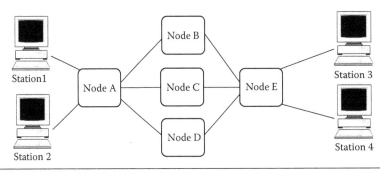

Figure 2.5 Example of a packet-switching network.

Synchronization issues are also dealt with by this layer. This includes the ability of a receiver to synchronize with a transmitter (start and end of bit instants) before bits are transferred.

This layer aims to optimize the channel capacity as defined by [Shannon 1948],* making use of encoding techniques (or modulation schemes), multiple transmitting and receiving antennas, regenerators, equalizers, and so on. Although the physical layer may use error control, the provision of reliability to the exchanged data is normally a functionality to be provided by the data link layer.

2.2.1.2 Data Link Layer This layer is responsible for providing reliability to the data exchanged by the physical layer. Note that the data link layer (and the physical layer)

---

* Shannon capacity is deÃned in Chapter 3.

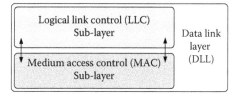

Figure 2.6 Data link layer and its sublayers.

focuses on the point-to-point exchange of data.* The data link layer is composed of two sublayers (see Figure 2.6):

- Logical link control (LLC), which deals with flow control and error control
- Medium access control (MAC), which determines when a station is allowed to transmit within a LAN. Note that this sublayer exists only in case of a LAN (and some types of metropolitan area network). When stations share the transmission medium in a LAN, it is said that the access method is with collisions (e.g., Ethernet). In this case, the MAC sublayer is responsible for defining when a station is allowed to transmit in such a way that collisions among transmissions from different stations are avoided (which originate errors). On the other side, when stations of a LAN do not share the transmission medium, it is said that the method is without collisions (e.g., token ring).

The exchange of bits performed by the physical layer is subject to noise, interferences, distortion, and so on. All these channel impairments may originate corrupted bits, which degrades the performance. The data link layer makes use of error control techniques to keep the errors at an acceptable level. Depending on the medium that is being used to exchange data, error control can be performed using either error detection or error correction techniques. In the case of error detection, codes such as cyclic redundancy checks (CRC) or parity bits are used to allow errors being detected at the receiver side, and the receiver may request the retransmission of the frame. However, if the medium is highly subject to noise and interferences (e.g., wireless medium), the choice is normally the use of error correction. In this latter case, the level of overhead per frame is higher, but it avoids successive repetitions, which also translates in a decrease of overhead. Note that, in both cases, the data link layer handles blocks of bits to which the corresponding overhead (redundant bits to allow error detection or error correction, as well as this layer address) is added. These blocks of bits, with a specific format depending on the sublayer protocols, are the previously mentioned LPDU, commonly known as frame.

Figure 2.7 shows the decomposition of an LPDU (frame), as composed of an NPDU (packet received from the upper layer at the transmitting side) plus this layer overhead. The LPDU overhead is the startup flag, the source and destination address, the

---

* For example, between adjacent routers or between a host and a router.

Start flag	Address	Packet	Redundant bits	End flag

Figure 2.7 Frame decomposition.

redundant bits for error control, and the end flag. Flags are used to allow synchronization, that is, for the receiver to identify the beginning and end of a frame. It is composed of a sequence of bits with low probability to occur in the information part of a frame.

In the example of Section 2.1, the letter follows different types of transportation corresponding to each point-to-point connection (data link layer). Peter walked from home to the post office, then a van took the letter into the airport, a flight was taken from the Boston airport to the Bristol airport, another van was taken to the Bristol post office, and so on. Each different type of transportation has its own protocol. Similarly, the end-to-end path is composed of a concatenation of links (data link layer). Each link may have a different transmission medium* and different data link layer protocol† running over it. Note that the data link layer also includes the interchange of data between hosts within the same LAN. This interconnection of hosts within a LAN is achieved using a hub (repeater of bits), a switch, or a bridge. Their function is to allow the distribution of data within a LAN. These devices are defined later in this book.

It is worth referring that another important functionality of the data link layer is flow control. Normally, the transmitter can transmit faster than the receiver is able to receive. To avoid the loss of bits, the receiver needs to send feedback (control data) to the transmitter about whether it is ready to receive more frames. This is achieved through flow control. The flow control protocol can be associated to the error control protocol as follow: when a receiver checks the existence of errors (e.g., using CRC or parity bits) and sends a feedback message to the transmitter informing that it is ready to receive the following frame (which means that the previously received frame was free of errors), the two protocols (error control and flow control) are working together.

2.2.1.3 Network Layer This layer relies mainly on routing of packets along the network as well as addressing issues. The network layer is the first one (from the bottom) that takes care of end-to-end issues.

Looking at Figure 2.5, end-to-end connection is the connection between station 1 and station 3, or between station 2 and station 4, and so on.

Let us focus again on the example of Section 2.1. The letter had to be sent from Boston to Bristol. The post office in Boston had to decide about the best way to make the letter arrive in Bristol. It could be by using a direct flight (in case there is one), or through London, and so on. This is the decision that has to be made by the network layer, allowing the NPDU to reach the destination through the best path. In case there is no direct flight, an intermediate "router" in London would have

---

* For example, wireless, twisted pair, satellite, optical fibers, and so on.
† For example, IEEE802.11, IEEE802.3, point-to-point protocol (PPP), and so on.

to read the destination address (layer 3 address, i.e., network service access point [NSAP]—country/city) and decide about the next hop to reach Bristol. Therefore, the network layer is responsible for the end-to-end routing of NPDU in the network. There are two different basic modes of routing as follows:

- Datagram*
- Virtual circuit[†]

In the datagram mode, each NPDU carries the destination address and each node (router) has to decide about the best way to forward the NPDU to reach the destination. On the other side, in the virtual circuit mode, each NPDU only has information about the virtual circuit to which it belongs. Several channels flowing in the network would belong to the same virtual circuit, and the node (router) only has to know about which output interface corresponds to a certain virtual circuit.

The virtual circuits are established in advance before the data is exchanged. In this case, all NPDU of a certain connection follow the same predefined path. Contrarily, in the datagram mode, each node decides the following path and different NPDU of the same connection may follow different paths.

In the virtual circuit mode, the routing tends to be faster as the amount of decision that has to be taken by routers is lower. Different packets with different destination addresses may belong to the same virtual circuit in a specific part of the path. Looking into the network depicted in Figure 2.5, let us consider that station 1 needs to send data to station 3. A possibility could be sending packets through node A, node B, and node E. The NPDU has an identifier that identifies the virtual circuit, and the node only needs to read this identifier to find the output interface to use, not having to know the final destination address of the packet. Note that the router does not make any decision. Routers only have to read the virtual channel identifier, which is shorter (in number of bits) than the destination address, and thus, the level of overhead is reduced.

In case the datagram mode is in use, node A receives packets from station 1, reads the destination address (a field within the packet whose length is longer than the virtual circuit identifier), and decides about the following node to use to make the packet arrive the final destination. Note that, in the datagram case, since the network changes dynamically in time, different packets may follow different paths, and the packets may reach the destination out of order. In this case, another layer would have to make the reordering of packets.[‡]

In both cases, routers make use of routing tables. In the datagram mode, a routing table stores information about the output interface to which packets should be sent in order to reach certain destination address. In the virtual circuit mode, a routing table

---

* For example, Internet protocol.
† For example, X.25 or MPLS.
‡ Normally, for services that require the reordering of packets, this is performed by the transport layer. Nevertheless, in some cases, this can be done by another upper layer (e.g., IP telephony service).

stores information about the output interface that corresponds to a certain virtual circuit. Note that the virtual circuit mode allows data to be forwarded quickly, but the construction of the routing table is more complex (and requires higher level of overhead) than is in the case of datagram. The IP is based on datagram mode.

2.2.1.4 Transport Layer   This layer is responsible for making sure that end-to-end data delivered by the network layer has the required quality of service (QoS) (reliability, delay, jitter, bandwidth etc.). In other words, it is responsible for providing the desired service to the upper layers. Depending on the classification of the service provided, there are two different types of connections, which influence the provision of this layer QoS:

- Connection-oriented
- Connectionless

Connection-oriented is a service provided by a layer that comprises three different phases: (1) connection setup, (2) data exchange, and (3) connection closure. A connection-oriented service assures that packets that reach the receiver follow the transmission order. In addition, it makes use of error control techniques to provide reliability to the delivered data.* Although connection-oriented services bring benefits in terms of data reliability, it demands more processing from both transmitter and receiver, which translates in additional resources and time (e.g., time to establish before sending data, delay due to request for repetition of packets, reordering of data at the receiver, etc.). On the other hand, a connectionless service is minimalist in terms of processing, but the delay is also minimized. It does not bring additional reliability to data and there is no need to make a connection setup before transmission. Reordering of packets is not performed at the receiver, and error control techniques are not adopted.† Let us consider the IP telephony service. As previously described in Section 1.2.1, an important characteristic of the voice communication service is that it is delay-sensitive, whereas not sensitive to loss (or corrupted) of data. Therefore, the use of a connection-oriented transport layer does not seem to be a good idea, as it may introduce delays.‡ Therefore, the IP telephony service is normally supported in connectionless mode as it minimizes the delay, whereas the errors that may occur are normally not critical for the message to be understood (typically, a voice service has an acceptable quality with a bit error rate of the order of 10^{-3}, whereas most of other data services require much more reliable data).

Let us now consider a file transfer between two terminals through the network. This service is highly sensitive to loss of data (otherwise the file would be corrupted),

---

* In the TCP/IP stack, the connection-oriented service of the transport layer is implemented using the TCP protocol.

† In the TCP/IP stack, the connectionless service of the transport layer is implemented using the user datagram protocol (UDP). Furthermore, the IP protocol (layer 3) is based on the connectionless mode.

‡ In fact, this delay is subject to fluctuations, which is called jitter. The level and variation of delay would depend on the amount of requests for repetition.

whereas not very sensitive to delay. It is important to make sure that packets arrive the destination in the correct order and free of errors. Therefore, it is clear that the service provided by the transport layer should be connection-oriented.

In addition to the above mentioned transport functions, this layer may also offer other functionalities: let us consider the case where the network layer has a 512 kbps connection established and where the application layer is requesting a 1024 kbps connection. In this situation, the transport layer may establish two 512 kbps network connections and, in a transparent manner, offer a 1024 kbps connection to the application layer [Monica 1998]. Similar function may be offered when the maximum NPDU length is lower than the NPDU length being requested by the application layer (APDU). In this situation, the transport layer performs the segmentation of the APDU into two (or more) NPDUs.

2.2.1.5 Session Layer This layer allows the mechanisms for setting up, managing, and closing down sessions between end-user application processes. While supported in the transport layer, it consists of requests and responses between applications. Logical sessions need to be properly managed in terms of which station can send data and when. As an example, although the lower layers can be full duplex, some applications are specified to communicate in half duplex. Another function of the session layer consists of the ability to reestablish the connection in case of failure. In this context, synchronism points are inserted such that, in case of failure, the session is reestablished from the last synchronism point correctly processed. Furthermore, the session layer may ask its counterpart (destination session layer) about whether the data received before a certain synchronism point has been properly processed.

2.2.1.6 Presentation Layer This layer is responsible for formatting and delivering of data to the application layer for processing. Different computers use internally different representations of data. However, the data exchanged along a network needs to follow a common representation; otherwise, the communication could not succeed. This definition is performed by the presentation layer. The interconnection of different networks that use different presentation layers requires a gateway. A gateway can be seen as a device that is able to understand two (or more) languages and is able to translate one language into another. It is still worth noting that, although it can be performed by other layers, since encryption can be viewed as a different way of representing data, it is typically performed by the presentation layer.

2.2.1.7 Application Layer This is the layer of the protocol architecture that is responsible for interfacing with the application program of the user. Note the difference between an application layer of a protocol architecture and an application program. The former has some attributes in the exchange of data in the network, whereas the latter is only a specific application resident in hosts. These attributes include the definition

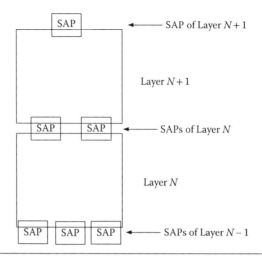

Figure 2.8 Service access point as the interface between layers.

of fields and rules of interpretation for these fields. It provides services specific to each kind of application (file transfer, web browsing, e-mail, etc.). As an example, the Microsoft Outlook is an e-mail application, whereas the CCITT X.400 is an e-mail application layer, which defines the way functionalities are carried out.

2.2.2 The Service Access Point

Each layer has its own addressing format. The purpose of an address is to allow the corresponding layer identify whether a certain host is the destination of a PDU. In addition, the source address allows the identification of who was the sender of such message that is circulating in the network.

The address of each layer is referred to as service access point (SAP) preceded by a letter corresponding to the layer and is part of the layer control data. As previously mentioned, a specific layer (N) communicates with the upper layer ($N + 1$) to offer services and communicates with the lower layer ($N − 1$) to use its services [Monica 1998]. This concept can be seen from Figure 2.8. An SAP can be viewed as the interface between adjacent layers. Note that a layer may communicate with adjacent layers using more than one SAP.

The SAP of layer N is the address of the interface between layer N and layer $N + 1$. The address between the application layer and the presentation layer is the presentation SAP. The address between the presentation layer and the session layer is the session SAP. The address between the session layer and the transport layer is the transport SAP (TSAP).* The address between the transport layer and the network layer is the NSAP. The same principle applies to the other layer SAP. Note that the

---

* In the TCP/IP stack, TSAP is known as port number, the NSAP is known as IP address, and LSAP is known as MAC address, hardware address, or physical address.

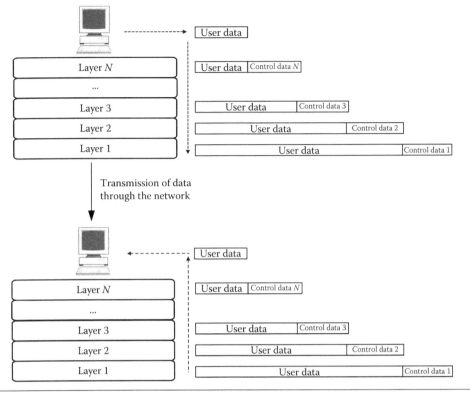

Figure 2.9   User data and control data added by several layers.

physical address is the data link layer address used in the interface between the data link layer and the network layer.

Returning to the example of Section 2.1, the two post office employees, respectively, in Boston and Bristol, have a virtual circuit between them (dashed lines in Figure 2.2), but this virtual circuit is supported by a lower layer, which is the airplane. Therefore, we may view the airplane as the physical layer (lower layer), being the unique that represents a real circuit [Forouzan 2007].

As can be seen from Figure 2.9, the communication between two terminals is performed by different layers. Each different layer will develop a specific function and have its own protocol.

Note that each upper layer uses the services made available by the lower layer to establish a virtual circuit with its counterpart layer at the destination address. Furthermore, on the transmitter side,* each layer adds specific control data (overhead), essential to allow the message (or part of it) being forwarded to its counterpart layer at the destination address. At the transmitter side, the lower layer receives the user

---

* Note that each station normally acts simultaneously as a transmitter and as a receiver. Nevertheless, for the sake of simplicity, in this description we assume that a station is acting as a transmitter and another one is acting as a receiver. In reality, their functions alternate in time (half duplex) or both stations may even act simultaneously as transmitter and receiver (full duplex).

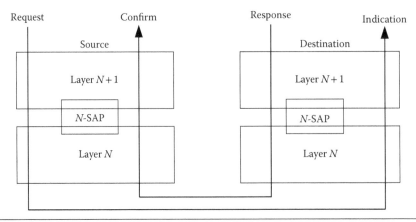

Figure 2.10   Service primitives.

data and control data from the upper layer and considers all these data as user data (this layer control data is also to be added). The control data of a specific layer is only of interest to the corresponding layer at the receiver and follows a specific protocol depending on the specifications of the layer. The receiver side removes the control data previously added by the corresponding layer at the transmitter and delivers the user data to the upper layer.

The service provided by a layer to its upper layer is defined as a group of elementary services. Each one of these elementary services is implemented using service primitives. As can be seen from Figure 2.10, the four basic service primitives considered by the OSI-RM are the following:

- Request
- Indication
- Response
- Confirm

The request is a service primitive sent by a source upper layer (N + 1-layer) to its adjacent lower layer (N-layer), being transmitted along the network until the destinations counterpart layer as an indication primitive (from the destination N-layer to the destination N + 1-layer). In case the service is confirmed,* the service primitive response is sent from the destination upper layer (N + 1-layer) to its adjacent lower layer (N-layer), meaning that such indication message was properly received by the destination layer, whereas in the source side this response is delivered from the lower layer (N-layer) to its adjacent upper layer (N + 1-layer) in the form of a confirm primitive.

As can be seen from Figure 2.11, a connection-oriented service comprises the exchange of the four service primitives for each of the following elementary operations:

---

* An example of a confirmed service is the TCP, whereas the UDP is not confirmed. Both these protocols are layer 4 protocols of the TCP/IP stack.

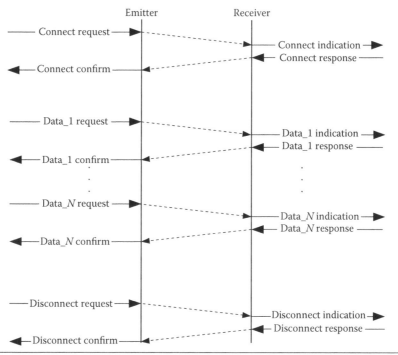

Figure 2.11 Service primitives of connection-oriented services.

- Connection establishment
- Exchange of each group of bits (frame or segment)
- Connection termination

As described in Chapter 1, a connectionless service can be confirmed or nonconfirmed. As can be seen from Figure 2.12, a confirmed connectionless service includes only the service primitives associated to the transmission of data, including response and confirm primitives, necessary for the confirmation of the service. Contrarily, as can be seen from Figure 2.13, a nonconfirmed connectionless service makes use of only the two first primitives (request and indication).

The reader should refer to Chapter 1 for the description of a nonconfirmed service.

2.3 An Overview of TCP/IP Architecture

As previously referred, the TCP/IP architecture adopted by the Internet,* is the most used real implementation of the OSI-RM. Nevertheless, while the basis is the same, there are some differences between these two architectures. While the OSI-RM is a seven-layer architecture, the TCP/IP model is composed of only five layers. This can be seen from Figure 2.14.

---

* The TCP/IP architecture is, sometimes, also known as the Internet model.

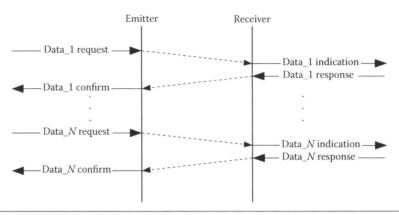

Figure 2.12 Service primitives of confirmed connectionless services.

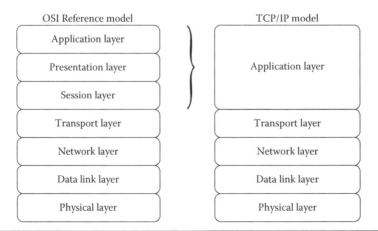

Figure 2.13 Service primitives of nonconfirmed connectionless services.

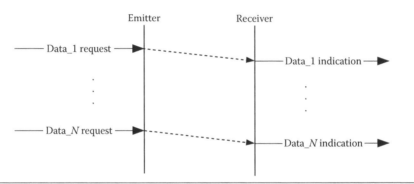

Figure 2.14 Comparison between open system interconnection reference model and the transmission control protocol/Internet protocol model.

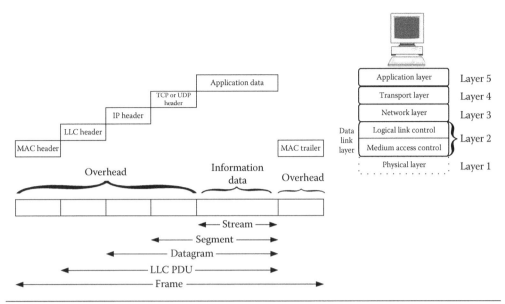

Figure 2.15 Description of overhead added by different transmission control protocol/Internet protocol layers and identification of message formats.

Similar to the OSI-RM, the layer N of the TCP/IP model uses the services made available by layer $N-1$ and provides services to layer $N+1$. In addition, layer N at the transmitting side has a virtual circuit with the layer N at the receiving side. Naturally, this virtual circuit is established making use of the services provided by the lower layers (the only real circuit is the one established by the physical layer). This can be seen from Figure 2.17. In addition, Figure 2.15 shows the control data added by each different layers of the TCP/IP in the transmitter side. Note that this control data is inversely removed at the receiver side.

TCP/IP application layer includes functionalities assigned to the application, presentation, and session layers of the OSI-RM (see Figure 2.14). In addition, some bibliographies refer to the two lower layers of the TCP/IP (physical layer and data link layer) as the "Network Access Layer." In fact, the functionality of these two layers is to provide access to the network.

The following subsections provide a generic description of each of the layers of the TCP/IP architecture. This description will start by the layer with a higher level of abstraction (application layer), as it is closer to the hosts that make use of the network protocol architecture to interchange data with a remote host.

The OSI-RM is defined such that each layer can only make use of services made available by the adjacent lower layer. In the TCP/IP architecture, the level of flexibility is much higher as any layer may invoke a service of any of the other layers, not only the lower layers but also the upper layers.*

---

* As an example, the open shortest path first (OSPF) of the TCP/IP architecture is a layer 3 protocol that is used to create routing tables. This protocol invokes the UDP, which is a layer 4 protocol.

2.3.1 Application Layer

The TCP/IP application layer incorporates most of the functions defined for the three upper layers of the OSI-RM (session, presentation, and application layers). In this case, the application layer deals with all the issues related to communication of user processes, whereas the OSI-RM splits these functionalities into three layers.

The reader should refer to the session, presentation, and application layers of the OSI-RM. In addition, the reader should refer to Chapter 13 of this book for a detailed description the TCP/IP application layer.

2.3.2 Transport Layer

As described for the OSI-RM, this layer is responsible for the provision of QoS. Such functionalities include the implementation and control of service requirements such as reliability of data, delay, jitter, low or high bit rate, constant or variable bit rate, and so on. From Figure 2.15, it is seen that the application layer generates data that is segmented and delivered to the transport layer. The message format of the TCP/IP is called segment and can be of two different types:

- UDP
- TCP

While the TCP is connection-oriented, it requires the setup of the connection before the data is exchanged. Making use of error detection and CRC, it provides reliability to the packets delivered. The provision of reliability is performed through the use of error detection (CRC codes) associated to the positive acknowledgement with retransmission (PAR) and the sliding window protocol.* In this sense, a receiving station acknowledges the good reception of packets. In case an error is detected, the transmitter "chronometer" reaches the time-out† without receiving the acknowledgement, and the packet is retransmitted. Note that this successive repetition of packets introduces variable delay (jitter) in signals.

In fact, the TCP performs other functions, namely it assures the following:

- Data is delivered with reliability.‡
- Packets are received in the correct sequence.
- Packet losses are detected and corrected.
- The duplication of packets is avoided.

---

* Refer to Chapter 10 for a detailed description of sliding window protocol.
† The transmitter starts the "chronometer" whenever a packet is transmitter.
‡ In fact, there are limits to reliability of data. TCP does not guarantee 100% of error-free packets, but keeps it at an acceptable level for the more demanding services in terms of error-sensitivity.

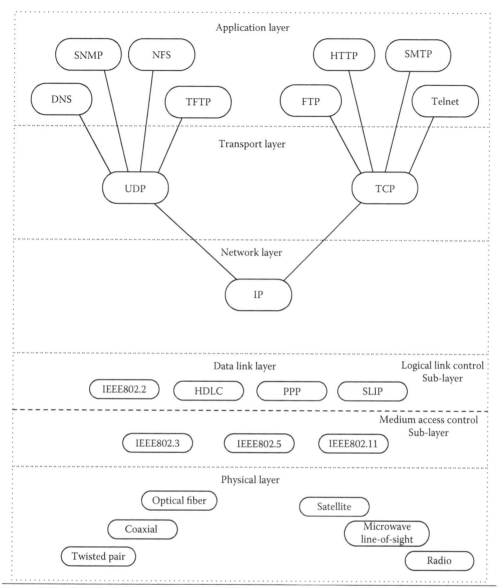

Figure 2.16 Example of some protocols and technologies used by different transmission control protocol/Internet protocol layers.

Therefore, this mode is ideal for services that require reliable data, without presenting sensitivity to delay or jitter. Figure 2.16 depicts several protocols and technologies that can be used by different TCP/IP layers. From Figure 2.16, it can be seen that some applications supported by TCP are file transfer protocol, telnet, simple mail transfer protocol, or hypertext transfer protocol.

The UDP transport protocol is connectionless and provides a nonconfirmed service, and therefore, the data is exchanged without requiring the preliminary setup of the connection. In this case, the delivery of data to the end station is based on the best effort. Although not providing data reliability, it presents the advantage of not introducing delay in signals. Therefore, this is the ideal mode for services that can resist to some level of errors. Such examples are the simple network management protocol (SNMP), the domain name server (DNS), or the network file system (NFS). Note that although UDP does not provide reliability, such reliability can be provided by the upper layers or by other means than error detection and retransmission. The SNMP uses UDP because transmission of data is very redundant in time (repeated from time-to-time). In the case of the IP telephony, ordering of packets is performed by the application layer.

The TSAP of the TCP/IP stack is the concatenation of the IP address with the port number. The port number is composed of a 16-bit address (between 0 and 65,535), referenced using the decimal notation. The reader should refer to Chapter 12 of this book for a detailed description the TCP/IP transport layer.

2.3.3 Network Layer

As described for the OSI-RM, the network layer deals with routing issues between different networks, which is an end-to-end issue. Nevertheless, actions are to be taken by each node (router) of a network.

The IP has been developed and standardized in different versions. This protocol started with version 1. Versions 2 and 3 were defined but were replaced by the version 4 (IPv4), which has been the most used version of the IP protocol. Version 5 was developed, being a specific protocol optimized for data streams (voice, video, etc.). Finally, the new IP, initially entitled IP next generation (IPng) during development phase, was standardized by the [RFC 2460; request for comments] as the IP version 6 (IPv6).

In the example of Section 2.1, where Peter sent a letter to Christine, the interconnection of different cities was viewed as the interconnection of LANs, as we can view a city as a LAN. Note that, between two cities, a letter may follow different paths (e.g., direct flight, through a third city, etc.). Furthermore, the forwarding of packets within a city (i.e., between houses) was viewed as the layer 2 switching.

We have seen in the definition of the network layer of the OSI-RM that it can be based on datagram or virtual circuit mode, depending on weather all packets between the origin and destination follow the same path (virtual circuit) or, eventually, different paths (datagram). The IP is based on datagram method, which is connectionless. Therefore, since the network changes dynamically in time, different packets may follow different paths, and the packets may reach the destination out of order. In case the reordering of packets is necessary for the service to be supported, another layer (normally this is performed by TCP in layer 4) would have to perform it.

In the IP, each node (router) has to decide about the best way to forward packets, and these packets transport information about the destination address. From

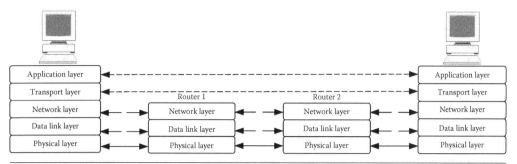

Figure 2.17 Example of a network using the transmission control protocol/Internet protocol.

Figure 2.15, we see that a datagram* is composed of the layer 4 data (segment) plus the IP header. As in the case of the OSI-RM, different control data is added at the source and removed at the destination. The exchange of data along the network depicted in Figure 2.17 is as follows:

- In the source host (leftmost host), the application layer of the TCP/IP stack receives the application data, formats it for transmission, establishes a session with the remote host (rightmost host), segments it, and transfers it to the transport layer.
- The transport layer receives the data from the application layer, adds the necessary overhead, including the source and destination port address (in case of TCP, this includes the addition of CRC redundant bits and the sequence number), and transfers it to the network layer.
- The network layer of the source host adds the layer 3 overhead, namely the source and destination IP address and passes it to the data link layer for point-to-point transfer.
- The data link layer receives the packet from layer 3 and adds the layer 2 overhead (source and destination MAC address, flags, error control redundant bits,[†] and other control data). Afterwards, the frame (layer 2 PDU) is transferred to layer 1.
- The physical layer of the source host receives the frame from the upper layer, formats the bits for transmission[‡] (type of modulation scheme or digital encoding, bit synchronization, etc.), and starts the transmission of symbols to the leftmost router (which is directly connected to the host).
- The physical layer of the router receives the symbols, performs equalization (if applicable), converts the symbols into bits, and transfers it to layer 2 of the router.

---

* Note that the IPv4 designates NPDU (packet) as datagram. The IPng, known as IPv6, returns the packet designation.

[†] The TCP/IP does not define the type of error control technique to be used by the data link layer. It depends on the data link protocol adopted in each point-to-point connection. It can be based on error detection (e.g., parity bits, CRC) or based on error correction (e.g., forward error correction [FEC]).

[‡] At the transmitter side, each bit, or group of bits, is encoded into one or more symbols. Conversely, at the receiver side, symbols are converted into bits. Symbols are transmitted, not bits. Symbols are generated using a modulator or digital encoder technique.

- The data link layer of the router groups the bits into the frames and checks for errors. If an error is detected in a frame there are two possibilities:
  - If error correction is used by layer 2, the frame is corrected by the layer 2 of the router.
  - If error detection and automatic repeat request is used by layer 2, a frame is sent from the router to the recipient (source) host, requesting for the retransmission of the corrupted frame. This frame has to go to layer 1 of the router for transmission. Afterwards, the corresponding bits are received by the host and passed to layer 2. The host layer 2 understands that this frame corresponds to a request for repetition and starts the retransmission of the frame, being then transferred into layer 1 for transmission. In the router, its layer 1 receives the bits and passes it again to layer 2 for another error check.
- Once the frame is assumed correctly received by the layer 2 of the router, this layer overhead is removed and the packet is transferred to layer 3.
- The router's layer 3 reads the destination address and consults its routing table. From this table, it extracts the output interface that should be used to send the packet. In the example of Figure 2.17, this is the next router. Therefore, the packet is transferred again to layer 2, where the new control data is added, and passed to layer 1 for transmission. This process is repeated until the packet reaches the destination host.
- At the destination host (rightmost host of Figure 2.17), the data goes up the several layers until the transport layer. This layer may have two difference procedures:
  - In case TCP is in use, the protocol checks for segment errors. If an error is detected, since the PAR is used by the TCP, it does not acknowledge the reception of the frame and the layer 4 of the recipient host reaches the time-out and starts the retransmission of such packet. Furthermore, once the packet is correctly received by the layer 4 of the destination host, it checks the correct sequence of packets (to avoid wrong sequence of packets, duplication of packets, or absence of packets), corrects it (if necessary), removes the layer 4 overhead, and transfers it to the application layer.
  - In case UDP is in use, it only removes the layer 4 overhead and transfers it to the application layer.
- The application layer reassemblies the several packets received from the source host (transferred from the transport layer), makes the necessary conversion of data, and transfers it to the application process.

The NSAP of the TCP/IP stack is the IP address. The Internet Assigned Numbers Authority (IANA) is responsible for the global coordination of the IP addressing, including the assignment of IP address groups. As can be seen from

Figure 2.18 An example of an Internet protocol version 4 address in both binary and dotted-decimal notation.

Table 2.1 Mapping Between the Address Class and the Leftmost Octet Value

CLASS	BINARY RANGE OF THE LEFTMOST OCTET
Class A	0XXXXXXX
Class B	10XXXXXX
Class C	110XXXXX
Class D	1110XXXX
Class E	11110XXX

Figure 2.18, an IPv4 address is composed of 32 bits, grouped into four octets,[*] that is, four groups of eight binary numbers. Nevertheless, for the sake of simplicity, it is normally displayed in four groups of decimal numbers, using the dotted-decimal notation.

IPv4 address space is divided into classes, from A to E (see Table 2.1). There are different possible ways to identify the class of an IPv4 address. Performing the conversion of the leftmost octet from decimal into binary and observing the position of the leftmost zero, from Table 2.1, the address class can be identified. Class A has the leftmost zero in the most significant bit (MSB). Class B has the leftmost zero in the second position and the MSB is 1. Class C has the leftmost zero in the third position and the two leftmost bits are 1.

A router is layer 3 device that is used to interconnect two or more networks. In addition, its two or more interfaces are normally of different types, that is, the data link layer protocols in each of the interfaces are different. As an example, a router is normally used to interconnect a domestic Ethernet LAN with a wide area network (WAN), which allows it to reach the Internet service provider (ISP).[†] The router is connected to adjacent devices using different links[‡] (at the data link layer level). In the example of Figure 2.19, the router has two network interface cards (NIC). Each NIC is connected to each of the networks to which the router is connected. Moreover, each NIC presents a different IP address.

Routing algorithms and IPv4 and IPv6 protocols are described in Chapter 11 of this book.

---

[*] The theoretical limit of the IPv4 address space is 2^{32}, corresponding to 4,294,967,296. Because of the rapid growth of the Internet, the available address space is depleted. IPv6 solves this problem, as its address is composed of 128 bits, which makes a much wider address space available for the Internet world.

[†] Using, for example, an asynchronous digital subscriber line or cable modem.

[‡] For example, it connects to a LAN using the IEEE802.3 protocol and it connects to the ISP through a WAN using, for example, the PPP protocol.

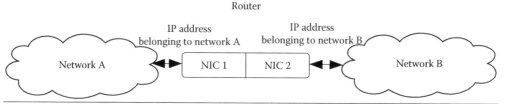

Figure 2.19   Example of a router with two network interface cards.

2.3.4 Data Link Layer

As already mentioned for the OSI-RM, this layer refers to point-to-point communication between devices. In the example of Section 2.1, the layer 3 switching (routing) is responsible for finding the best path to forward the letter between cities. This could be a direct flight, through London, and so on. On the other hand, each elementary connection of the full path between the origin and the destination can be seen as a data link layer. Each one of the connections may use a different mean of transportation: walking, car, van, bus, plane, ship, and so on. Moreover, we have seen that a city can be viewed as a LAN, and the distribution of letters within the city can be viewed as a layer 2 switching. Note that end-to-end (layer 3) forwarding is based on the country/city part of the address, whereas the distribution within a city is based on the street/number part of the address. Similarly, data link layer is responsible for the connection between two successive routers or between a router and a host (even though if it is performed through a switch or a hub). Each connection may use a different layer 2 protocol (e.g., high level data link control, PPP, IEEE802.3,* etc.) and a different communication medium (e.g., satellite, optical fiber, twisted pair, etc.).†

To allow a better understanding of the differences between the layer 2 (data link layer) and the layer 3 (network layer), let us analyze Figure 2.20. The connection between router 1 and router 2 refers to layer 2 (point-to-point connection). The same applies to the connection between router 1 and the hosts in its LAN (197.139.18.0). If the host with the IP address 193.139.18.2, in the LAN connected to router 1, needs to exchange data with a host connected to router 2 (e.g., 197.139.18.2), the layer 3 protocol is used to forward the packets between the origin and the destination. It deals with routing of packets in intermediate nodes (router 1 and router 2), based on the destination IP address (197.139.18.2). Packets use different point-to-point connections, with different layer 2 protocols. The LAN connected to router 1 may use the IEEE802.3/ IEEE802.2 protocols,‡ the connection between the two routers may use the PPP,

---

* The IEEE802.3 corresponds to a standardization of the widely used Ethernet technology. This technology was developed by the consortium Digital, Intel, and Xerox. For this reason, Ethernet was also known as DIX. The standard IEEE802.3 presents some variations to the Ethernet, being for this reason also known as Ethernet II or DIX II. This standard defines the physical (type of cable, etc.) and the MAC sublayer. This standard is detailed in Chapter 10.

† In this case, the communication mediums refer to the physical layer.

‡ The IEEE802.2 is a LLC protocol, whereas the IEEE 802.3 consists of a MAC protocol.

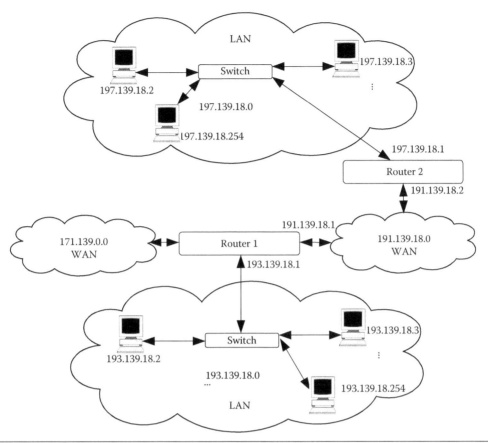

Figure 2.20   Layer 2 and layer 3 switching.

and the connection between router 2 and hosts may be supported on the IEEE802.5 (Token Ring)/IEEE802.2 protocols.

A switch is responsible for the distribution of frames (instead of packets) within a LAN, based on the destination MAC address* (instead of IP address). Note that the MAC address is composed of 48 bits and represented by six groups of two hexadecimal digits (e.g., 00-1f-33-ac-c5-bb). In fact, a switch learns the MAC address of each device connected to each interface. Every time a frame is received in a certain interface, if such interface is not associated to any MAC address, the mapping is added to the table. Such switching table maps interfaces to MAC addresses.

In addition, before a host or a router sends frames to a certain destination, such station needs to find the MAC address that corresponds to the destination IP address included in the packet. Such mapping is carried out by the address resolution protocol (ARP) [RFC 826], whose procedure is as follows: a station has an ARP table (see Figure 2.21), which maps IP addresses into MAC addresses. When a host has a packet to send or to relay, it tries to find the destination IP address in the ARP

---

* This is also called physical address or hardware address.

Internet address	Physical address	Type
192.168.0.1	00-1f-33-ac-c5-bb	Dynamic
192.168.0.255	ff-ff-ff-ff-ff-ff	Static
224.0.0.23	01-00-5e-00-00-18	Static
239.255.255.251	01-00-5e-00-00-fd	Static
239.255.255.249	01-00-5e-7f-ff-fh	Static
255.255.255.255	ff-ff-ff-ff-ff-ff	Static

Figure 2.21   Example of an address resolution protocol table.

table to extract the corresponding MAC address. In case there is no entry table corresponding to such address, it broadcasts (in the LAN) an ARP packet that contains information about a desired IP address. The station with such IP address answers with a hello packet and the station extracts its MAC address, inserting a new line into the ARP table with the mapping. This entry to this table is kept for a certain period of time. After a certain period without traffic to be passed to (or received from) this station, the entry to this table is removed and the procedure is restarted, when required. Figure 2.21 shows an example of an ARP table.

Although most of the LANs being implemented nowadays use a switch as the central node, the IEEE802.3 protocol was standardized for use with a hub as a central node.* While the switch performs the layer 2 switching based on the destination MAC address, the hub simply repeats in all other outputs the bits received in one input. Therefore, the medium becomes common, and when a station sends data, all other stations in the same LAN receive such data.

The carrier sense multiple access with collision detection (CSMA-CD)† is used to establish when a host is allowed to transmit within a LAN.

In terms of performance, since the hub broadcasts data through the LAN, the amount of collisions tends to be high, for medium-to-high traffic load (typically above 20%). On the other hand, the switch mitigates this problem as it allows having half of the stations transmitting to the other half of the stations in the LAN (considering a half-duplex network).

The data link layer is composed of two sublayers: the LLC sublayer, which deals with error control and flow control, and the MAC sublayer whose responsibility relies on the coordination of when a station is allowed to transmit and using which format. Note that MAC sublayer exists only in LAN. In the example of Figure 2.20, the LAN connected to router 1 could be an IEEE802.3 LAN. In the interconnection between the two routers, since there is only one pair of stations (router 1 and router 2), there is no need to coordinate authorizations to transmit, and therefore, the MAC sublayer does not exist.

---

* This corresponds to the worst case scenario.
† This protocol is defined in detail in Chapter 11.

The reader should refer to Chapter 10 for a detailed description of several data link layer and sublayer protocols.

2.3.5 Physical Layer

This is the lowest layer of the TCP/IP stack and, as the data link layer, also refers to point-to-point interchange of data. As defined for the OSI-RM, this is the only layer where data is physically moved across the nodes. On the other hand, the other layers only create messages that are passed to lower (at the transmitter side) or to upper (at the receiver side) layers. The type of data interchanged by the physical layer* is bits. This layer deals with all impairments of the transmission medium such as interference, noise, distortion, and attenuation. In addition, this layer also deals with transmission parameters (which may mitigate the above mentioned channel impairments) such as modulation schemes or digital encoding techniques, bandwidth, transmission rate, transmission power, equalization, advanced receivers to mitigate channel impairments, and so on.

The physical layer is only responsible for the transmission of bits, whereas error control is typically provided by the layer 2. Nevertheless, in some cases, the physical layer may also adopt some error control techniques. Such example is the transmission of bits through a wireless medium. In this case, because of the high probability of error, this layer may adopt the FEC,[†] which is a type of error correction technique.

The important functionalities of the physical layer include the following:

- Encoding of signals: the way bits are sent over the network. Such functionalities include the decision about the type of modulation scheme or digital encoding technique to use, the voltages and powers, and so on.
- Transmission and reception of bits: the choice of the bandwidth to use, the transmission rate, whether an equalizer is adopted at the receiver side, whether regenerators are necessary in the transmission path, decision about the use of multiple transmitting and receiving antennas, and so on.
- Mechanical specifications: the definition of the type of connectors (such as RJ45, RJ11, BNC) and cables to use (e.g., unshielded twisted pair, shielded twisted pair, coaxial cable, optical fiber).
- Physical topology of the network: the definition of a physical topology to use within a network such as star, ring, tree or mesh. It also includes the definition about whether cabling will be half (e.g., one cable pair) or full duplex (e.g., two cable pairs).

For a detailed description of the physical layer, the reader should refer to Chapters 3 through 7.

---

* Physical layer is also referred to as PHY.
[†] The reader should refer to Chapter 10 for a description of the FEC.

End of Chapter Questions

1. How is it possible to implement a logical bus topology and a star topology?
2. Enumerate the differences between a hub and a switch.
3. What is the different between the physical and the logical network topology?
4. To which layer of the TCP/IP model belong the following protocols: TCP and UDP?
5. Which of the OSI-RM layers is responsible for forwarding packets along the several nodes (routers) of the network?
6. Let us consider that a router needs to send packets to a host in a LAN to which it is connected to and that the corresponding MAC address is not known. Which protocol is used and what is the sequence of packets expected to be exchanged?
7. Which layer of the OSI-RM ensures that a connection is previously established before data is sent and ensures that the correct sequence of packets is maintained at the receiver side?
8. How can a switch make a better usage of the LAN bandwidth?
9. What are routing tables used for?
10. What is the difference between a router and a switch?
11. For which purpose is the ARP used for?
12. What is the difference between the UDP and TCP? Enumerate services which use either protocol?
13. Which of the OSI-RM layer is responsible for end-to-end forwarding of data?
14. What are the advantages of using a network architecture model based on several layers? What is the most implemented network architecture model based on layers?

3

CHANNEL IMPAIRMENTS

In a communication system, signals are subject to a myriad of impairments that accumulate over the path between the transmitter and receiver (see Figure 3.1). These signals are used to allow the exchange of messages between these two parties. For this message to be properly extracted, the received signal must have a signal-to-noise plus interference ratio (SNIR) higher than a certain threshold. Otherwise, the message cannot be properly understood by the receiving party.

There are several impairments that degrade the SNIR. The SNIR degradation occurs in two different ways: (1) by decreasing the signal level (S), and (2) by increasing the noise (N) and interference (I) levels.

The attenuation is the factor that causes a decrease in the signal level, whereas the increase of noise and interference levels is caused by different factors, namely:

- Different noise sources
- Distortion
- Other interferences

3.1 Shannon Capacity

In analog communications, the degradation of a signal is approximately linear to the decrease in the SNIR level, whereas in the case of digital signals, the bits degrade heavily below a certain threshold, originating an abrupt increase in bit error rate (BER). This can be seen in Figure 3.2 for the binary phase shift keying (BPSK) and quadrature phase shift keying (QPSK) modulation schemes.*

The acceptable BER threshold depends on the service under consideration (the threshold level for voice is different than that for file transfer). For voice, the tolerated bit error probability is approximately 10^{-3}, meaning that the BPSK or QPSK modulation requires a minimum of 7 dB of bit signal-to-noise ratio (SNR) (see Figure 3.2).

Depending on the source of impairments and whether the signal is analog or digital, there are different measures that can be used to mitigate it. Since the currently most used type of communication is digital, this chapter description focus on this type of transmission.

---

* As detailed in Chapter 6, the BER for BPSK is the same as for QPSK. However, this is an exception, as for M-QAM modulation schemes, increasing the modulation order M leads to a degradation in the BER. This occurs because the Euclidian distance between constellation points decreases and the modulation becomes more subject to noise and interferences.

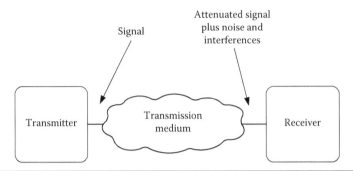

Figure 3.1   Generic chain of a communication system.

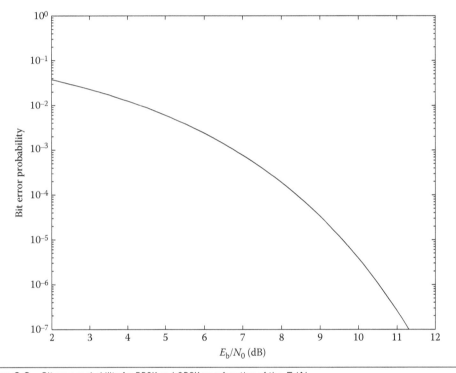

Figure 3.2   Bit error probability for BPSK and QPSK as a function of the E_b/N_0.

The capacity limit of any telecommunication system is taken to be the resulting throughput obtained through the full usage of the allowed spectrum.

Claude Shannon derived the following capacity formula in 1948, for an Additive White Gaussian Noise (AWGN) channel [Shannon 1948]:

$$C = W \log_2\left(1 + \frac{S}{N}\right)(bit/s) \tag{3.1}$$

This equation provides information about maximum theoretical rate at which the transfer of information bits[*] can be achieved, with an acceptable quality, in a certain

---

[*] Note that the Shannon capacity refers to the maximum rate of information bits (i.e., overhead and control bits are not included).

transmission medium that has a channel bandwidth W (in Hz), a noise power of N, and assuming a transmission power S (both in Watts).

Examining the voice-graded twisted pairs, which have a typical channel bandwidth W of 3.1 kHz, and assuming a SNR* of 3.7 dB,[†] from Equation (3.1), we conclude that the maximum speed of information bits is 38.103 kbps. Therefore, the solution to accommodate higher transmission rates must correspond to either an increase in the available medium channel bandwidth or an increase in the signal power, or a decrease in the power of noise and interferences.

If multiple transmitting and receiving antennas are employed,[‡] the capacity may be raised. If there are a sufficient number of receiving antennas, it is possible to resolve all messages as long as the channel correlation between the antennas is not too high. The pioneer work of Foschini [Foschini 1998] established the mathematical roots from the information theory field that, with multipath propagation, multiple antennas at both the transmitter and the receiver can establish essentially multiple parallel channels that operate simultaneously on the same frequency band and at the same time. In a more practical case of a time variable and randomly fading wireless channel, the capacity is written as

$$C = W \log_2 \left[1 + \frac{S}{N} \cdot |\mathbf{H}^2| \right] \tag{3.2}$$

where $\mathbf{H}^2$ is the normalized channel power transfer function. $\mathbf{H}$ is a $M \times N$ power complex Gaussian amplitude of the $M \times N$ channel, where M stands for the number of transmit and N for the number of receiving antennas. Multiple antenna systems are described in Chapter 7.

The minimum bandwidth B_{min} of the transmission medium is related to the symbol rate R_S[§] through the following equivalence [Proakis 1995]

$$B_{min} = \frac{R_S}{2} \tag{3.3}$$

where it is assumed that the generated signal is transmitted in baseband. Note that this equivalence corresponds to the maximum symbol rate that can be accommodated in the aforementioned bandwidth. In the case of a bandpass (carrier-modulated) signal, the minimum bandwidth becomes $B_{min} = R_S$.

---

* Equation (3.1), while generic, establishes a relationship between capacity and *SNR* (as it corresponds to *S/N*). However, the generic noise designation corresponds, in the sense of this equation, to noise plus interference (including distortions, interferences, etc.). In other words, the *S/N* of (3.1) should be understood as $S/(N + I)$.

[†] Since we are dealing with logarithmic units, it is calculated using $SNR_{dB} = 10 \log_{10} \left(\frac{S}{N} \right)$.

[‡] Multiple antennas are adopted by the air interface of IEEE802.11n standard, WiMAX standard, long-term evolution (LTE) cellular systems, and so on.

[§] Symbol rate is also referred to as transmission rate, being expressed in symbols per second (symb/s) or in baud.

As an example, let us consider a satellite link with a 2-MHz bandwidth. From Equation 3.3, we conclude that the maximum symbol rate that can be transmitted within such bandwidth is 4 Msymbol/s. If one needs to transmit 4 Mbps, the symbol constellation should be such that each symbol transports one bit.* Alternatively, if one intends to transmit 8 Mbps in the referred satellite link bandwidth, a symbol constellation which accommodates 2 bits per symbol is the solution (e.g., QPSK).

Returning to the noise power N defined in Equations 3.1 and 3.2, it is worth noting that, here, the letter N refers generically to all sources of noise and interference (not only the noise). Therefore, it is important to define all of the impairments that influence this parameter. Depending on the system and transmission medium, there are different factors, which are defined in the following subsections.

3.2 Attenuation

The propagation of a signal in any transmission medium is subject to attenuation. This effect corresponds to a decrease in the signal strength. The level of attenuation depends on the medium, but as a rule of thumb it increases with the distance at a variable scale. Figure 3.3 shows an example of transmitted signal and the corresponding attenuated signal, as a result of the propagation losses through a medium.

As an example, let us consider the propagation of electromagnetic waves in the free space. In this case, the attenuation is due to the path loss, being known as free space path loss (FSPL). Therefore, the propagation can be viewed as a sphere, whose surface area increases with the increase of the distance from the transmitting antenna. While the power can be considered as constant, since it spreads out over the surface area of the sphere and its radius corresponds to the propagation distance, its power spatial density decreases with the distance.

From Chapter 5, it can be seen that the FSPL is given by [Proakis 1995] $FSPL = (4\pi d f / c)^2$. Therefore, in the special case of electromagnetic waves and assuming free space propagation, the attenuation increases with the square of the distance. In case of a coaxial cable, the dependence of the distance is different, as the attenuation tends to increase with increase of the logarithm of the distance. Furthermore, it increases with the frequency of the signal being transmitted. This latter rule of thumb is generically applicable to most of the transmission mediums.

Note that the attenuation depends on the type of propagation between the transmitter and the receiver. Electromagnetic propagation is normally composed of several paths, namely a direct line-of-sight, and several reflected, diffracted, and scattered waves (e.g., in buildings, trees, etc.). As a result, in real propagation scenarios (others than free space), the attenuation depends on distance with a higher rate than its square. Normally, an exponent between three and five is experienced in scenarios

---

* Note that modulation schemes are dealt with in Chapter 6.

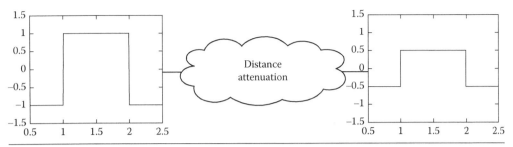

Figure 3.3 Transmitted signal and attenuated version caused by the propagation channel.

subject to shadowing and multipath.* Consequently, we conclude that the attenuation experienced in real scenarios is higher the FSPL. This results from the three basic propagation effects [Theodore 1996]: (1) reflection; (2) refraction; and (3) scattering. These propagation effects are characterized in Chapter 5.

Having a signal subject to attenuation, there is the need to increase its level such that detection is possible at the receiver. For detection to be possible, the received signal needs to be above the receiver's sensitivity threshold. The increase in signal level is performed by an amplifier at the receiver side. Note that, similarly, the signal at the receiver's input must be above the **amplifier** sensitivity threshold, or otherwise the signal cannot be amplified. Therefore, for long distances, there is the need to use amplifiers in the transmission path, before the distance attenuation is too high, and before the signal is below the amplifier's and receiver's sensitivity threshold.

Since this device amplifies its input signal, which is composed of signal plus noise and interferences, amplifying such signal does not bring any added value in terms of SNR gain.† Moreover, an amplifier also introduces additional noise, leading to a noise factor‡ higher than one. This means that the SNR suffers degradation after the amplification process.

One important advantage of digital signals consists of the ability to implement regeneration, which is a more effective process than amplification, as it allows a gain in terms of SNR. A regenerator§ includes a detector followed by an amplifier. Therefore, inserting regenerators along the transmission chain allows recovering the original signal (performed by its detector) before the signal is amplified. This enables removing the negative effects of noise and interferences, and thus, improves the SNR value. In addition, as in the case of the amplifier, a regenerator has to be placed in locations along the transmission chain such that the distance attenuation does not degrade the signal level below the regenerator's sensitivity threshold. As an example, note that synchronous digital hierarchy (SDH) makes use of regenerators, typically every 60 Km of optical fibers.

---

* Shadowing effect is detailed in Chapter 5. Multipath effect is detailed later in subsection 3.5.1.

† It amplifies the signal, the noise, and the interferences present at the input.

‡ Noise factor is defined by Equation 3.9.

§ A regenerator is also known as a repeater.

3.3 Noise Sources

One of the most important impairments of a telecommunication medium is the noise. The noise is always present, with higher or lower intensity. A receiver detects the desired (attenuated) signal superimposed with the noise and interferences. Therefore, the noise can be defined as unwanted impairments superimposed on a desired signal, which tends to obscure its information content. As previously described, a signal needs to be received with a SNR higher than a certain threshold to allow a good service quality.

Figure 3.4 shows an attenuated signal (on the left), due to the distance attenuation, equal to the one considered in Figure 3.3. Since this signal does not present any kind of noise or interferences, its SNR is infinite. This signal is then received together with the noise present at the antenna's location (signal on the right). The resulting SNR is now degraded. Although the noise has been added to the signal, since its power is not too high, we observe that the envelope shape is similar to the original signal without noise.

A receiver uses a certain instant within the pulse duration to perform the sampling. Based on the sampled signal at the sampling instant, a decision is made about whether the received signal is assumed as a symbol +1 or −1.* In this case, the hard decision should be as follows: if the sampled signal is above 0, it is assumed a +1; otherwise (with the sampled signal below 0), it is assumed that the received symbol is a −1 (as the estimated transmitted symbol).

Figure 3.5 presents the same signals but the plot on the right includes a signal subject to a stronger noise power. Note that, although the transmitted signal between instants 1 and 2 was +1, depending on the exact sampling instant, the sampled received signal can be considered as −1 (because the sampled value can be below zero, which is assumed as a decision threshold), that is, a symbol error may occur. This is because, for the same signal level as the one in Figure 3.4, the noise power is much more intense, resulting in a lower SNR value. In the case of digital signals, the resulting bit error probability becomes higher, which results in a degraded signal.

The most important types of noise can be grouped as:

- External noise
  - Atmospheric
  - Human
- Extraterrestrial noise
- Internal noise
  - Thermal
  - Electronic

---

* In this example, we assume the use of amplitude shift keying (ASK) as defined in Chapter 6. A +1 level may correspond to a logic value +1 and a −1 level may correspond to a logic value 0. In fact, the +1 and −1 levels can be any value, depending on the transmitting power.

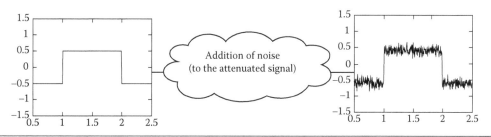

Figure 3.4   Addition of low-power noise to an attenuated signal (possible received signal).

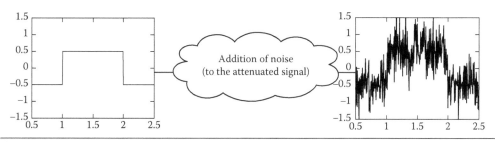

Figure 3.5   Addition of high-power noise to an attenuated signal (possible received signal).

The total noise power, resulting from all different noise sources, is summed and seen at the receiver's detector. Different types of noise are defined in the following sections.

3.3.1 Atmospheric Noise

This noise is electromagnetic, being caused by natural atmospheric phenomenon, namely lightning discharges in thunderstorms. It consists of cloud-to-ground and cloud-to-cloud flashes. While more intense in tropical regions, it consists of a high-power and low duration current discharge, which results in a high-power electromagnetic impairment. These flashes occur approximately 100 times a second, on a worldwide scale, and the sum of all these flashes results in the random atmospheric noise.

In an area surrounding the thunderstorms, the noise presents an impulsive profile (i.e., low duration but very intense). Since the pulse is very narrow, its bandwidth is very wide. This means that the noise is experienced by many nearby communication systems that make use of different parts of the electromagnetic spectrum.

The combination of all distant thunderstorms (low-duration pulses) results in white noise,* which is felt with continuity over time but with a lower power level. Its power varies with the season and proximity of the thunderstorm centers. In addition, since this phenomenon is more frequent in tropical regions, atmospheric noise tends to decrease with the increase of the latitude.

---

* White noise present a constant PSD.

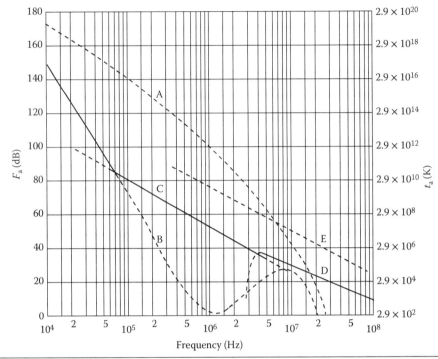

Figure 3.6 Comparison of noise figure (F_a) and temperatures (t_a) between different types of electromagnetic noise: (A) percentile 0.5 of atmospheric noise, (B) percentile 0.5 of atmospheric noise, (C) Median man-made noise for Business, (D) Median man-made noise for Galactic, and (E) Median man-made noise for Rural. [From CCIR 322. With permission.]

As described by the FSPL equation (Section 3.2), electromagnetic attenuation increases with frequency. Consequently, the higher frequency components of the noise are subject to higher attenuation levels. This is the reason why this wideband noise is felt at long distance with higher power at lower frequencies, and with lower power at higher frequencies. Consequently, the atmospheric noise dominates at very-low-frequency (VLF) and low-frequency (LF) bands. Note that the frequency bands are defined in Table 3.1. This can be seen in Figure 3.6.

In addition, as explain in Chapter 5 for the groundwave propagation, at low frequencies, waves with horizontal polarization experience higher attenuation levels than vertically polarized waves. Consequently, the vertically polarized atmospheric noise tends to be more intense than horizontally polarized noise.

3.3.2 Man-Made Noise

This noise is electromagnetic and caused by human activity, namely the use of electrical equipment, such as car ignitions, domestic equipment, vehicles, and so on. The intensity of this kind of noise varies substantially with the region. Urban man-made noise tends to be more intense than rural noise. This noise is characterized by the emission of low duration and high-power pulses, when the corresponding source is activated (e.g., when the car ignition is activated). Figure 3.7 shows the typical man-made noise

Table 3.1 Definition of Frequency Bands and Their Limits

FREQUENCY BAND	FREQUENCY BANDS							
	VLF	LF	MF	HF	VHF	UHF	SHF	EHF
Designation	Very low frequency	Low frequency	Medium frequency	High frequency	Very High frequency	Ultra high frequency	Super high frequency	Extremely high frequency
Lower limit	3 kHz	30 kHz	300 kHz	3 MHz	30 MHz	300 MHz	3 GHz	30 GHz
Upper limit	30 kHz	300 kHz	3 MHz	30 MHz	300 MHz	3 GHz	30 GHz	300 GHz
Applications	Navigation, maritime and submarines communications	Navigation, maritime and submarines communications	Radio broadcast and maritime	Radio broadcast and maritime	Radio and TV broadcast and maritime	TV broadcast, maritime and cellular communications	Satellite communications	Satellite and microwave communications

along the frequency spectrum, for different environments (business, residential, rural, and quiet rural). While very intense at high-frequency (HF) band, its noise figure (F_{am}) decreases at higher frequencies.

3.3.3 *Extraterrestrial Noise*

This type of electromagnetic noise comes from certain limited zones of the cosmos and galaxies. Extraterrestrial noise is also known as galactic noise and solar noise. An antenna directed towards certain regions of the sky, such as the sun or other celestial objects, may experience powerful wideband noise. Note that this type of noise depends on the relative orientation of the antenna's radiation pattern. This orientation varies along the day due to the earth's rotation, and therefore, attention needs to be taken about a sudden increase of noise experienced by some type of stations (e.g., by a satellite earth station). The pattern of extraterrestrial noise can be seen in Figure 3.7.

Figure 3.6 shows the different electromagnetic noise contributions in different environments. From Figure 3.6, it is noticeable that atmospheric noise dominates at VLF and LF bands, whereas man-made noise is more intense in the HF band.

Since the noise is a random process, its measure is normally performed using statistical tools. Therefore, the noise is normally expressed in percentile. As an example, percentile 10 is a value characterized by having 90% of the sample values above this percentile 10 value and having 10% of the sample values below the percentile 10 value.

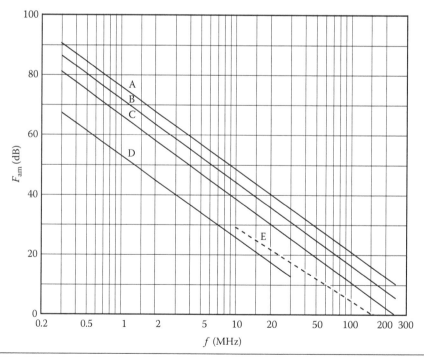

Figure 3.7   Noise factor for (A) Business, (B) Residential, (C) Rural, (D) Quiet Rural, and (E) Galactic. [From CCIR 322.]

The noise may also be expressed in median. The median corresponds to the percentile 50 (50% of the samples are above the median value, and the other 50% of the samples are below).

3.3.4 Thermal Noise

This type of noise occurs inside electrical conductors (wire, electrolyte, resistors, etc.), being caused by the thermal agitation of charges at the amplifier's input resistance. In case of radio communications, this is also generated by external sources, presenting a wide variation of amplitude, depending on the temperature viewed by the receiving antenna.

This type of noise is experienced even without any applied voltage. The frequency profile of the thermal noise presents a power spectral density (PSD) approximately constant along the frequency spectrum, that is, it is approximately white.

The noise power P_n captured by its amplifier's input resistance is given by [Carlson 1986] [CCIR 1995]

$$P_n = k_B T_n B \tag{3.4}$$

where k_B is Boltzmann's constant, with $k_B = 1.3806503 \times 10^{-23} \, JK^{-1}$ (expressed in joules per kelvin degree), T_n is the resistor's absolute temperature (expressed in kelvin degrees), and B is the receiver's bandwidth (expressed in Hertz) [Burrows 1949]. Note that, in statistical terms, the noise power P_n corresponds to the noise variance σ^2.

The PSD N_0 is given by:

$$
\begin{aligned}
N_0 &= \frac{P_n}{B} \\
&= k_B T_n
\end{aligned}
\tag{3.5}
$$

The root mean square (RMS) voltage of thermal noise generated in a given amplifier's input resistance R (expressed in ohm) is given by the following:

$$\upsilon_n = \sqrt{4 k_B T_n R B} \tag{3.6}$$

In case of electromagnetic waves, the value of T_n captured by the amplifier's input resistance depends on the orientation of the antenna's radiation pattern. Consequently, the thermal noise of a satellite earth station pointing towards the sky is typically very low, as the temperature of the sky is also low ($200°K > T > 80°K$). On the other hand, the thermal noise of a satellite transponder is typically high, as it is pointing toward the earth, whose temperature is also high ($T > 300°K$).

Another way to express the noise level is by using the noise factor f_a coefficient. This is defined by

$$f_a = \frac{P_n}{k T_o B} \tag{3.7}$$

The noise factor is defined as the relationship of the received noise power with the noise power delivered by a charge with the reference noise temperature of 300°K (T_o). Expressing this value in logarithmic units leads us to the noise figure* defined by

$$F_a = 10 \log f_a \tag{3.8}$$

3.3.5 Electronic Noise

This internal noise is generated in the active elements (e.g., transistors) in the interior of equipment such as amplifiers. As a consequence, the SNR at the output of active equipment (e.g., amplifier, active filter, etc.) is lower than the SNR at its input. Similar to the thermal noise, the electronic noise level is typically quantified by the noise factor as

$$f_a = \frac{\text{SNR}_{\text{IN}}}{\text{SNR}_{\text{OUT}}} \tag{3.9}$$

or in logarithmic units, as expressed in Equation 3.8. Using such equivalence, from f_a, and knowing the received SNR at the input, of a satellite transponder (SNR_{IN}), it is straightforward the computation of the SNR at its output (SNR_{OUT}).

In case of a cascade of N electronic devices (e.g., amplifier and filter), the resulting noise factor f_{OUT} becomes

$$f_{\text{OUT}} = f_1 + \frac{f_2 - 1}{g_1} + \frac{f_3 - 1}{g_1 g_2} + \frac{f_4 - 1}{g_1 g_2 g_3} + \dots + \frac{f_N - 1}{\prod_{i=1}^{N-1} g_i} \tag{3.10}$$

where g_i stands for the gain and $f_i : i = 1 \dots N$ for the noise factor of the ith electronic device.

Still in the case of a cascade of N electronic devices (e.g., amplifier, filter), the resulting SNR_{OUT} becomes

$$
\begin{aligned}
SNR_{\text{OUT}} &= \frac{S_{\text{OUT}}}{N_{\text{OUT}}} \\
&= \frac{g_{\text{OUT}} \cdot S_{\text{IN}}}{g_{\text{OUT}} \cdot f_{\text{OUT}} \cdot N_{\text{IN}}} \\
&= \frac{S_{\text{IN}}}{f_{\text{OUT}} \cdot N_{\text{IN}}}
\end{aligned}
\tag{3.11}
$$

where $g_{\text{OUT}} = \prod_{i=1}^{N} g_i$ and f_{OUT} as defined in Equation 3.10. Note that Equation 3.11 is equivalent to Equation 3.9, but considers a cascade of electronic devices.

---

* Note that Figure 3.6 and Figure 3.7 express the noise using the noise figure notation.

Note that other types of noise may also be viewed as the noise generated in electronic devices, but where the device gain should correspond to, for example, the path loss attenuation and the noise figure of the device is, for example, the noise figure of the thermal noise. Therefore, with such approach one could use Equation 3.10 to compute f_{OUT}, and then, use Equation 3.11 to compute SNR_{OUT} from the initial SNR_{IN}, that is, without computing all intermediate SNR.

3.4 The Influence of the Transmission Channel

An ideal transmission channel would be such that the received signal would equal the transmitted one. Nevertheless, it is known that real channels introduce attenuation and phase shifts to signals. When such attenuation is constant over the signal's frequency components (i.e., over the entire signal's bandwidth), the channel does not introduce amplitude distortion to the signal. Similarly, when the phase shift is linear over the signal's frequency component, the channel does not introduce phase distortion to the signal. In these cases, only amplification process may be necessary before detection.

Nevertheless, it is very often that the channel introduces different attenuations and nonlinear phase shifts for different frequency components of the signal. This is more visible when the signal bandwidth is relatively high. In this case, equalization process may also be required at the receiver side, before amplification and detection.

The channel's frequency response $H(f)$, also known as the channel's transfer function, translates mathematically the way the channel processes different frequency components of the signal which crosses it, in terms of both signal's attenuation and phase shift. The signal at the output of the channel is given by

$$V_R(f) = V_E(f) \cdot H(f) \tag{3.12}$$

where $V_E(f)$ stands for the Fourier transform of the time-domain transmitted signal $V_E(t)$ and $V_R(f)$ for the Fourier transform of the time-domain received signal $V_R(t)$ (see Figure 3.8). Annex A presents a short description of the Fourier transform theory.

3.4.1 Delay and Phase Shift

Let us consider a carrier-modulated signal in the carrier frequency f_c, which propagates through the transmission medium between a transmitter and a receiver. The phase θ of the signal varies by 2π radians for every propagation distance corresponding to a wavelength λ (see Figure 3.9). Similarly, the propagation delay τ increases

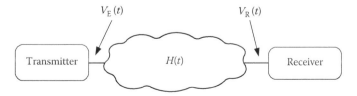

Figure 3.8 Generic communication system with the signals depicted in the time domain.

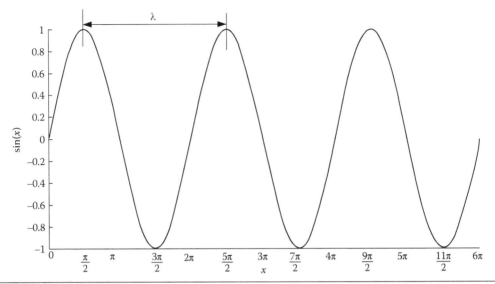

Figure 3.9   Sinusoidal wave and its characteristics.

$\tau = T_C = 1/f_c$ for every propagation distance corresponding to a wavelength λ (f_c is the carrier frequency and T_C is the carrier period). Moreover, the phase shift in the instant $t = \tau$ is given by $\theta = 2\pi f_C \tau$.

As the propagation distance increases, the phase shift and the delay vary accordingly. In fact, the received signal is composed of a superposition of components, including direct, reflected, diffracted, and scattered components.

Since each of these elementary waves experience a different distance,* each component arrives at the receiving antenna with a different phase shift and delay, before they are superimposed. As can be seen from Figure 3.10, the ideal[†] phase shift response of a channel corresponds to slop, with a constant gradient across the different frequencies.[‡] Note that time delay is related with the phase shift by [Marques da Silva et al. 2010]:

$\tau(f) = -\dfrac{1}{2\pi}\dfrac{d\theta(f)}{df}$, where both τ and θ are a function of the frequency f. As a result,

the superposition of waves from different paths results in either constructive or destructive interference, amplifying[§] or attenuating the signal power seen at the receiver (relating to the single path propagation). The variation of the signal level received from one or more of the multipaths causes a variation in the resulting superimposed signal level, which is known as fading.

Assuming a variation of distance between the transmitting and receiving antennas, it is observed that the envelope of the received signal level presents a cyclic period for every $\lambda/2$ distance variation. In fact, this fluctuation of signal level across distance can be viewed in the frequency spectrum as distortion, being explained in the following.

---

* Note that reflected waves experience longer paths.
† Which does not distort signals.
‡ A phase shift profile defined by a curve introduces distortion.
§ Relating to the line-of-sight version of the signal.

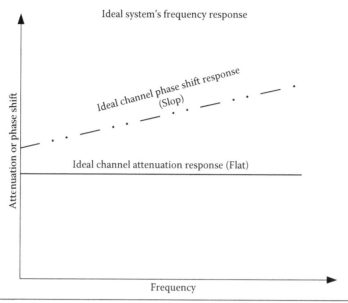

Figure 3.10 Ideal channel's frequency response (distortionless).

3.4.2 Distortion

Transmitted signals are not composed of a single frequency. On the contrary, signals are composed of a myriad of frequency components, presenting a certain bandwidth. Note that the audible spectrum spans from around 20 Hz up to around 20 kHz.

As previously described, transmission mediums tend to introduce different attenuations at different frequencies. As a rule of thumb, the attenuation level tends to increase with an increase in the frequency. Moreover channels tend to introduce different delays* and nonlinear phase shifts at different frequency components. Therefore, a signal, after propagation through a medium, is subject to different attenuations and nonlinear phase shifts at different frequency components, and hence, the received signal is different from the transmitted one. This effect is known as distortion.[†]

3.4.3 Equalization

A postprocessing which aims to introduce the opposite effect of that introduced by the channel (ex: higher attenuations at lower frequency components and lower attenuations at higher components), results in a signal equally attenuated after the combined system composed of the propagation path plus this postprocessing. Such postprocessing is called equalization, whose purpose is to mitigate the distortion introduced by the channel.

The equalizer is normally part of the receiver, being especially important for signals with higher bandwidths. Note that, from Equation 3.3, higher transmission rates

---

* As a result of different delays at different frequency components.

[†] Note that distortion may also be introduced by devices, such as amplifiers, filters, and so on.

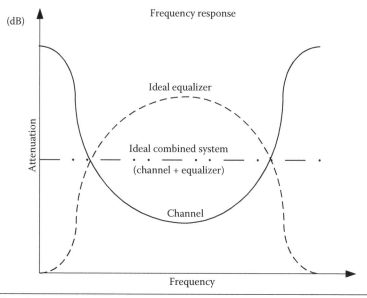

Figure 3.11   Example of frequency response of a channel and its corresponding ideal equalizer.

correspond to higher bandwidths. A frequency response* of a channel is approximately constant for a very narrowband signal (low transmission rate). In the case of very narrowband signal, its frequency span tends to zero. In this case, the curve of the channel's frequency response tends to its tangent in the reference point. On the other side, a wideband signal (high transmission rate) suffers heavily from distortion, as the signal spans over a wide bandwidth. Therefore, increasing the transmission rate demands a higher processing from the receiver in order to perform the required equalization, and this processing is never optimum, resulting in some residual level of distortion.

Figure 3.11 shows an example of the attenuation introduced by a channel, as a function of the frequency. Ideally, the equalizer frequency response should be such that the attenuation of the combined system that results from cascading the channel and the equalizer is a straight line, that is, continuous attenuation over the bandwidth of interest (i.e., absence of distortion). Note that Figure 3.11 only depicts the attenuation as a function of the frequency, but the same principle is applicable to phase shift, where the ideal phase shift response is a straight line (see Figure 3.10).

Taking the signal's description from Section 3.4 and observing Figure 3.12, the resulting signal at the output of a zero force (ZF) equalizer is given by

$$\mathrm{Eq}(f)_{\mathrm{ZF}} = \frac{1}{H(f)} \tag{3.13}$$

---

* System's frequency response is a graphic which shows the attenuation and phase shift introduced by the system as a function of the frequency.

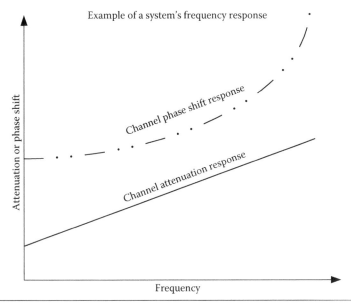

Figure 3.12 An example of channel's frequency response which introduces distortion.

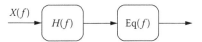

Figure 3.13 Generic communication including the channel and the equalizer at the receiver.

where $H(f)$ stands for the channel transfer function. Similarly, the signal at the output of a minimum mean square error (MMSE) equalizer becomes

$$Eq(f)_{MMSE} = \frac{[H(f)]^*}{1/SNR + [H(f)]^2} \tag{3.14}$$

where $Eq(f)$ stands for the equalizer's transfer function, that is, the Fourier transform of the equalizer's impulse response $eq(t)$. Figure 3.13 depicts the concatenation of the channel and the equalization (part of the receiver). Note that $X(f)$ represents the Fourier transform of the transmitted signal $x(t)$.

An equalizer requires being tuned to the channel. This is normally performed using pilots or training sequences. It consists of a predefined signal (sequence of known symbols), which is periodically transmitted. While the receiver has information about the transmitted signal, it computes the difference between the transmitted signal and the received one. This difference is a function of the distortion introduced by the channel. From this received pilot sequence, the receiver may extract the channel coefficients (in terms of attenuations and phase shifts at different frequencies) that are then utilized to implement the equalization process.

As previously described, a wireless channel is typically subject to fluctuations (fading). This signal variation is the result of variations of the distance between the transmitting and receiving antennas, variation of the environment surrounding the

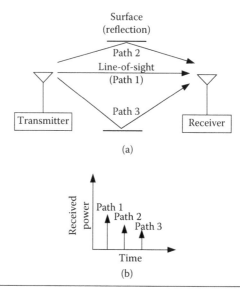

Figure 3.14 Propagation of a signal in a multipath environment: (a) diagram with multipaths; (b) average received power by each path.

receiver, variations of the refraction index, and so on. In the case of intense fading (frequency-selective fading), this corresponds to a variation in time of the attenuation levels and nonlinear phase shifts at different frequencies (distortion).* In this case, the equalizer should be able to follow the channel parameters, requiring the periodic transmission of pilots in order to allow the implementation of the equalization process performed adaptively. This process is called adaptive equalization. Note that some advanced receivers may also perform channel estimation directly from the received modulated signal [Proakis 1995].

3.5 Interference Sources

The main sources of interferences can be grouped as

- Intersymbol interference (ISI)
- Multiple access interference (MAI)
- Co-channel interference (CCI)
- Adjacent channel interference (ACI)

3.5.1 Intersymbol Interference

This source of interference occurs in digital transmissions of symbols when the channel is characterized by the existence of several paths (i.e., multipaths, as can be seen from Figure 3.14), where some paths reach the receiver's antenna with delay higher than the RMS delay spread† of the channel. In this case, this effect can be viewed in the

---

* Note that frequency-selective fading corresponds to a type of distortion.

† See Chapter 5 for the definition of delay spread.

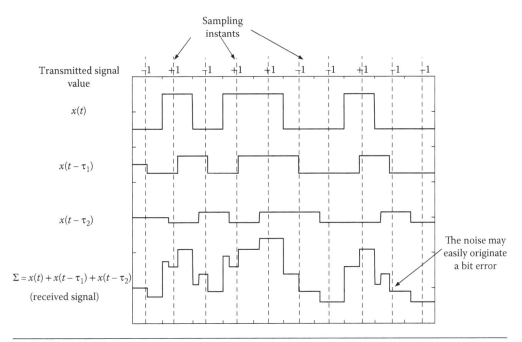

Figure 3.15 Plot of received signal through different paths and the resulting received signal.

frequency domain as having two sinusoids with frequency separation greater than the channel coherence bandwidth, being affected differently by the channel (in terms of attenuation and delay/phase shift) [Benedetto 1987].

The channel coherence bandwidth is the bandwidth above which the signal presents frequency-selective fading, that is, different attenuations, different delays, and nonlinear phase shifts for different frequencies (the signal is severely distorted by the channel). In the case of frequency-selective fading, and observing this effect in the time domain, it can be concluded that different digital symbols suffer from interference from each other, whose effect is usually known as intersymbol interference. This can be viewed as a type of distortion, applicable to digital transmissions. This effect can be seen from Figure 3.15. It tends to increase with the increase in the signal's bandwidth (increase of data rates, according to the Nyquist theorem).

On the other hand, if the signal's bandwidth is within the channel coherence bandwidth, the channel is said to be frequency nonselective and the type of fading is characterized as flat fading. In this latter case, all frequencies fade in unison (i.e., different signal frequency components present the same attenuations and linear phase shifts). In this case, the channel does not originate ISI.

Since in real radio propagation scenarios there are always some level of multipaths, the decision to evaluate the type of channel depends on the average power of the received multipaths. When the channel profile has a normalized average power below 30 dB for nondirect path replicas, although these multipaths may present a delay higher than the RMS delay spread, it is normally assumed that we are in the presence of a single-path channel (frequency nonselective fading or flat fading).

On the other hand, when the channel profile has an average power of nondirect multipaths higher than this threshold, it is normally assumed that the channel presents selectivity in frequency. In this case, ISI is experienced in digital transmission of symbols when the symbol rate is sufficiently high. Note that no serious ISI is likely to be experienced if the symbol duration is longer than a several times the delay spread. This is the reason why higher symbol rate correspond to higher levels of ISI.

Figure 3.15 depicts the received signal without noise (represented by $\Sigma = x(t) + x(t-\tau_1) + x(t-\tau_2)$) being composed of the cumulative sum of the signal received through direct path (represented by $x(t)$) plus the signals received through two multipaths (represented by $x(t-\tau_1)$ and $x(t-\tau_2)$, respectively). Each multipath consists of a reflected ray in a certain surface, and the corresponding signal level depends on the reflection index $\Gamma(0 < \Gamma < 1)$. In the example of Figure 3.15, it was assumed $\Gamma = 0.5$ for the path 2 and the reflection coefficient assumed for path 3 was $\Gamma = 0.3$. Furthermore, the delay of the first multipath τ_1 corresponds to the delay between the first multipath and the direct path, whereas the delay of the second multipath τ_2 corresponds to the delay between the second multipath and the direct path.

A decision is to be taken by the receiver at sampling instants. As can be seen from Figure 3.15, in some cases, the resulting signal at sampling instants has a level above the one received through direct path, which results in constructive interference of the multipaths (e.g., the first symbol). Nevertheless, in other cases, the resulting signal has a level below the one received through direct path, corresponding to a destructive interference caused by ISI (e.g., second symbol). In these cases, a low noise power may be enough to make this sample resulting in erroneous symbol estimation.

There are measures that can be implemented to mitigate the effects of ISI, namely the use of equalization, channel coding with interleaving, antennas diversity, and frequency diversity [Proakis 1995].

In code division multiple access (CDMA) networks, the presence of multipaths is used by the receiver (RAKE receiver*) to explore multipath diversity.

Finally, it is worth noting that the variation of distance between transmitter and receiver also originates the Doppler effect, which results in a variation of the received carrier frequency relating to the transmitted one. The Doppler frequency is given by $f_D = d\theta/dt$ (variation of the wave's phase).

3.5.1.1 Nyquist Intersymbol Interference Criterion   As stated before, ISI can be caused by the frequency-selective channel. Nevertheless, ISI is also caused by nonoptimum sampling instant of the detector.

Typically, the transfer function of the channel and the transmitted pulse shape are specified, and the problem is to determine the transfer functions of transmit and receive filters so as to reconstruct the original data symbol. The receiver performs this by extracting and then decoding the corresponding sequence of channel coefficients

---

* The RAKE receiver is described in Chapter 7.

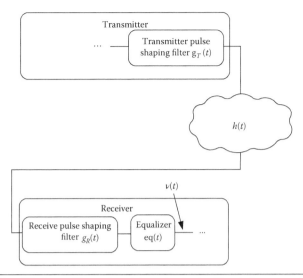

Figure 3.16 Positioning of the transmitting and receiving pulse shaping filters in the communication chain.

from the output $v(t)$ (see Figure 3.16). The extraction involves sampling the output $v(t)$ at time $t = iT_S$, where T_S stands for the symbol period. The decoding requires that the weighted pulse contribution $h(t)p(iT_S - kT_S)$ for $k = i$ be free from ISI due to the overlapping tails of all other weighted pulse contributions represented by $k \neq i$. This, in turn, requires that it controls the overall pulse $p(t)$,

$$p(iT_S - kT_S) = \begin{cases} 1, & i = k \\ 0, & i \neq k \end{cases} \tag{3.15}$$

where $p(0) = 1$ by normalization. If $p(t)$ satisfies the condition of Equation 3.15, the receiver output $v(t_i)$ (with $t_i = T_S$) implies zero ISI, that is, $v(t_i) = \mu(h)_i$ (for all i).

Since it is assumed that the pulse $p(t)$ is normalized such that $p(0) = 1$, the condition for zero ISI is satisfied if [Proakis 1995]

$$\sum_{n=-\infty}^{+\infty} P(f - nR_S) = T_S \tag{3.16}$$

Therefore, the Nyquist criterion for distortionless baseband transmission in the absence of noise can be stated: the frequency function $P(f)$ eliminates ISI, namely that caused by nonoptimum sampling instant, for samples taken at intervals T_S provided that it satisfies Equation 3.16. Note that $P(f)$ refers to the overall system, incorporating the transmit filter, the channel, the receive filter, and so on, such that

$$\mu P(f) = G_T(f)H(f)G_R(f) \tag{3.17}$$

In Equation 3.17, $G_T(f) = F[g_T(t)]$ and $G_R(f) = F[g_R(t)]$, where $g_T(t - nT)$ stands for the transmitting pulse shaping filter and $g_R(t - nT)$ for the receiving pulse shaping filter ($F(x)$ is the Fourier transform of x). $H(f)$ stands for the channel frequency response. The above system and signal description refers to Figure 3.16, where the carrier modulator and demodulator, as well as the symbol modulator and demodulator, were not depicted for the sake of simplicity.

Besides nonoptimum sampling instant, the ISI generated due to the frequency-selective fading (i.e., multipath channel) is usually not completely removed by pulse shaping but by different techniques such as equalization.

In order to assure that ISI is not present at the receiver due to nonoptimum sampling instant, the Fourier transform of the signal at the equalizer's output $V(f)$ must be described by a function that satisfy the Nyquist ISI criterion. In other words, if a communication channel satisfies the Nyquist ISI criterion, the received signal is free of ISI originated by nonoptimum sampling instant. A possibility to follow the Nyquist ISI criterion consists of assuring that the signal $V(f)$ presents a pulse shaping filter which follows the raised cosine function, defined as follows [Proakis 1995]:

$$P(f) = \begin{cases} T_S & |f| \le f_N(1-\alpha) \\ \dfrac{T_S}{2}\left[1 - \sin\left(\dfrac{2\pi f_N}{\alpha}|f| - \dfrac{\pi}{2\alpha}\right)\right] & f_N(1-\alpha) \le |f| \le f_N(1+\alpha) \quad (3.18) \\ 0 & |f| \ge f_N(1-\alpha) \end{cases}$$

In Equation 3.18, α stands for the roll-off factor, taking values between 0 and 1; it indicates the excess bandwidth over the ideal solution, $f_N = 1/(2T_S)$, and is usually expressed as a percentage of the Nyquist frequency f_N. Specifically, the baseband transmission of absolute values B_T is defined by $B_T' = (1+\alpha)/2T_S$. However, in the case of bandpass (carrier modulated) transmissions, earlier negative baseband part of the spectrum becomes positive, being also transmitted. Therefore, the transmitted bandwidth is $f \in [-W + W]$, which is defined by $B_T = 2 \cdot B_T'$, that is,

$$B_T = 2 \cdot \left[\frac{1+\alpha}{2T_S}\right] \qquad (3.19)$$

For the special case of single side band (SSB) transmissions, only the positive (or negative) part of the baseband spectrum is transmitted, making $B_T = B_T'$.

As can be seen from Figure 3.17, for roll-off factor $\alpha = 0$ and SSB transmission, we obtain the minimum bandwidth capable of transmitting signals with zero ISI defined by

$$\begin{aligned} B_{min} &= f_N \\ &= \frac{1}{2T_S} \\ &= \frac{R_S}{2} \\ &= \frac{R_b}{2}\log_2 M \end{aligned} \qquad (3.20)$$

where R_S stands for the symbol rate and R_b stands for the bit rate. Furthermore, M stands for the symbols constellation order and $\log_2 M$ for the number of bits transported in each symbol.

Note that Equation (3.20) corresponds to Equation (3.3). Furthermore, the spectrum of the pulse-shaping filter depicted in Figure 3.17 corresponds to Fourier transform of its impulsive response depicted in Figure 3.18. As can be seen from Table A.2, the Fourier transform of the sinc function corresponds to the rectangular pulse, that is, $F[\text{sinc}(2Wt)] = \dfrac{1}{2W}\Pi\left(\dfrac{f}{2W}\right)$ (valid for $\alpha = 0$). Since, for $\alpha \neq 0$, the pulse in the time domain is a variation of the sinc function, its Fourier transform may be viewed as a variation of the rectangular pulse.

It is also worth defining the spectral efficiency. Assuming a baseband signal of a M-ary constellation, the spectral efficiency becomes

$$\begin{aligned} \varepsilon &= \frac{R_b}{B_T} \\ &= \frac{2\log_2(M)}{1+\alpha} \end{aligned}$$

(3.21)

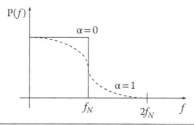

Figure 3.17 Raised cosine pulse in the frequency domain for $\alpha = 0$ and $\alpha = 1$.

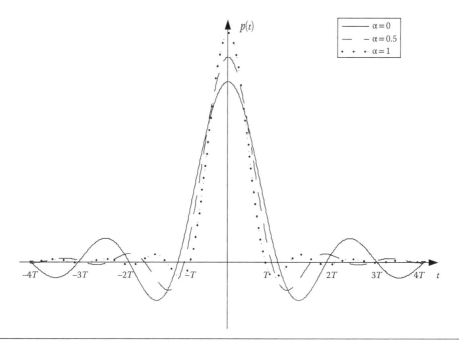

Figure 3.18 Raised cosine pulse in the time domain for $\alpha = 0$, $\alpha = 0.5$, and $\alpha = 1$.

Naturally, for the bandpass (carrier modulated) signal, the spectral efficiency becomes $\varepsilon = \dfrac{\log_2(M)}{1+\alpha}$.

In the case of binary transmission (i.e., a symbol such that each symbol transports a single bit), the minimum channel bandwidth is $R_b/2$. Naturally, the transfer function that leads to the minimum bandwidth with $\alpha = 0$ cannot be physically implemented. Consequently, the transmission bandwidth is always higher than the minimum bandwidth B_{min}.

The function $p(t)$ consists of the product of two factors: the factor $\operatorname{sinc}(\pi t/T_S)$ characterizing the ideal Nyquist channel and a second factor that decreases as $1/|t|^2$ for large $|t|$. The first factor ensures zero crossing of $p(t)$ at the desired sampling instants of time $t = iT_S$ with i an integer (positive and negative). The second factor reduces the tails of the pulse considerably below that obtained from the ideal Nyquist channel, so that the transmission of binary waves using such pulses is relatively insensitive to sampling time errors. In fact, for $\alpha = 1$, this leads to the most gradual roll-off in that the amplitudes of the oscillatory tails of $p(t)$ are smallest. Thus, the amount of ISI resulting from timing error decreases as the roll-off factor α increases from zero to unity. The special case with $\alpha = 1$ is known as the full-cosine roll-off characteristic. This response exhibits two interesting properties:

- At $t = \pm T_S/2 = \pm 1/4W$, we have $p(t) = 0.5$; that is, the pulse width measured at half amplitude is exactly equal to the bit duration T_S.
- There are zero crossings at $t = \pm 3T_S/2, \pm 5T_S/2, \ldots$ in addition to the usual zero crossings at the sampling times $t = \pm T_S, \pm 2T_S, \ldots$

These two properties are extremely useful in extracting a timing signal from the received signal for the purpose of synchronization. However, the price paid for this desirable property is the use of channel bandwidth double than that required for the ideal Nyquist channel corresponding to $\alpha = 0$.

Due to the smooth characteristics of the raised cosine spectrum, it is possible to design practical filters for the transmitter and the receiver that approximate the overall desired frequency response. In the special case of an ideal channel, that is, $H(f) = 1, |f| \le W$, it leads to

$$P(f) = G_T(f)G_R(f) \tag{3.22}$$

where $G_T(f)$ and $G_R(f)$ are the frequency responses (transfer function) of the transmit and receive filters, respectively, and $P(f)$ is the frequency response of the raised cosine pulse. Assuming that the receive filter is matched to the transmit filter, it leads to $P(f) = G_T(f)G_R(f) = |G_T(f)|^2$. Ideally,

$$G_T(f) = \sqrt{|P(f)|}\, e^{-j2\pi f t_0} \tag{3.23}$$

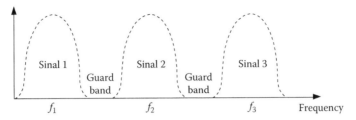

Figure 3.19 Adequate frequency control using guard bands between different channels.

and $G_R(f) = G_T^*(f)$, where t_0 is some nominal delay that is required to ensure physical implementation of the filter. Thus, the overall raised cosine spectral characteristic is split evenly between the transmitting and receiving filters. Note also that an additional delay is necessary to ensure the physical implementation of the receive filter.

Moreover, note that the PSD is proportional to $|G_T(f)|^2 = P(f)$, that is, $PSD \propto |G_T(f)|^2 = P(f)$.

The pulse shaping filter in the transmitter has the main function to allow the symbols formatting (in order to avoid ISI, as previously described) and to limit the spectrum inside the desired band, whereas the receive filter intends not only to contribute to format the symbols jointly with the transmitting pulse shaping filter but also to eliminate the noise outside the signal bandwidth, allowing only the reception of the noise inside the signal's bandwidth.

3.5.2 Multiple Access Interference

MAI occurs in networks that make use of multiple access techniques. This type of interference is experienced when there is no perfect orthogonality between signals from different users, viewed at the receiver's antenna of a certain user. In time-division multiple access (TDMA) networks, this orthogonality is normally assured through guard periods, which avoids the overlapping of signals transmitted in different time slots (from different users). In frequency-division multiple access (FDMA) networks, this orthogonality is assured through the use of guard bands (see Figure 3.19) and through the use of filters that reject undesired in-band interferences. In CDMA networks, this kind of interference is normally present in real scenarios and represents the main limitation of CDMA networks. MAI exists in CDMA networks due to the following:

- The use of spreading sequences that are not orthogonal (nonzero cross-correlation between different spreading sequences).
- Even using orthogonal spreading sequences, the orthogonality between spreading sequences is not assured when the network is not synchronized.*

---

* The uplink of a cellular network is normally asynchronous, that is, the symbols' transmissions from different mobiles do not start at the same instant.

For this reason, the use of quasi-orthogonal spreading sequences is often preferable,* when in the presence of an asynchronous network.

- Even in the downlink of a cellular network and even with the use of orthogonal spreading sequences, the multipath channel profile (frequency-selective fading) originates a relative level of asynchronism in the network. This originates nonzero cross-correlation values between superimposed signals received from different multipaths. Consequently, MAI is also present in this scenario.

In CDMA networks, the MAI is directly related to the received power from different users. A certain user, with an excessive power, originates a higher level of MAI than others. Therefore, in CDMA networks, the use of an effective power control to mitigate the fading effect as well as the near–far problem is essential.† The MAI can be reduced by using multiuser detection (MUD), power control, as well as sectored/adaptive antennas [Marques da Silva 2010].

3.5.3 Co-Channel Interference

This type of interference occurs when two different communications using the same channel interfere with each other. In a cellular environment, this occurs when a communication is interfered by another communication being transmitted in the same carrier frequency but in an adjacent cell. In cellular networks using TDMA/FDMA, this type of interference can be mitigated by avoiding the use of the same frequency band in adjacent cells, introducing the concept of frequency reuse factor higher than one.‡ In CDMA networks, this kind of interference is always present because the whole spectrum is reused in all cells, making the reuse factor as one. This reuse factor is normally adopted in CDMA networks because, although co-channel interference occurs, leading to a decrease of performance, the gain in capacity is higher than the decrease of performance. Furthermore, co-channel interference can be mitigated through the use of MUD and adequate power control. In CDMA networks, this kind of interference is also known as MAI, being however generated by users located in adjacent cells.

3.5.4 Adjacent Channel Interference

This type of interference consists of an inadequate bandwidth overlapping of adjacent signals. This is due to inadequate frequency control, transmission, broadband noise, intermodulation distortion (IMD), transmission with a bandwidth greater than the

---

* Quasi-orthogonal spreading sequences present some level of orthogonality. However, they present better autocorrelation properties in asynchronous networks than orthogonal spreading sequences (ex: Gold sequences).
† Near–far problem: a received signal originated from a transmission in the neighborhood is much more powerful than a received signal originated from a transmission made at a long distance.
‡ See Chapter 8.

one to which the operator is authorized, and so on. As depicted in Figure 3.19, the uses of guard bands are measures that are normally used to minimize the inadequate frequency control.

Since any transmitter's oscillator presents a certain level of broadband noise, direct interference may be generated from this source. The power generated by an oscillator presents typically a Gaussian shape around the desired transmitting frequency. Therefore, the energy out of the desired signal's spectrum is considered as broadband noise. Broadband noise consists of an unwanted signal transmitted in a frequency adjacent to the desired transmitted signal. This noise has a power very much below the transmitted signal (typically −155 dBc/Hz @ 3 MHz offset from carrier). Nevertheless, receiving a signal at short distance and very close in frequency to the transmitted signal may result in direct interference.

Spurious is another type of direct interference, being normally generated in transmitters. A receiver located at short distance from such transmission with spurious may result in a high-power interfering signal that may block one or more channels. The measure that can be used to mitigate this direct interference consists of keeping transmitting and receiving antennas sufficiently spaced apart to assure the required isolation. Furthermore, spurious transmission filtering is normally mandatory from frequency management regulators. Pre and postselector filters may also mitigate the negative effects of spurious.

IMD is another type of interference, consisting of undesired signal generated within nonlinear elements (transmit amplifier, receiver, low noise amplifier [LNA], multicoupler, etc.). Two or more signals present at this nonlinear element are processed, and an additional signal is generated, at sum and difference frequencies. Note that IMD can be of third order, fifth order, seventh order, and so on. Assuming that two isolated carriers f_a and f_b are present at a nonlinear element, the generated third-order intermodulation product becomes $(2f_a - f_b)$ and $(2f_b - f_a)$.

Finally, it is worth noting that IMD is considered as indirect interference, and its effect can be worse than direct interference. This can be mitigated with the use of postselector filters (at the transmitter) as well as preselector filters (at the receiver).

End of Chapter Questions

1. Which kind of channel impairments do you know?
2. What are the effects of channel impairments in either analog or digital signals?
3. Which kind of noise do you know?
4. What does ISI stands for?
5. What is the difference between adjacent channel interference and co-channel interference?
6. What is MAI?
7. Assuming baseband transmission, what is the minimum bandwidth necessary to accommodate a digital signal with a symbol rate of R_S?

8. Assuming passband (carrier modulated) transmission, what is the minimum bandwidth necessary to accommodate a digital signal with a symbol rate of R_S?

9. What does distortion stands for?

10. Which types of distortion do you know?

11. What does thermal noise stands for?

12. How can we quantify the FSPL?

13. What is the meaning of the Shannon capacity?

14. Assuming a twisted pair with a 1-MHz bandwidth, and a SNR of 5 dB, according to the Shannon capacity limit, what is the maximum speed of information bits that can be transmitted?

15. What is the difference between atmospheric noise and human noise? Characterize these two types of noise.

16. What does noise factor stands for?

17. How can we compute the noise factor of a system composed of a cascade of N electronic devices?

18. What is the ideal frequency response of a system, in terms of phase shift and attenuation?

19. How can we mitigate the effects of nonideal frequency response of a system, in terms of phase shift and attenuation?

4

CABLE TRANSMISSION MEDIUMS

The transmission medium that still dominates houses and offices is the twisted pair. In the past, local area networks (LAN) were made of coaxial cables, which were also used as a transmission medium for medium- and long-range analog communications. Although, its use in LAN was replaced by the twisted pair, the development of the cable television made the coaxial cable reused. With the improvement of isolators and copper quality and with the development of shielding, the twisted pair became widely used for providing high-speed data communications, in addition to the initial use for analog telephony. Currently, most of the companies use Internet protocol (IP) telephony, which represents a convergence between voice and data, using the same physical infrastructure. We are observing an increase in the demand for optical fibers, in both LAN, metropolitan area networks (MAN), and wide area networks (WAN) segments, due to its immunity to electromagnetic interferences and extremely high bandwidth.

As a rule of thumb, the attenuation of cable transmission mediums increases with the increase of the distance, but at a different rate for different transmission mediums (i.e., it is different for twisted pair, coaxial cable, optical fiber, etc.). Moreover, the attenuation and phase shift also tend to increase with the increase of the frequency, whose effect is more visible at longer distances. This results in distortion and, in the case of digital communications, it is viewed as intersymbol interference (ISI).* Consequently, it can be stated that the available bandwidth decreases with the increase of the link distance. Decreasing the link distance, the attenuation and phase shift at limit frequencies also reduces, which results in a higher throughput supported by the cable. Table 4.1 shows the typical bandwidths for different cable transmission mediums.

As described in Section 3.2, when the transmitter and the receiver are sufficiently far apart, amplifiers (used for analog signals) or regenerators (used for digital signals) need to be used at regular intervals to improve the signal-to-noise ratio, as well as to allow keeping the signal strength above the receiver's sensitivity threshold. The maximum distance where regenerators need to be placed depends on the cable transmission medium and on the bandwidth under consideration. Higher bandwidths require regenerators at shorter distances.

The following sections describe each of these important transmission mediums, in terms of use, bandwidth, attenuation, distortion, resistance to interference, and so on.

---

* As previously described, a way to mitigate this effect is by using an equalizer at the receiver.

Table 4.1 Bandwidths of Different Cable Transmission Mediums

	VOICE-GRADED TWISTED PAIR (GRADE 1)	TWISTED PAIR CATEGORY 6	COAXIAL	OPTICAL FIBER
Bandwidth	3.4 kHz	250 MHz	500 MHz	150 THz

4.1 Twisted Pairs

Low bandwidth twisted pairs, normally referred to as voice-graded twisted pairs, have been widely used for decades, at home and in offices, for analog telephony. Twisted pairs were also widely used to link houses and offices with local telephone exchanges. To reduce distortion, inductors (load coils) can be added to voice-graded twisted pairs, at certain distance intervals. This results in a flatter frequency response of the twisted pair, over the analog voiceband (300 Hz–3.4 kHz), which translates in lower attenuation level. Note that twisted pair with loading cannot be used to flatten the frequency response of twisted pair cables used for data, as the bandwidth of data communications (several megahertz or even gigahertz) is typically much higher than that of analog voice.

Due to the pre-existence of voice-graded twisted pairs in houses and offices, their use for data communications became a viable and inexpensive solution. Nevertheless, since they are very susceptible to noise, interferences, and distortion, these cables could not allow the data rates in use by most of LAN. The improvement of twisted pair's technology (shielding, twisting length, cable materials, etc.) increased the resistance to these impairments, which resulted in higher data rates. Consequently, these improved twisted pair characteristics, added to its reduced cost, resulted in a massive use of this physical infrastructure, instead of the previously used coaxial cable.

Nevertheless, comparing the twisted pair with coaxial and optical fiber, the distances and bandwidths reached with the twisted pair are below those obtained with coaxial and optical fiber.

4.1.1 Characteristics

As can be seen from Figure 4.1, a twisted pair is considered as a transmission line, being composed of two isolated and twisted conductors in a spiral pattern. The proper selection of the twisting length of these conductors leads to a reduction of low frequency interferences and cross talk.

Cross talk consists of an electromagnetic coupling of one conductor into another (wire pairs or metal pins in a connector). The electromagnetic field received by an adjacent conductor generates an interfering current, being superimposed to the signal's current. This originates a degradation in the signal-to-noise plus interference ratio.

The material used in conductors is normally the copper, whereas the polietilene is normally used as isolators. To improve the cross talk properties, twisted pairs are normally twisted and bundled in two pairs (four wires): one pair for transmission and another pair for reception (full duplex). To further improve the cross talk properties and to optimize

Figure 4.1 Twisted pair as composed of two isolated copper wires properly twisted.

the cables, two groups of two pairs (i.e., eight wires) are twisted and wrapped together, using a protective sheath. This results in cables composed of four pairs.

The quality of the twisted pair depends on several factors, such as the material and width of the isolator, the copper wire purity and width (typically between 0.4 and 0.9 mm), the twist length, the type of shielding (when used), the number of pairs twisted together, and so on. All these parameters define the impedance of the twisted pair, which results in a certain attenuation coefficient (dB per kilometer) and phase shift coefficient, both as a function of the signal's frequency. The maximum bandwidth and distance supported by a certain type of cable depends on these parameters. Naturally, due to increased resistance to interference, multipair cabling presents a bandwidth higher than single pair.

4.1.2 Types of Protection

Twisted pairs are normally grouped as unshielded twisted pairs (UTP), foiled twisted pairs (FTP), shielded twisted pairs (STP), or screened STP (S/STP) [ANSI/TIA/EIA-568].

As the name refers, UTP cabling is not surrounded by any shielding, whereas STP presents a shielding with a metallic braid or sheathing, which protects wires from noise and interferences.

The attenuation coefficient of a 0.5 mm copper wire UTP cable can be seen from Figure 4.2 (as a function of the frequency).

The resulting attenuation, expressed in decibel, is given by

$$A_{dB} = l \cdot \alpha(f) \tag{4.1}$$

where l stands for the cable length (in km) and $\alpha(f)$ for the attenuation coefficient (in dB/km).

UTP cabling is used in Ethernet and telephone networks, being normally installed during building construction. Since it does not present any shielding, UTP cabling is very subject to noise and external interferences, presenting typically an impedance of 100 ohm. UTP cabling is typically less expensive than STP and less expensive than coaxial and fiber optic cables, as well. Furthermore, STP and FTP are more difficult to handle than UTP. Consequently, a cost-benefit analysis needs to be made, before a decision is made about the type of cabling to use.

STP cabling includes a metal grounded shielding surrounding each pair of wires, presenting a typical impedance of 150 ohm. STP supports up to 10 Gbps, being considered in 10GBASET technology used to implement the IEEE802.3 LAN.

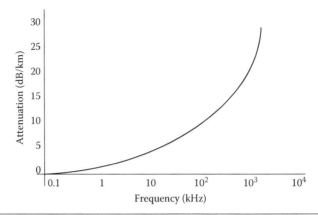

Figure 4.2 Attenuation coefficient of a twisted pair with 0.5 mm copper wires as a function of frequency. (From Keiser, G. 1991. *Optical Fiber Communications*. 2nd ed. McGraw-Hill. With permission.)

Table 4.2 Protection Type for Different Twisted Pair Cablings

	SHIELDING	SCREENING
UTP	No	No
FTP	No	Yes
STP	Yes	No
S/STP (S/FTP)	Yes	Yes

When the shielding is applied to multiple pairs, instead of a single pair of wires, it is referred to as screening. This is the case of FTP, being also referred to as screened UTP (S/UTP). It consists of a UTP cabling whose shielding surrounds the cable (screened), not presenting shielding in each pair of copper wires. Consequently, while this cabling presents good resistance to interferences originated from outside of the cable, the cross talk properties (interference between different pairs of the cabling) is typically poorer than STP.

Finally, screened FTP (S/FTP) cabling, also referred to as S/STP cabling, presents a shielding surrounding both individual pairs and the entire group of copper pairs, and therefore, it is both externally and internally protected from adjacent pairs (cross talk). The above description is summarized in Table 4.2.

4.1.3 Categories

Another way to characterize twisted pair cabling is grouping them into different categories, from 1 to 7. As can be seen from Table 4.3, increased cabling category results in a higher bandwidth and data rate. The better performance is achieved at the cost of better and thicker copper wires, isolation, improved shielding, or improved twisting. Consequently, higher bandwidths tend to correspond to higher costs. With the exception of the voice-graded twisted pair, cables of the other categories comprise four pairs of conductors.

Table 4.3 Characteristics of Different Twisted Pair's Categories (@ 100 meters)

	CATEGORY 1	CATEGORY 2	CATEGORY 3	CATEGORY 4	CATEGORY 5	CATEGORY 6	CATEGORY 7
Cable type	Voice-graded TP	UTP (multipairs voice graded)	UTP (multipairs voice graded)	UTP	UTP/FTP	UTP/STP/FTP	S/STP/S/FTP
Copper connector/ termination	Not specified	8P8C/T568A	8P8C/T568B	Not specified	8P8C/T568B	8P8C/T568A or 8P8C/ T568B	8P8C/T568B
Bandwidth	3.4 kHz	6 MHz	16 MHz	20 MHz	100 MHz	250 MHz	600 MHz
Maximum data rate	Analog voice	4 Mbps	12 Mbps	16 Mbps	100 Mbps	1 Gbps	40 Gbps @ 40m

Table 4.4 T568A Termination of 8P8C Modulator Connectors ("RJ-45") [ANSI/TIA/EIA-568-A]

PIN	PAIR	COLOR
1	3	White/green
2	3	Green
3	2	White/orange
4	1	Blue
5	1	White/blue
6	2	Orange
7	4	White/brown
8	4	Brown

Although the standards [ANSI/TIA/EIA-568-A] and [ISO/IEC 11801] only recognize categories 3 up to 6, categories (or grade) 1, 2, and 7 are also listed in Table 4.3, as these designations are normally assigned to cabling configurations. Table 4.3 lists the cabling and connectors, as well as the corresponding bandwidths and maximum data rates. Note that the copper connectors listed in Table 4.3 are defined in Tables 4.4 and 4.5, respectively, for T568A* and T568B† terminations of 8P8C modular connectors (often referred to as "RJ-45").

As previously described, the maximum bandwidth supported by a cable transmission medium depends on the link distance. Shorter distances allow accommodating higher bandwidths and reciprocally. This results from the fact that different frequencies present different attenuation coefficients and different delays, whose effect is more visible at longer distances. In fact, the attenuation of a twisted pair increases approximately exponentially with the increase of the frequency. The effect that results from this impairment is known as distortion (attenuation and/or phase distortion) and, in the case of digital communications, results in ISI. Improved twisted pair quality results in longer distances for the same bandwidth or higher bandwidth for the same distances, as compared to lower quality twisted pair. As a rule of thumb, for digital signals, there is the need to use regenerators at a distance interval of 2–3 km of twisted pair cable.

---

* T568A is the designation of the 8P8C termination standardized by [ANSI/TIA/EIA-568-A].
† T568B is the designation of the 8P8C termination standardized by [ANSI/TIA/EIA-568-B].

Table 4.5   T568B Termination of 8P8C Modulator Connectors ("RJ-45") [ANSI/TIA/EIA-568-B]

PIN	PAIR	COLOR
1	2	White/orange
2	2	Orange
3	3	White/green
4	1	Blue
5	1	White/blue
6	3	Green
7	4	White/brown
8	4	Brown

The bandwidths listed in Table 4.3, for different categories, are those specified for 100 m of distance (90 m of cable plus 10 m of patch cord). These values may be exceeded for shorter distances.

Category 1 UTP consists of low quality twisted pairs specified for analog voice only (without external isolation). The twisting of multiple pairs, previously individually twisted, into the same cable originates the category 2 UTP. Some authors also consider category 2 and 3 as voice-graded twisted pair. Nevertheless, the exceeding bandwidth, besides that of the analog voice, is used for data communications. Category 3 and 4 UTP cabling is similar to category 2, but with improved copper and isolation, as well as using a twisting length that improves the resistance against noise and interferences. This results in the ability to support a bandwidth of 16 MHz for category 3 and 20 MHz for category 4 twisted pair cabling. Category 3 is considered in the LAN standard IEEE 802.3 (at 10 Mbps). Moreover, categories 3 and 4 are considered in the LAN standard IEEE 802.3u at 100 Mbps, using several parallel pairs.

Category 5, while consisting of either UTP or FTP, is currently installed during construction in most offices. It supports a bandwidth of 100 MHz, being considered by the LAN standard IEEE 802.3u (at 100 Mbps) and by IEEE 802.3ab (at 1 Gbps), in this latter case using four negotiated parallel pairs for transmitting or receiving.

Category 6 supports 250 MHz of bandwidth and a data rate of 1 Gbps, being based on UTP (adopted by the LAN standard IEEE 802.3ab), STP, or FTP cabling (IEEE 802.3z). There is a variation, referred to as Category 6a, which allows twice the bandwidth of category 6, that is, 500 MHz.

Finally, category 7 is defined to support 600 MHz of bandwidth and data rates as high as 40 Gbps (contrarily to the other categories, the listed value refers to 40 meters). This is achieved using S/STP, which makes use of double shielding, resulting in a high level of immunity to noise and interferences. There is a variation, referred to as category 7a, defined to support frequencies up to 1 GHz.

Tables 4.4 and 4.5 list the T568A and T568B terminations layout for 8P8C modulator connectors.

The IEEE 802.3 cabling, also referred to as Ethernet cabling, may have two different basic configurations as follows:

- Straight cable: typically used to interconnect devices of different types, namely
  - A switch/hub to a computer or network printer
  - A router to a modem
- Crossover cable: typically used to interconnect devices of same types, namely
  - Two computers
  - Two switches/hubs
  - Two routers
  - A router to a switch/hub

The visual identification of straight and crossover cables is simple. The wire arrangement of both cable terminals of straight cables is the same, whereas the wire arrangement of the two sides of crossover cables is different. Pins 1 and 2* are used for transmit, whereas pins 3 and 6† are used for receive. The transmit pins in one terminal of a crossover cable becomes the receive pins in the other terminal, and reciprocally. Consequently, pins 1 and 2 in terminal A of a crossover cable becomes pins 3 and 6 of the terminal B, and pins 3 and 6 of terminal A becomes pins 1 and 2 of terminal B.

4.2 Coaxial Cables

Coaxial cables were the main transmission medium of long-range analog transmission, namely between local telephones exchange. They supported high capacity communications as defined by [ITU-T G.333] recommendation (10800 telephone channels, with a maximum frequency of 60 MHz). The utilization of coaxial cables in telephone networks has been replaced by optical fibers. The initial use of coaxial cables as LAN infrastructure was also replaced, in this case by twisted pair cables. Currently, coaxial cables are used for either analog or digital signals. Its use in cable television networks (analog and digital) is still of high importance. Furthermore, this physical infrastructure is currently used for video distribution between devices, namely between an antenna and a television receiver or a DVD recorder, and so on. Its use in LAN is limited to special applications. Nevertheless, the current trend is to use more and more optical fibers.

4.2.1 Characteristics

A coaxial pair consists of two concentric conductors, an inner conductor and an outer conductor. As can be seen from Figure 4.3, the inner conductor is isolated and centered with the help of a dielectric material or isolator. The outer conductor is cylindrical

---

* Pair 3 of T568A termination or pair 2 of T568B termination.
† Pair 2 of T568A termination or pair 3 of T568B termination.

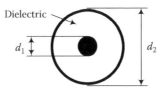

Figure 4.3   Coaxial cable as composed of two isolated and concentric conductors.

shape, being covered with a shield or jacket. Coaxial pairs are grouped in low number, being wrapped into a cable, protected from traction forces.

The design of coaxial cable leads to good electromagnetic isolation, which translates into good resistance properties against noise and interferences. Consequently, the bandwidth of a coaxial cable is typically higher than that of a regular twisted pair. The characteristic impedance of coaxial cables is 75 ohm.

The structure of coaxial cables almost eliminates the possibility of coupling between parallel coaxial pairs of the same cable. Albeit galvanic coupling may exist between external conductors, the cross talk properties of this transmission medium are excellent for operating frequencies above 60 kHz. Below this frequency, coaxial cables cannot be used as it captures high level of external interferences (high cross talk) and suffers heavily from distortion.

It can be shown that the attenuation is minimized for a relationship between the outer conductor diameter d_2 and the inner conductor diameter d_1 equal to 3.6. As can be seen from Table 4.6, coaxial cables normalized by ITU-T present a relationship close to that value.

The attenuation coefficient of a 2.6/9.5 mm coaxial cable, standardized by [ITU-T G.623], is approximated by

$$\alpha \cong 0.01 + 2.3\sqrt{f} + 0.003\,f \tag{4.2}$$

where the attenuation coefficient α is expressed in dB/km and the frequency f is expressed in MHz. Note that Equation 4.2 is an approximation valid for frequencies above 1 MHz. The resulting function is plotted in Figure 4.4.

Having as a reference an attenuation coefficient variation from very low values up to 30 dB/km, we observe from Figure 4.4 that this corresponds to an approximate bandwidth of 400 MHz, whereas in the case of the twisted pair cable (Figure 4.2), this corresponds to about 2 MHz. From these results, we conclude that the level of distortion typically introduced by a coaxial cable is much lower than that of a twisted pair. Consequently, the bandwidth available in a coaxial cable is higher than that of a twisted pair. The typical bandwidth provided by a coaxial cable is 500 MHz. Note that since the attenuation coefficient of a coaxial cable carrying a high bandwidth is high (30 dB/km or higher), the distance between adjacent repeaters or regenerators is reduced. Conversely, carrying a lower bandwidth in a coaxial cable leads to lower attenuation levels, which translates in a longer distance between adjacent repeaters.

Table 4.6 Dimensions of Normalized Coaxial Cables

ITU-T RECOMMENDATION	[ITU-T G.623]	[ITU-T G.623]	[ITU-T G.623]
d_1 (mm)	2.6	1.2	0.7
d_2 (mm)	9.5	4.4	2.9
d_2/d_1 (mm)	3.65	3.67	4.14

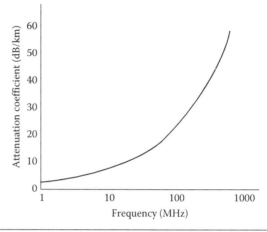

Figure 4.4 Attenuation coefficient of 2.6/9.5 mm coaxial cable as a function of frequency.

4.3 Optical Fibers

Since the seventies, optical fibers have become one of the most important transmission mediums used in medium- and long-range communications. Their use at short range has also taken place, instead of the previously used coaxial cables and twisted pairs. Typical use of optical fibers includes WAN and MAN, namely as the transmission medium of synchronous digital hierarchy (SDH), synchronous optical network (SONET), or asynchronous transfer mode (ATM). Their applications also include the interconnection between different local telephone exchanges, as well as the interconnection of local telephone exchanges and cable television with a street cabinet that is typically located within few hundred meters from homes (fiber to the curb), with a patch cord box just outside of a home (fiber to the home), or even with a home itself or a small business (fiber to the premises).

Optical fibers have been of high importance to implement the collaborative era of telecommunications principle defined in Chapter 1. In the LAN environment, we have observed the gradual replacement of twisted pairs by optical fibers. IEEE 802.3z (namely the 1000Base-SX and 1000Base-LX) is an example of LAN standard, which includes optical fibers as transmission medium. In fact, optical fibers have already been considered in 100 Gbps standards, and its use in 400 Gbps networks are under consideration [Winzer 2010]. Finally, optical fibers are starting to be used to interconnect different modules of electronic equipment (e.g., of a radio transmitter or receiver). This results in high performance equipment.

4.3.1 Characteristics

The emergence of optical fibers, as compared to other cable transmission mediums, was due to the following reasons:

- Extremely high bandwidth, which translates in high throughputs (typically 150 THz).
- Low attenuation coefficient, as compared to twisted pairs, or even to coaxial cables. Moreover, as opposed to other cable transmission mediums, the attenuation coefficient of an optical fiber varies very smoothly with the frequency, which translates in low level of distortion.
- Longer repeating distances, which results from the low level of attenuation. As an example, SDH networks include typically regenerators only every 60 km distance, whereas the regeneration distance of twisted pairs and coaxial cables is about 1–2 km.
- Immunity to electromagnetic interferences, which results from the fact that optical fibers are typically made of fiber of silica glass (S_iO_2). Since this substance is not an electrical conductor, optical fibers do not suffer from cross talk.
- Small dimensions and low weight: an optical fiber presents typically a diameter 10 times smaller than that of a coaxial cable and its weight is typically 30 times lower than that of a coaxial cable.
- Greater capacity: this results from the fact that a duct previously used for coaxial cables can accommodate 10 times more optical fibers, and because the capacity of an optical fiber is much higher than that of a coaxial cable (typically 300,000 times higher).
- Reduced cost: silica is one of the most abundant materials in the earth, which makes it much less expensive than copper.

An optical fiber is made of a dielectric guide capable of guiding an optical ray, composed of a core made of glass or plastic with refraction index n_1 and a cladding with refraction index n_2, where $n_2 > n_1$ [Keiser 1991] (see Figure 4.5). For the sake of protection, this dielectric guide is surrounded by a jacket that is normally made of plastic. A light pulse is propagated along the core of a fiber, between a light emitter and a light receiver.

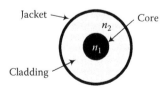

Figure 4.5   Optical fiber as composed of a core and a cladding, surrounded by a jacket.

4.3.2 Categories

There are two main types of optical fibers: single mode and multimode. The former type of fibers supports a single propagation mode, whereas the latter supports several modes. As can be seen from Figure 4.6, in the single mode, a single path exists over the fiber length, between the emitter of the light pulse and the corresponding receiver. In this case, there are no rays propagated in any direction different from the normal to the fiber section. Conversely, in the multimode fiber, several paths exists between the light emitter and the receiver. As can be seen from Figure 4.6, multimode fibers have several rays propagated with different angles, from the normal to the fiber section until a certain critical angle above which reflection does not occur. Different rays are reflected at the border between the core and the clading, following the reflection principles described in Chapter 5, with a reflected angle equal to the incident angle. Consequently, an emitted pulse arrives at the receiver through several different paths, with different delays. This results in a pulse dispersion, which can be viewed as ISI. As the level of dispersion increases with the link distance, multimode fibers are normally limited to short range applications. This reduced range is also a result of the higher attenuation of multimode fibers, as compared to single mode (see Figure 4.7).

There are two different types of multimode fibers: step index and graded index. The core of multimode step index fiber presents a single refraction index, whereas the graded index fiber presents an index that decreases from the center to the extremities of the core. The refraction index profile is depicted in Figure 4.6. In the graded index fiber, different paths still exist, but they converge to the same point. Therefore, as in the case of a step index fiber, the receiver of a graded index fiber detects a signal that is composed of the superimposition of several replicas

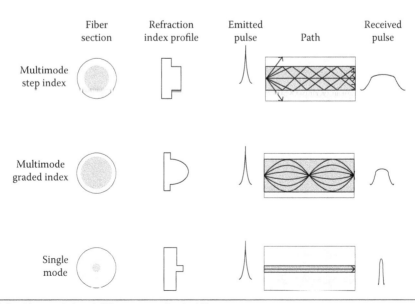

Figure 4.6 Characteristics of different types of optical fibers.

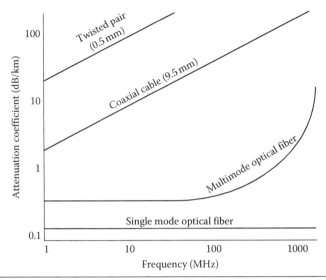

Figure 4.7   Attenuation coefficient of a typical optical fiber versus twisted pair and coaxial cable.

(i.e., the same signal from different paths), but these replicas tend to be aligned in time, that is, they tend to be synchronized. The level of alignement is never perfect, which results in a certain level of pulse dispersion. However, the level of pulse dispersion is smaller than in the case of step index fiber, which results in an increased bandwidth. Note that, as a result of the gradual refraction index profile, the rays suffer a refraction or bending effect, as they propagate from the center to the extremities of the fiber's core. The refraction phenomenon is modeled by the Snell's law, defined by

$$n_1 \sin \theta_1 = n_2 \sin \theta_2 \tag{4.3}$$

where n_1 and n_2 stands for the refraction index of the core and the cladding of the optical fiber, respectively. Moreover, θ_1 and θ_2 stands for the wave's direction, measured from the normal to the border between the two materials (see Figure 4.8). A wave that crosses a border between the medium n_1 into the medium n_2 suffers a refraction, that is, a deviation or bending in the wave's direction. After refraction, the new angle of the wave's direction becomes

$$\theta_2 = \arcsin\left(\frac{n_1 \sin \theta_1}{n_2}\right) \tag{4.4}$$

Assuming that $\theta_2 = \theta_1 - \delta$, with δ the deviation angle, we have

$$\delta = \theta_1 - \arcsin\left(\frac{n_1 \sin \theta_1}{n_2}\right) \tag{4.5}$$

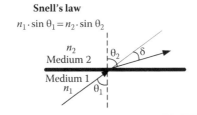

Figure 4.8 Refraction effect.

Single mode fibers present a core diameter smaller than that of multimode fibers, which makes it more difficult to handle than multimode fibers. Moreover, as can be seen from Figure 4.7, the attenuation coefficient in any of the optical fiber's transmission windows presents a very low variation, as compared to the attenuation coefficient variation of twisted pairs or coaxial cables.*

This results in a low level of distortion, which translates in a high bandwidth. Furthermore, since a single mode is propagated over a single path, the level of dispersion is very low, confirming the enormous bandwidth made available by this type of optical fibers. This makes single mode fibers well fitted for long-range links. Note that the bandwidth of the single mode fibers is much higher than that of multimode. Nevertheless, single mode fibers are more difficult to handle as the core diameter is smaller.

The core diameter varies from around 10 μm for single mode optical fibers up to about 200 μm for multimode step index fibers. The cladding diameter varies from about 125 μm for single mode fibers up to about 400 μm for the multimode step index fibers.

As previously described, the received pulse is typically wider than the emitter pulse. This is due to dispersion, which can be classified as intermodal and intramodal. Intermodal dispersion results from the fact that different propagation modes present different propagation times, whereas the intramodal is due to the fact that different wavelengths present different propagation times. Single mode optical fibers present only intramodal dispersion, whereas multimode fibers experience both types of dispersion.

The devices used to emit a beam of light in optical fibers can be either the light emitting diode (LED) or the injection laser diode (ILD). In high data rate links, the ILD is normally adopted, but it is more expensive. The modulation of these devices is achieved by varying the polarization current at their terminals. The variation of the light's intensity is directly proportional to the current variation. The data exchange is achieved in optical fibers with a variation of the light intensity. The most used modulation scheme is the amplitude shift keying, where the different amplitudes correspond to different light intensities. Note that the emitted light is not

---

* Note that the scale of the ordinate of Figure 4.7 is logarithmic.

monochromatic, presenting a certain spectral width σ_λ, whose value is lower for ILD than for LED.

A photodiode is normally used as a detector to convert light energy into electrical energy. There are two main types of photodiodes: P intrinsic N (PIN) photodiode and Avalanche photodiode (APD). PIN photodiodes are less sensitive and less expensive than APD.

The intramodal dispersion is a consequence of this spectral width (σ_λ). The resulting optical pulse dispersion is given by [Keiser 1991]

$$\sigma_t = |D_\lambda| \cdot l \cdot \sigma_\lambda \tag{4.6}$$

where D_λ is a dispersion parameter characteristics of the fiber under consideration and l is the optical fiber length.

The optical bandwidth of a fiber at –3 dB becomes [Keiser 1991]

$$B_0 = \frac{0.187}{\sigma_t} \tag{4.7}$$

Since, from the Nyquist ISI criterion, the minimum electrical baseband bandwidth B_e that can accommodate the symbol rate R_S is given by $B_e = \frac{R_S}{2}$, and once the electrical bandwidth is related to the optical bandwidth by $B_0 = \sqrt{2} \cdot B_e$, the relationship between the symbol rate and the optical pulse dispersion becomes

$$R_S = \frac{0.264}{\sigma_t} \tag{4.8}$$

Figure 4.9 shows the attenuation coefficient for different wavelengths of a typical fiber. Depending on the application and fiber's type, there are different wavelengths used in data transmission. Note that these transmission windows were selected based on two important requirements: flatness of the attenuation profile over the different wavelengths and low level of attenuation that translates in high ranges without the need to use regenerators.

The first transmission window uses a wavelength surrounding the 850 nm, being typically considered by multimode fibers in LAN applications. The second window, also referred to as S band, uses a wavelength around 1300 nm, being typically considered by single mode fibers in various applications. The third and fourth windows (C and L bands) use a wavelength around 1550 and 1600 nm, being normally considered by single mode fibers in long-range links [Stallings 2010]. Note that 1300 and 1550 nm windows correspond to the infrared light.

It is worth noting that a carrier wave transmitted in optical fibers is identified by its wavelength, instead of its frequency. In electric and electromagnetic propagation,

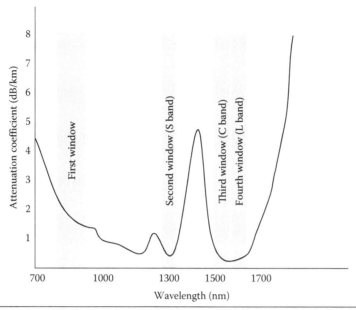

Figure 4.9 Attenuation coefficient of a typical optical fiber as a function of the wavelength.

we typically assume that the propagation speed is $v = c$, that is, we assume the propagation speed of the wave in the medium under consideration is the same as the propagation speed of the wave in the vaccum. Consequently, the approximation that relates wavelength and frequency becomes $c = \lambda \cdot f$, with $c = 3 \times 10^8$ m/s. In fact, the true relationship between the wavelength λ and the frequency f is $v = \lambda \cdot f$, where $n = c/v$. Since the refraction index n of an optical fiber is sufficiently different from 1 (the refraction index in the vaccum), the propagation speed of the light in an optical fiber v is sufficiently different from the propagation speed of the light in the vaccum c. In fact, different fibers have different refraction index. Consequently, the equivalence that should be used is $v = \lambda \cdot f$. Therefore, referring to frequency does not have the same correspondency as in the case of electric or electromagnetic waves, having been adopted the wavelength unit in optical fibers. Note that a certain frequency value has different meanings in two optical fibers with different refraction indexes. Contrarily, a certain wavelength value is always applicable, regardless the refraction index.

The aforementioned reasons also justify the fact that the multiplexing performed in optical fibers is known as wavelength division multiplexing (WDM) instead of frequency division multiplexing (FDM).* Note that WDM is the multiplexing technique normally used in MAN/WAN networks.

The reader should refer to Chapter 10 for the description of the physical layer used in different LAN and MAN technologies.

---

* The reader should refer to Chapter 6 for a description of FDM.

End of Chapter Questions

1. What are the advantages of the optical fibers relating to the twisted pairs?
2. Which kind of cable transmission mediums do you know? What are their main characteristics?
3. In terms of isolation, which types of twisted pair cabling do you know?
4. Which categories of twisted pair cabling exist?
5. What is the difference between the FTP and the STP cablings?
6. What is the difference between the STP and the S/STP cablings?
7. What are the key parameters that allow different twisted pair cabling presenting different bandwidths and attenuations?
8. What is the relationship between range and distortion of cabling transmission mediums?
9. Which types of optical fibers do you know?
10. How does a single mode optical fiber achieve a higher bandwidth than multimode?
11. Which types of multimode optical fibers do you know? Which one present higher bandwidth and how is it possible?
12. What are the typical applications of single mode and multimode optical fibers?

5

WIRELESS TRANSMISSION MEDIUMS

A typical radio communication system is composed of a transmitter and a receiver, which use antennas to convert electric signals into electromagnetic waves and vice versa. Those electromagnetic waves are propagated over the air.

In case the communication system is digital, a modem needs to be added at each end of the link. The modem is responsible for modulating and demodulating the bits using a certain modulation scheme (e.g., amplitude shift keying, frequency shift keying, phase shift keying, quadrature amplitude modulation). In addition, the modem is normally responsible for the implementation of error correction techniques and bit and frame synchronization. The output of the modem in the transmission chain consists of a certain bandpass (carrier modulated) signal (analog), which serves as input signal to the radio transmitter. The radio transmitter is responsible for centering this signal around a certain carrier frequency and amplifying it before delivery to the transmitting antenna. The reverse of these operations is performed in the receiving chain.

Figure 5.1 shows a conventional client/server communication system established over a radio channel (i.e., over the air). The propagation channel can be of any type, namely using direct wave propagation, surface wave propagation, or ionospheric propagation. Moreover, the communication system can be of any type, namely a radio broadcast, a terrestrial microwave system, a satellite communication system, and so on. This chapter deals with different types of electromagnetic propagation and different types of radio communication systems.

5.1 Wireless Propagation

5.1.1 Direct Wave Propagation

5.1.1.1 Free Space Path Loss   Free space path loss (FSPL) is defined as the loss in the electromagnetic signal strength due to free space propagation (line of sight) between a transmitting and a receiving antenna. Before FSPL is deducted, it is worth defining the power spatial density S as a function of the distance d and of the transmitting power P_{E}.[*]

Assuming the spherical propagation of electromagnetic waves, the power spatial density[†] becomes [Carlson 1986]

$$S = P_{\mathrm{E}} \frac{1}{4\pi d^2} \qquad (5.1)$$

---

[*] P_{E} is expressed in Watt.
[†] S is expressed in Watt/m².

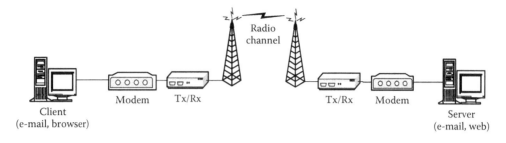

Figure 5.1   Conventional client/server communication system established over a radio channel.

Note that $4\pi d^2$ is the surface area of the sphere. It is viewed from Equation 5.1 that the transmitting power is spread out over the space. Moreover, this space (surface area of the sphere) increases with the distance d.

A common definition normally used in link budget calculations is the equivalent isotropic radiated power (EIRP). It is defined as the amount of power captured by an isotropic receive antenna* at a distance d, assuming a transmit power P_E and a transmitting antenna gain g_E (in the direction of maximum antenna gain):

$$
\begin{aligned}
\text{EIRP} &= g_E S \frac{\lambda^2}{4\pi} \\
&= P_E g_E \left(\frac{\lambda}{4\pi d}\right)^2
\end{aligned}
\tag{5.2}
$$

where the FSPL becomes

$$
\text{FSPL} = \left(\frac{4\pi d}{\lambda}\right)^2
\tag{5.3}
$$

and where $\lambda^\dagger$ stands for the wavelength.

Note that $A_R(r)$ (with $A_R(r) = g_R(r)\dfrac{\lambda^2}{4\pi} \iff g_R(r) = 4\pi\dfrac{A_r(r)}{\lambda^2}$) corresponds to the receiving antenna aperture in the direction of the emitter (r), relating to an isotropic antenna, whereas $g_R(r)$ corresponds to the receiving antenna gain in the direction of the emitter (r), relating to an isotropic antenna.

Using the approximation that electromagnetic waves propagate at the speed of light c in vacuum ($c \approx 3 \times 10^8\,\text{m/s}$) and using the equivalence $c = \lambda f$, the FSPL becomes [Carlson 1986]

$$
\text{FSPL} = \left(\frac{4\pi df}{c}\right)^2
\tag{5.4}
$$

---

* With antenna aperture $A_R = \dfrac{\lambda^2}{4\pi}$ in all directions.

$\dagger$ λ is expressed in m.

From Equation 5.4, we conclude that the path loss is proportional to the square of the distance d and to the square of the frequency f.

Since link budget calculations are normally performed in logarithmic units (decibel), instead of linear units, it is worth expressing FSPL in decibel as

$$FSPL_{dB} = 10\log_{10} FSPL$$
$$= 10\log_{10}\left[\left(\frac{4\pi df}{c}\right)^2\right]$$
$$= 20\log_{10}\left(\frac{4\pi df}{c}\right) \tag{5.5}$$
$$= 20\log_{10}\left(\frac{4\pi}{c}\right) + 20\log_{10}(d) + 20\log_{10}(f)$$
$$= -147.55 + 20\log_{10}(d) + 20\log_{10}(f)$$

Expressing d in kilometers and f in megahertz, Equation 5.5 becomes

$$FSPL_{dB} = 32.45 + 20\log_{10}(d) + 20\log_{10}(f) \tag{5.6}$$

5.1.1.2 Link Budget Calculations   The received power is defined by the Friis formula

$$P_R = EIRP \cdot g_R$$
$$= g_E S \frac{\lambda^2}{4\pi} g_R \tag{5.7}$$
$$= P_E g_E \left(\frac{\lambda}{4\pi d}\right)^2 g_R$$

where g_R stands for the receiving antenna gain [Carlson 1986]. Alternatively, we could express Equation 5.7 as a function of the frequency as

$$P_R = P_E g_E \left(\frac{c}{4\pi f d}\right)^2 g_R \tag{5.8}$$

Another way to compute the received signal power is in logarithmic units as

$$(P_R)_{dBW} = 10\log_{10}\left[P_E g_E \left(\frac{c}{4\pi f d}\right)^2 g_R\right] \tag{5.9}$$
$$= (P_E)_{dBW} + G_E + G_R + 147.56 - 20\log_{10} f - 20\log_{10} d$$

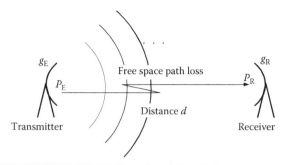

Figure 5.2   Generic diagram of a communication system with a line-of-sight propagation.

where $(P_E)_{dBW} = 10\log_{10}(P_E)$ and $(P_R)_{dBW} = 10\log_{10}(P_R)$, with P_E and P_R expressed in Watt.*

Furthermore, in Equation 5.9, G_E and G_R the transmitting and receiving antenna gains are expressed in decibel, that is, $G_E = 10\log_{10}(g_E)$ and $G_R = 10\log_{10}(g_R)$, respectively (Figure 5.2).

Expressing d in kilometers and f in megahertz, and using Equation 5.6, Equation 5.9 becomes

$$
\begin{aligned}
(P_R)_{dBW} &= (P_E)_{dBW} + G_E + G_R - \text{FSPL} \\
&= (P_E)_{dBW} + G_E + G_R - 32.45 - 20\log_{10}(d)_{km} - 20\log_{10}(f)_{MHz}
\end{aligned}
\tag{5.10}
$$

Observing Equation 5.6, it is clear that the increase in the carrier frequency leads to a higher path loss. From Equation 5.10, it can be seen that the higher path loss results in a weaker received signal. This limitation may be overcome by using transmitting and receiving antennas with higher gains.

The link budget calculation of a real system should also include additional losses (as cable losses, connector losses, etc.). Taking these parameters into account, Equation 5.10 becomes

$$
\begin{aligned}
(P_R)_{dBW} &= (P_E)_{dBW} + G_E + G_R + \text{Att}_{dB} \\
&= (P_E)_{dBW} + G_E + G_R + \left[\underbrace{-32.45 - 20\log_{10}(d)_{km} - 20\log_{10}(f)_{MHz} + \text{Att}_{add\,dB}}_{\text{Att}_{dB}} \right]
\end{aligned}
\tag{5.11}
$$

where Att_{dB} is a negative value, defined as the sum of different attenuations in decibel, namely the path loss and the additional attenuations, and where $\text{Att}_{add\,dB}$ is also a negative value, which stands for the additional attenuations (cable losses plus connectors losses, rain attenuation, etc.) in decibel. Moreover, note that the FSPL does not take into account any effect such as shadowing or multipath (i.e., there is no

---

* A dBW is a decibel relating to the Watt. Similarly, a dBm is a decibel relating to the milliWatt, that is, $(P_E)_{dBm} = 10\log_{10}P_E$, with P_E expressed in milliWatt.

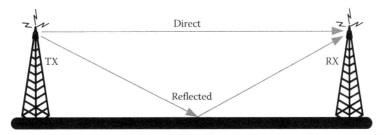

Figure 5.3 Received signal composed of a direct component and a reflected component.

diffraction, reflection, or scattering effects), as it refers to free space propagation. The FSPL has a distance decay rate 2,* whereas in real scenarios this value varies between 3 and 5. Rural scenarios present typically a decay rate of the order of 3, whereas decay rate varies between 4 and 5 for urban scenarios, depending on the shadowing effects, multipath propagation scenario, and so on. Therefore, depending on the propagation environment, the received power calculation in Equation 5.11 should take this modified path loss parameter into account.

The previous description was made taking into account the free space propagation, that is, including only direct path and that reflected, diffracted, and scattered wave components were not present. In addition, antenna heights were not taken into account in the calculations. Therefore, assuming the flat earth (i.e., neglecting earth curvature) and that the received signal is composed of a direct path to which one reflected path in the ground is summed (as depicted in Figure 5.3) and taking into account the antennas heights, Equation 5.11 becomes [Parsons 2000]

$$
\begin{aligned}
\left(P_R\right)_{dBW} = \left(P_E\right)_{dBW} &+ G_E + G_R + 20\log_{10}(h_E)_m + 20\log_{10}(h_R)_m \\
&- 120 - 40\log_{10}(d)_{km} + \text{Att}_{add\_dB}
\end{aligned}
\tag{5.12}
$$

where $(h_E)_m$ and $(h_R)_m$ stands for the transmitting and receiving antenna heights (in meters), respectively. As in Equation 5.11, Att_{add_dB} is a negative value, which stands for the additional attenuations (cable losses, connectors losses, rain attenuation, etc.) in decibel.

Alternatively, we may express Equation 5.12 in linear units as

$$
P_R = P_E g_E g_R \text{Att}_{add}(h_E h_R)^2/d^4
\tag{5.13}
$$

Note that the flat earth model used for Equations 5.12 and 5.13 can be considered for short distances d such that $\dfrac{10 \cdot h_E \cdot h_R}{\lambda} < d < d_h$, where the radio horizon d_h is approximated by $d_h \simeq \sqrt{2 r_E h}$ and where r_E stands for the earth's radius ($r_E = 6373$ km). For distances $d > d_h$, the earth curvature needs to be taken into account in the calculations and, therefore, Equation 5.12 loses validity.

---

* That is, the FSPL increases linearly with the increase of the square of the distance.

It is worth noting that the received power strength presents a decay rate 4 with the distance (see Equation 5.13), whereas in the free space model the received power presents a decay rate of 2 (see Equation 5.8). Consequently, one may conclude that the presence of reflected waves presents a negative effect in the received signal strength. Establishing a link over a ground that presents bad reflection properties makes this effect less visible, reducing the decay rate of the received power strength with the distance.

For a receiver to be able to decode a signal, two important conditions must be achieved as follows:

- The received power strength (defined by Equations 5.8 or 5.11) must be higher than the receiver's sensitivity threshold;
- The E_b/N_0 of the received signal must be higher than that required for the service being transported. As an example, for voice, the bit error rate (BER) should be lower than 10^{-3}. From the graphic of Figure 3.2 (Chapter 3), we extract that the E_b/N_0 level should be higher than 7 dB.

5.1.1.3 Carrier-to-Noise Ratio Calculations   Assuming the free space propagation, from Equation 5.8, and taking the noise power into account, as defined in Chapter 3, the carrier-to-noise ratio (C/N) becomes

$$C/N = \text{EIRP} \cdot g_R \cdot A_{tt} \cdot \frac{1}{k_B T_n B} \tag{5.14}$$

where k_B is the Boltzmann's constant, T_n is the resistor's absolute temperature (expressed in kelvin degrees), and B is the receiver's bandwidth [Carlson 1986]. In Equation 5.14, the carrier-to-noise ratio is $C/N = \dfrac{P_R}{P_N}$, with the power of the carrier $C = P_R = \text{EIRP} \cdot g_R \cdot A_{tt}$, and where $N = P_N = k_B T_n B$ stands for the power of noise at the receiver (see Chapter 3). Note that the carrier-to-noise ratio C/N differs from the signal-to-noise ratio (SNR) because the former refers to the power of the modulated carrier, whereas the latter refers to the signal power after carrier demodulation. The conversion between C/N and SNR depends on the modulation scheme [Carlson 1986]. In fact, the SNR is normally the performance measure adopted in analog communications, whereas in digital communications the performance measure considered is the E_b/N_0.

For the free space propagation, we may express Equation 5.14, in logarithmic units, as

$$\begin{aligned}(C/N)_{dB} &= 10\log_{10}\left(\frac{P_R}{P_N}\right)\\&= \text{EIRP} + (G_R/T_n)_{dB} - 10\log_{10}(k_B B) + A_{ttdB}\end{aligned} \tag{5.15}$$

where $\left(G_R/T_n\right)_{dB}$ in Equation 5.15 is expressed in decibel, which stands for the receiver's merit factor, defined as [Carlson 1986] [Ha 1990]

$$\left(G_{\mathrm{R}}/T_{\mathrm{n}}\right)_{\mathrm{dB}} = G_{\mathrm{R}} - 10\log_{10} T_{\mathrm{n}} \qquad (5.16)$$

Note that, in Equation 5.15, the attenuation A_{ttdB} is a negative value. Since A_{tt} is a coefficient lower than 1, its logarithmic value A_{ttdB} becomes negative.

Entering with the Boltzman constant $k_{\mathrm{B}} = 1.3806503 \times 10^{-23}\,\mathrm{J\cdot K^{-1}}$, as defined in Chapter 3, Equation 5.15 can be rewritten as

$$\left(C/N\right)_{\mathrm{dB}} = \mathrm{EIRP} + \left(G_{\mathrm{R}}/T_{\mathrm{n}}\right)_{\mathrm{dB}} - 10\log_{10}\left(B\right) + A_{\mathrm{ttdB}} + 228.6 \qquad (5.17)$$

where the antenna gain is defined as

$$g = \eta D \qquad (5.18)$$

and where η stands for the antenna performance and D for the antenna directivity [Burrows 1949]. In case of a parabolic, D becomes

$$D = \left(\frac{\pi d_{\mathrm{a}}}{\lambda}\right)^2 \qquad (5.19)$$

with d_{a} the parabolic antenna (dish) diameter [Burrows 1949]. In this case, the antenna gain becomes [Burrows 1949]

$$g = \eta\left(\frac{\pi d_{\mathrm{a}}}{\lambda}\right)^2 \qquad (5.20)$$

Taking the C/N values expressed by Equation 5.14, we can calculate the bit error probability (P_{e})* for M-ary PSK (M-PSK) modulation,[†] valid for $M > 2$, as

$$P_{\mathrm{e}} = \frac{1}{\log_2 M}\, erfc\left(\sin\left(\frac{\pi}{M}\right)\sqrt{\frac{C}{N}}\right) \qquad (5.21)$$

where $erfc$ is the complementary error function [Proakis 1995].

We may also express Equation 5.21 for M ary PSK (valid $M = 2$ or 4, i.e., for BPSK or QPSK) as a function of E_{b}/N_0, making [Proakis 1995]

$$P_{\mathrm{e}} = Q\left(\sqrt{\frac{2E_{\mathrm{b}}}{N_0}}\right) \qquad (5.22)$$

where E_{b} stands for the bit energy and N_0 for the power spectral density of noise.[‡]

---

* Bit error probability P_{e} is also known as BER.
† The modulation schemes are detailed in Chapter 6.
‡ In fact, N_0 in Equation 5.22 refers to $N_0 + I_0$, that is, the sum of power spectral density of noise with the power spectral density of interferences, as long as the power spectral density of interference has a Gaussian behavior. For the sake of simplicity, in the system description of this chapter, it is only assumed to be N_0.

Similarly, the bit error probability for M-ary PSK, valid for $M > 2$, is approximated by [Proakis 1995]

$$P_e \approx \frac{2}{\log_2(M)} Q\left(\sqrt{2\log_2(M)\sin^2\left(\frac{\pi}{M}\right)\frac{E_b}{N_0}}\right) \qquad (5.23)$$

and for M-QAM (quadrature amplitude modulation) or M-PAM (pulse amplitude modulation) as [Proakis 1995]

$$P_e \approx \frac{2}{\log_2(M)}\left(1 - \frac{1}{M}\right) Q\left(\sqrt{\frac{6\log_2(M)}{M^2-1}\cdot\frac{E_b}{N_0}}\right) \qquad (5.24)$$

In Equations 5.22 through 5.24, the bit energy E_b becomes [Carlson 1986]

$$\begin{aligned} E_b &= P_R \cdot T_B \\ &= \frac{P_R}{R_B} \end{aligned} \qquad (5.25)$$

where R_B is the transmitted bit rate and T_B is the transmitted bit period. In the case of M-ary modulation, this corresponds to $R_B = R_S \cdot \log_2 M$, where R_S stands for the transmitted symbol rate and $\log_2 M$ for the number of bits transported in each symbol. The power spectral density of noise is [Carlson 1986]

$$\begin{aligned} N_0 &= \frac{P_N}{B} \\ &= k_B T_n \end{aligned} \qquad (5.26)$$

From Equations 5.25 and 5.26 and knowing that $C = P_R$ and $N = P_N$, we deduct the relationship between E_b/N_0 and C/N as

$$\begin{aligned} \frac{E_b}{N_0} &= \frac{C \cdot T_b}{\dfrac{N}{B}} \\ &= \frac{C \cdot T_b \cdot B}{N} \\ &= \frac{C}{N} \cdot \frac{B}{R_b} \end{aligned} \qquad (5.27)$$

where B/R_b stands for the inverse of the minimum spectral efficiency.*

---

* From the Nyquist ISI criterion, the spectral efficiency is $\varepsilon = \dfrac{R_b}{B} = \dfrac{2\log_2 M}{1+\alpha}$. The minimum spectral efficiency is achieved for $\alpha = 0$, leading to $\varepsilon_{min} = 2\log_2 M = \dfrac{1}{\dfrac{1}{2T_S}T_b} = \dfrac{2T_S}{T_b}$.

5.1.2 Wireless Propagation Effects

As depicted in Figure 5.4, a received electromagnetic wave may be the result of several propagation effects, namely reflection, diffraction, and scattering. Moreover, when a line-of-sight component is present, all these components are summed together. In the following, each one of these propagation effects is characterized.

5.1.2.1 Reflection   It consists of a change in the wave's propagation direction as a result of a collision into a surface. Electromagnetic waves are typically reflected in buildings, vehicles, streets, and so on.

As can be seen from Figure 5.5, the reflected wave presents the same angle as the incident wave (relating to the normal of the surface).

An electromagnetic wave progresses in three perpendicular axes. The electric field progresses in one of the axis and the magnetic field progresses in another, perpendicular to the first. Finally, the third axis relates to the direction of wave's propagation. When the electric field has vertical polarization (and consequently, the magnetic field is horizontal), it is said that the wave presents vertical polarization. On the other hand, when the electric field has horizontal polarization (and consequently, the magnetic field is vertical), it is said that the wave presents horizontal polarization. This can be seen from Figure 5.6. Finally, a wave may be oblique, having both electric and magnetic fields with vertical and horizontal components.

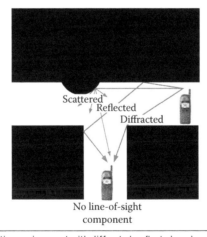

Figure 5.4   Example of propagation environment with diffracted, reflected, and scattered waves.

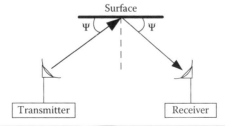

Figure 5.5   Reflection effect.

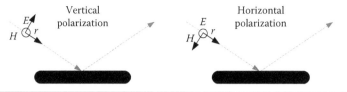

Figure 5.6 Vertical and horizontal polarizations.

The intensity of the reflected wave ($\vec{E}_{\text{REFL}}$ and $\vec{H}_{\text{REFL}}$) depends on the intensity of the incident wave ($\vec{E}_{\text{INCID}}$ and $\vec{H}_{\text{INCID}}$) and of the Fresnel coefficient (Γ_{H} and Γ_{V}) as

$$\vec{E}_{\text{REFL}} = \vec{E}_{\text{INCID}} \Gamma_{\text{H}} \tag{5.28}$$

for the horizontally polarized electromagnetic waves with $\vec{E}_{\text{INCID}} = E_{\text{TX}} \dfrac{e^{-jkd}}{d}$* and as

$$\vec{H}_{\text{REFL}} = \vec{H}_{\text{INCID}} \Gamma_{\text{V}} \tag{5.29}$$

for the vertically polarized electromagnetic waves with $\vec{H}_{\text{INCID}} = H_{\text{TX}} \dfrac{e^{-jkd}}{d}$ (with the propagation constant k defined by $k = 2\pi/\lambda$).

The power spatial density of an electromagnetic wave depends on both the electric and magnetic fields as

$$
\begin{aligned}
S &= \frac{1}{2}|E||H| \\
&= P_{\text{E}} g_{\text{E}} \frac{1}{4\pi d^2}
\end{aligned}
\tag{5.30}
$$

As incident components of both electric and magnetic fields depend inversely on the distance d, the power spatial density depends inversely on the square of the distance d^2. It is worth noting that the amplitudes of electric and magnetic fields are related through

$$E = Z \cdot H \tag{5.31}$$

where Z stands for the wave's impedance. The wave's impedance in the vacuum is quantified as $Z_0 = 120\pi\ \Omega$.

Let us define the Fresnel coefficients as a function of the reflection index n of the surface (ground). The Fresnel coefficient for the horizontal polarization becomes (NBS 1967)

$$
\begin{aligned}
\Gamma_{\text{H}} &= \frac{\vec{E}_{\text{REFL}}}{\vec{E}_{\text{INCID}}} \\
&= \frac{\sin\Psi - \sqrt{n^2 - \cos^2\Psi}}{\sin\Psi + \sqrt{n^2 - \cos^2\Psi}}
\end{aligned}
\tag{5.32}
$$

---

* Spherical wave is considered in this formulation.

and the Fresnel coefficient for the vertical polarization is given by (NBS 1967)

$$\Gamma_V = \frac{\vec{H}_{REFL}}{\vec{H}_{INCID}}$$

$$= \frac{\vec{E}_{NORMAL\_REFL}}{\vec{E}_{NORMAL\_INCID}} \qquad (5.33)$$

$$= \frac{n^2 \sin \Psi - \sqrt{n^2 - \cos^2 \Psi}}{n^2 \sin \Psi + \sqrt{n^2 - \cos^2 \Psi}}$$

where it was assumed that the reflected direction is in the same plane as the incident direction.

The reflection index n of the surface (ground) is given by NBS (1967) as follows:

$$n = \sqrt{\frac{\varepsilon_G'}{\varepsilon_0}} \qquad (5.34)$$

where ε_G' stands for the dielectric constant of the ground defined as $\varepsilon_G' = \varepsilon_G - j\frac{\sigma_G}{\omega}$, and ε_0 stands for the dielectric constant of the air. Furthermore, ε_G is the ground permittivity (real part of the dielectric constant), σ_G stands for the ground conductivity (imaginary part of the dielectric constant), and ω is the angular speed defined as $\omega = 2\pi f = \frac{2\pi}{T}$ (f is the electromagnetic wave's frequency and T its period). Note that the real part of the reflection index n is responsible for the reflection, whereas its imaginary part is responsible for the absorption of electromagnetic waves by the ground surface. Therefore, the part of the energy subject to absorption is the amount not subject to reflection, and vice versa.

We may also express Equation 5.34 as

$$n = \sqrt{\frac{\varepsilon_G - j\frac{\sigma_G}{\omega}}{\varepsilon_0}}$$

$$= \sqrt{\frac{\varepsilon_G\left(1 - j\frac{\sigma_G}{\omega\varepsilon_G}\right)}{\varepsilon_0}} \qquad (5.35)$$

$$= \sqrt{\frac{\varepsilon_G}{\varepsilon_0}\left(1 - j\frac{\sigma_G}{\omega\varepsilon_G}\right)}$$

$$= \sqrt{\varepsilon_R\left(1 - j \cdot tg(\delta)\right)}$$

where ε_R is the relative permittivity of the ground (relating to the air) and δ is the component responsible for the phase shift [Burrows 1949].

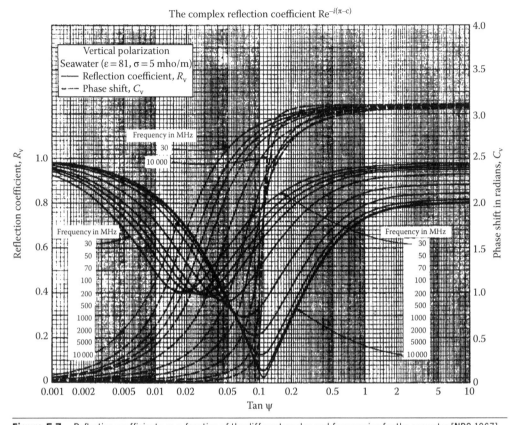

Figure 5.7   Reflection coefficients as a function of the different angles and frequencies for the seawater [NBS 1967].

From Equations 5.35, 5.32, and 5.33, we conclude that the Fresnel coefficients are a function of the incident angle ψ, of the surface's characteristics (i.e., dielectric constant), and of the frequency f. Figure 5.7 depicts the reflection coefficient as a function of the incidence angle Ψ and for different frequencies f, for the seawater. Note that, in Figure 5.7, the value $R_V = |\Gamma|$ and $\arg(\Gamma) = \pi - C_V <=> C_V = \pi - \arg(\Gamma)$. Different curves for different reflection materials can be found in NBS (1967).

Focusing on Figure 5.3, where the flat earth model was considered, the received electric field E_{REC} is composed of the sum of the direct component added to a reflected component as

$$
\begin{aligned}
\vec{E}_{REC} &= \vec{E}_{DIR} + \vec{E}_{REFL} \\
&= \vec{E}_{DIR} + \vec{E}_{DIR} \cdot \Gamma
\end{aligned}
\tag{5.36}
$$

Assuming that the distance between the two antennas is sufficiently high, Equation 5.36 becomes

$$
\begin{aligned}
\vec{E}_{REC} &\simeq E_{TX} \frac{e^{-jkr_{DIR}}}{r_{DIR}} + E_{TX} \frac{e^{-jkr_{REFL}}}{r_{REFL}} \Gamma \\
&\simeq E_{DIR} \left(1 + \Gamma \cdot e^{-jk\Delta r}\right)
\end{aligned}
\tag{5.37}
$$

where the propagation constant is $k = 2\pi/\lambda$, and where Δr is the difference between the reflected path distance r_{REFL} and the direct path distance r_{DIR}, that is, $\Delta r = r_{\text{DIR}} - r_{\text{REFL}}$. Moreover, in Equation 5.37, the direct wave electric field is $E_{\text{DIR}} = E_{\text{TX}} \dfrac{e^{-jkr_{\text{DIR}}}}{r_{\text{DIR}}}$ and E_{TX} stands for the electric field measured at 1 m from the transmitting antenna defined as $E_{\text{TX}} = \sqrt{30 P_{\text{E}} G_{\text{E}}}$. Note that we have assumed into Equations 5.36 and 5.37 that the transmitting antenna gain G_{E} is the same in the direction of the direct path and of the reflected path.

For distances sufficiently high and assuming that the refraction index is approximated by $\Gamma = -1$, Equation 5.37 can be approximated by Parsons (2000)

$$\vec{E}_{\text{REC}} \simeq 2 E_{\text{DIR}} \left| \sin(k \cdot h_{\text{E}} \cdot h_{\text{R}}/d) \right| \qquad (5.38)$$

Note that Equation 5.38 corresponds to Equation 5.13 with the difference that Equation 5.38 refers to the electric field strength, whereas Equation 5.13 corresponds to the received power strength. Using Equation 5.30 one can easily convert one into another.

Figure 5.8 depicts the received electric field strength using Equation 5.38 as being composed of the sum of the direct component and one reflected component, for distances between the transmitting and the receiving antenna between 1 and 10 m, considering the following parameters: 60 MHz ($\lambda = 5$ m); $\Gamma = -1$ (this parameter is already included in the deduction of Equation 5.38); $h_{\text{E}} = h_{\text{R}} = 10$ m. As can be seen, the resulting received electric field strength fluctuates with the distance as a function of the interference between the direct and reflected components. Note that the type

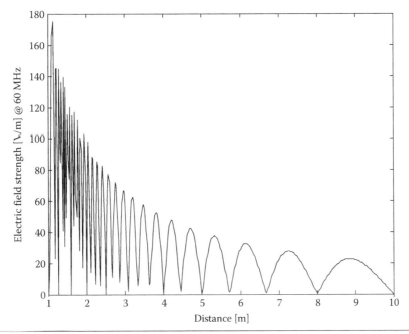

Figure 5.8 Plot of the received field strength for distances between 1 and 10 m ($\Gamma = -1$ and $h_{\text{E}} = h_{\text{R}} = 10$ m, at 60 MHz).

of interference between direct and reflected waves alternates between constructive and destructive. This type of interference is also known as multipath interference. It is seen that the difference between two consecutive maximums (constructive interference) and consecutive minimums (destructive interference) corresponds to $\Delta r = r_{REFL} - r_{DIR} = n\lambda$. In fact, consecutive maximums and minimums occur at path variation $\Delta r = n/2$, where Δr depends on the distance d and on the antennas height (h_E and h_R). For even values of n, destructive interference occurs between direct and reflected waves and, for odd values of n, constructive interference occurs between those component waves. Moreover, as can be seen from Equation 5.38, increasing the antenna's height, or decreasing the distance, the amplitudes of the maximums and minimums increases, that is, the link becomes more subject to multipath interference.

The decay rate of the electric field strength envelope with the distance is two. This corresponds to the decay rate of the received power envelope with the distance four, as viewed from Equation 5.13.

Figure 5.9 depicts the received electric field strength using Equation 5.38, in the same conditions as for Figure 5.8, but for distances between 10 and 500 m. As before, the field strength decreases with the distance, but the signal fluctuations stop for distances beyond a certain value (in this scenario, beyond around 50 m).

The level of interference generated by the reflected signal depends on several factors such as the antenna directivity. The use of an antenna with low antenna gain in the direction of the reflected wave reduces the level of interferences and, consequently, the signal fluctuations caused by fading, as well as the decay rate of the

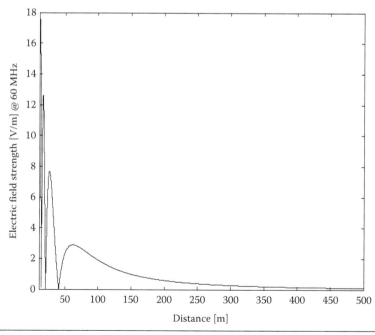

Figure 5.9   Plot of the received field strength for distances between 10 and 500 m ($\Gamma = -1$ and $h_E = h_R = 10$ m, at 60 MHz).

received signal strength with the distance. Alternatively, the selection of a path that blocks the reflected wave also leads to a reduction in the level of interference, reducing the decay rate, as well. Moreover, for long distances between the transmitting and the receiving antenna, the signal fluctuations tend to decrease, but the level of attenuation tends to be higher than that in free space propagation (i.e., only direct path). Transmitting electromagnetic waves over a soil that presents low refraction index also leads to low level of interference between the direct and reflected path and lower decay rate. Finally, using diversity such as multiple input multiple output (MIMO) systems avoid the fading effects, improving very much the performance of communications.

5.1.2.2 Diffraction   It occurs when a wave faces an obstacle, which does not allow it reaching the receiveing antenna in a direct path. In this case, even in the absence of direct path, a bending effect of waves is experienced, allowing the waves to reach the receiving antenna, but properly attenuated (Figure 5.10).

This phenomenon is normally quantified using the knife edge model. Such model considers a semi-infinite plan, located between a transmitting and a receiving antenna, in a certain position relating to this plan (Figure 5.11).

To calculate the level of attenuation introduced by a semi-infinite plan, let us focus on the geometry depicted in Figure 5.12.

First, the value $\bar{x}$ is defined as [Parsons 2000]

$$\bar{x} = \frac{x_E d_R + x_r d_E}{d_R + d_E} \tag{5.39}$$

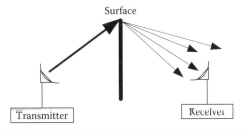

Figure 5.10   Diffraction effect.

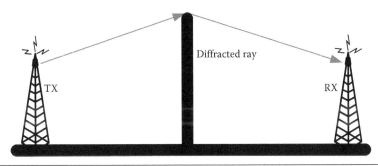

Figure 5.11   Propagation path between a transmitting and a receiving antenna achieved through diffraction.

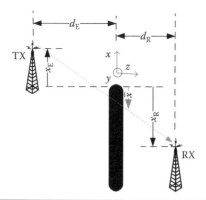

Figure 5.12   Geometry of the knife edge model.

Note that, in the example depicted in Figure 5.12, $\bar{x}$ is placed below the semi-infinite plan. This means that $\bar{x}$ is negative.

The equivalent height h_E is defined as [Parsons 2000]

$$
\begin{aligned}
h_E &= \sqrt{k \frac{(d_E + d_R)}{\pi d_E d_R}} \, \bar{x} \\
&= \sqrt{\frac{2\pi}{\lambda} \frac{(d_E + d_R)}{\pi d_E d_R}} \, \bar{x} \\
&= \sqrt{\frac{2\pi f}{c} \frac{(d_E + d_R)}{\pi d_E d_R}} \, \bar{x}
\end{aligned}
\tag{5.40}
$$

Taking the value for h_E, we are now in the position to calculate the level of attenuation using the Euler formula as

$$
A\left(h \frac{-b \pm \sqrt{b^2 - 4ac}}{2a} \right) = \frac{1}{2}\left[\frac{1}{2} + C(h_E)\right]^2 + \frac{1}{2}\left[\frac{1}{2} + S(h_E)\right]^2
\tag{5.41}
$$

where $S(x)$ and $C(x)$ stand for the Fresnel sine integral and Fresnel cosine integral functions [Parsons 2000].

We may conclude that the received signal level increases with the decrease of the carrier frequency, the increase of the horizontal distance between the receiving antenna and the semi-infinite plan (which represents the obstacle), and with the decrease in the depth of the receiving antenna.

This model is very useful in many different scenarios. One common use of this model is to quantify the attenuation introduced by an obstacle or by the earth curvature in a microwave line-of-sight link.

5.1.2.3 Scattering   It occurs when a wave is reflected by an obstacle that is not flat. Since the incident wave covers a certain area (a group of points in the surface) and

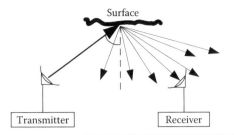

Figure 5.13 Scattering effect.

since each point of such area has a different normal to the obstacle, the scattering effect corresponds to an amount of successive reflections, each one in each point of the surface's obstacle covered by the incident wave (Figure 5.13).

Due to the high complexity of this phenomenon, its characterization is not dealt with here. Nevertheless, a detailed description of such phenomenon can be found in Parsons (2000).

5.1.3 Fading

In mobile communications, the channel is one of the most limiting factors for achieving a reliable transmission. The different types of fading are characterized by random variation in the received signal level. This is caused by several factors, such as atmospheric turbulence, movement of the receiver or the transmitter, movement of the environment that surrounds the receiving antenna, variation in the atmospheric refraction index, and so on.

Because of the mobility of the transmitter, receiver, or both, the resulting channel affects the received signal that basically suffers from two effects: slow fading (shadowing) and fast fading (multipath fading).

Slow fading is mainly caused by the terrain contour between the transmitter and receiver, being directly related to the presence of obstacles in the path of the signal. This effect can be compensated for with power control schemes.

Fast fading is caused by the reflection of the signal in various objects (buildings, trees, vehicles, etc.), which originate in multiple replicas of the signal reaching the receiving antenna through different paths. These replicas arrive with different delays and attenuations, superimposed in such a way that they will interfere with each other, either constructively or destructively. Because of the mobility of the transmitter or receiver and of the surrounding objects, the replicas are subject to variations on their paths, and hence in their delays and attenuations, leading to great oscillations on the envelope of the received signal.

Since the multiple replicas of the signal arrive with different delays, there will be temporal dispersion of the received signal. This means that if a Dirac impulse is transmitted, the received signal will have a non-infinitesimal duration, that is, the received signal shape will not be of impulsive type. This temporal dispersion can be represented using a power delay profile (PDP), $P(\tau)$, which represents the average received power as a function of the delay τ. Figure 5.14 shows an example of a PDP. There are two main types of fading as follows:

- Shadowing fading
- Multipath fading

Depending on the depth of received power fluctuations and fluctuations rate, there are two statistical models and types of fading, defined in the following section [Fernandes 1996].

5.1.3.1 Shadowing Fading   This type of fading is characterized by slow variation (slow fading) in the received signal level. This is caused by an obstruction to the line of sight caused by an object.

The factors that influence the depth of this slow signal strength variation are as follows:

- The movement of the receiver (although in a lower scale than the variation caused by the multipath)
- The nature of the terrain
- The nature, density, and orientation of the buildings, as well as the width and orientation of the streets

This effect is experienced when there is no direct line of sight between the transmitter and the receiver, and therefore, the propagation is characterized by diffraction.* Figure 5.15 shows the shadowing effect caused by a building between the transmitter and the receiver. The average value of the received signal level follows a lognormal distribution† (the logarithm of the amplitude of the field follows a normal distribution).

Higher attenuations have been experienced in urban zones with higher building densities. The standard deviation σ increases with

- Increase of the considered area
- Increase of the building proportions
- Increase of the frequency

Typical values of σ for cellular environments are between 6 and 18 dB.

It is worth noting that variations in the refraction index, ducts, rain, and fog may create similar effects as those described earlier.

5.1.3.2 Multipath Fading   This type of fading is characterized by fast changes in the received signal level. This is caused by variations such as turbulence of the local atmosphere, variation of the distance between the transmitter and the receiver, variation of

---

* The diffraction effect is normally quantified using the knife-edge model previously introduced.

† A log-normal distribution is defined by [Proakis 1995] $f(x) = \dfrac{1}{\sigma\sqrt{2\pi}}\dfrac{1}{x}\exp\left[-\dfrac{1}{2}\left(\dfrac{\ln(x)-\mu}{\sigma}\right)^2\right]$.

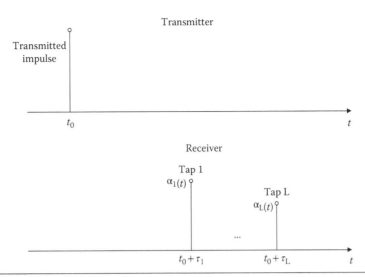

Figure 5.14 Discrete impulsive response of a multipath channel.

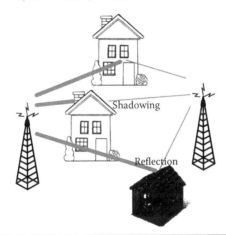

Figure 5.15 Shadowing and multipath effects.

the environment surrounding the receiver, and so on. This type of fading depends on the carrier frequency, the environment surrounding the antennas, and so on.

The above mentioned causes of fast fading originate variations of the interferences (constructive/destructive) between different propagation paths (line of sight, reflected, diffracted, scattered).*

When the receiver moves about one wavelength or when the environment surrounding the receiver moves, the intensity of the received signal experiences a deep fading of the order of 30–40 dB.

Figure 5.14 shows an example of an impulsive response of a multipath channel. In Figure 5.14, t_0 corresponds to the instant of transmission (of a pulse) and $t_0 + \tau_1$ and $t_0 + \tau_L$ correspond to instants at which the transmitted pulse was received, after being propagated and reflected in the environment.

---

* This is caused by a variation in amplitude or delay (or both) of one or more received multipaths.

In case of digital transmission, if the delay spread of the channel defined by Equation 5.45, caused by the multipath environment, is greater than the symbol period, this means that the signal bandwidth is greater than the channel coherence bandwidth (defined by Equation 5.44). In this case the channel presents frequency selectivity and the receiver experiences intersymbol interference.

Assuming a discrete multipath propagation channel with L paths, the complex equivalent low pass of the channel impulse response becomes[*]

$$h(\tau,t) = \sum_{i=0}^{L-1} a_i(t)e^{j\theta_i(t)}\delta(t-\tau_i)$$ (5.42)

where $a_i(t)$, $\theta_i(t)$, and τ_i stand for the attenuation, phase shift, and delay of the ith multipath. $\delta(t)$ is the Dirac function. The frequency response of the channel becomes

$$H(f,t) = \sum_{i=0}^{L-1} a_i(f)e^{j\theta_i(f)}e^{-j2\pi f\tau_i}$$ (5.43)

Note that the time delay is related to phase shift by $\tau(f) = -\dfrac{1}{2\pi}\dfrac{d\theta(f)}{df}$, both parameters being a function of the frequency f [Marques da Silva 2010].

Depending on the depth of the fast fading, there are two statistical models characterizing these effects [Fernandes 1996]:

- Rayleigh model—Fast and deep variation: it is typically experienced when there is no line of sight between the transmitter and the receiver, that is, there is only interference between the several reflected, diffracted, and scattered multipaths. This is normally experienced in urban environments. Considering that for each delay, τ_i, a large number of scattered waves arrive from random directions, in accordance with the central limit theorem, $a_i(t)$ can be modeled as a complex Gaussian process with zero mean. This means that the phase $\theta_i(t)$ will follow a uniform distribution in the interval $[0\ 2\pi]$, and the fading amplitude $|h(\tau,t)|$ will follow a Rayleigh distribution.[†]
- Ricean model—Fast but low deep variation: it is typically experienced when in the presence of a line of sight between the transmitter and the receiver, to which several multipaths are added at the receiver. In this case, there is interference between the line of sight and the several reflected paths. This effect is defined by a Ricean distribution. It consists of a sum of a constant component (direct path, i.e., line-of-sight component) with several reflected paths (defined by a

---

[*] This equation is generic, being valid for both Rayleigh or Ricean models.

[†] A Rayleigh distribution is defined by [Proakis 1995] $f(x) = \dfrac{x}{\sigma^2}\exp\left[-\left(\dfrac{x^2}{2\sigma^2}\right)\right]$, whose probability density function (PDF) is expressed by [Proakis 1995] $p_R(|\alpha|) = \dfrac{2|\alpha|}{E[|\alpha|^2]}\exp\left(-\dfrac{|\alpha|^2}{E[|\alpha|^2]}\right)$.

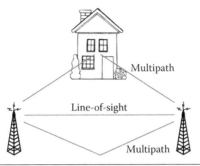

Figure 5.16 Propagation environment with line-of-sight and multipath components.

Rayleigh distribution). This effect is typical of rural or indoor environments. Assuming the presence of a line-of-sight component with amplitude A arriving at the receiver, then $a_i(t)$ will be a complex Gaussian process with nonzero mean and thus the fading amplitude $|h(\tau,t)|$ will follow a Ricean distribution.[*]

The rapid movement between the transmitter and the receiver creates, with the variation of the propagation path distance, a change in the corresponding interferences (see Figure 5.16). In addition, this movement may create a change in the environment surrounding the receiving antenna, which also creates a fast fading. Finally, the turbulence of the local atmosphere is also a cause of the multipath fading, as it consists of the fast and random variation in the refraction index, creating similar variation in the interference between the multipaths.

A cellular architecture includes three different types of cells: macro-cells (rural and urban), micro-cells (of lower dimensions), and pico-cells (coverage of an office, a room, etc.).

An urban macro-cell is the environment where the shadowing effect is experienced with higher intensity. This is caused by two main reasons as follows:

- There is no line-of-sight component between the transmitter and receiver and, therefore, the link is established through reflected, diffracted, and/or scattered rays.
- The high rate of constructions.

The absence of line of sight between the transmitter and the receiver also shows that the multipath fading of urban macro-cells is characterized by a Rayleigh distribution. On the other hand, due to the presence of line of sight, rural macro-cells and pico-cells tend to be characterized by a Ricean distribution.

---

[*] A Ricean distribution is defined by [Proakis 1995] as $f(x) = \dfrac{x}{x_{ef}^2}\exp\left[-\left(\dfrac{x^2 + A^2}{2x_{ef}^2}\right)\right]\cdot I_0\left(\dfrac{x\cdot A}{x_{ef}^2}\right)$, whose PDF is

expressed by [Proakis 1995] as $p_R(|\alpha|) = \dfrac{2|\alpha|}{E[|\alpha|^2]}I_0\left(\dfrac{2|\alpha|A}{E[|\alpha|^2]}\right)\exp\left(-\dfrac{|\alpha|^2 + A^2}{E[|\alpha|^2]}\right)$, where I_0 is the modified

Bessel function of zero order and where x_{ef} stands for the root mean square value of x.

The coherence bandwidth is defined as the bandwidth above, which the signal starts presenting a frequency selective fading. In other words, a signal with a bandwidth higher than the coherence bandwidth presents different attenuations and nonlinear phase shifts* at different frequencies. This effect is known as distortion. As exposed in Chapter 3, in the case of digital transmission, the distortion is viewed in the time domain as creating intersymbol interference.

The coherence bandwidth is obtained by

$$(\Delta f)_C = \frac{1}{2\pi S} \tag{5.44}$$

where S is the RMS delay spread defined as

$$S = \sqrt{\frac{\int_0^\infty (\tau - D)^2 P(\tau) d\tau}{\int_0^\infty P(\tau) d\tau}} \tag{5.45}$$

and where $P(\tau)$ is the power delay profile and D is the average delay defined as [Marques da Silva 2010]

$$D = \frac{\int_0^\infty \tau P(\tau) d\tau}{\int_0^\infty P(\tau) d\tau} \tag{5.46}$$

There are different measures that can be adopted to combat the fading effects, such as the use of multiple spaced antennas (spatial diversity), sectored antennas, matched filter equalizer, channel coding with interleaving,[†] or the use of frequency diversity.[‡]

5.1.4 Groundwave Propagation

There are three basic propagation modes: direct wave, ionospheric wave, and groundwave.

Direct coverage was already dealt with in the last section. Groundwave propagation can be used to cover areas that go beyond the direct line-of-sight coverage.

Figure 5.17 depicts the coverage by groundwave and ionospheric wave. Groundwave coverage may extend up to about 400 kilometers from the transmitting antenna.

---

* That is, the phase shift response as a function of the frequency is nonlinear (it is a curve).
† To avoid bursts of errors and allow the channel coding to correct a certain number of corrupted bits per frame.
‡ Transmit the same signal in different frequency bands, with a separation higher than the channel coherence bandwidth, to behave as uncorrelated.

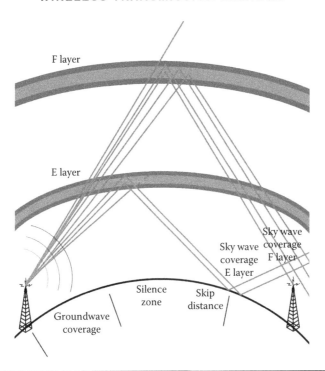

Figure 5.17 Propagation of electromagnetic waves using the ionospheric layers.

The coverage depends on the carrier frequency, characteristics of the terrain, polarization, absent of obstacles between the transmitting and receiving antennas, and so on.

Groundwave propagation is the sum of several elementary waves: (a) the surface wave, whose electromagnetic waves are guided over the earth's surface; (b) the direct wave; and (c) the reflected wave in the ground.

Surface wave can be viewed as the result of diffraction of low frequency electromagnetic waves by the earth's surface. As known from the knife edge model, diffraction effect is experienced with higher intensity at lower frequencies (as lower frequencies are less subject to attenuation by objects).

The surface wave propagates mainly using the vertical polarization, as horizontal polarization experiences high attenuation levels [Burrows 1949]. With regard to direct and reflected waves, as previously described, they are present in line-of-sight, reflected, or scattered paths between transmitting and receiving antennas. Since the reflection in the ground at short distance tends to originate a phase inversion, the combination of these two components in line-of-sight coverage is normally destructive at low frequencies. At long range, the groundwave is normally only composed of the surface wave, as the other two components are not present.

The groundwave attenuation corresponds approximately to the FSPL added by a 20 dB attenuation per decade. This decay becomes exponential with the increase of the distance d after a critical distance d_c, expressed in kilometers by

$$d_c = \frac{80}{\sqrt{f_{MHz}}} \tag{5.47}$$

The terrain permitivity and conductivity is determinant for the surface's wave propagation. Lower losses are achieved above surfaces with higher conductivity. Note that the seawater is highly favorable for the surface's wave propagation due to the high rate of salinity, which improves the conductivity.

Figure 5.18 shows the field strength curve as a function of the distance for several different frequencies, for seawater with $\sigma = 5$ S/m and $\varepsilon = 70$. Similar curves for different terrains can be obtained from (ITU-R Recommendation P.368–7 1992). As can be seen from Figure 5.18, the groundwave propagation is normally achieved with frequencies that span from few kilohertz up to around 3 MHz. Frequencies higher than this upper limit are subject to high attenuations, and therefore, their range becomes limited.

Using the graphic of Figure 5.18, we can calculate the received field strength expressed in dBuV/m (abscissa) at a certain distance from a 1 kW transmitter, for different frequencies.

Alternatively, we may extract the range obtained with a certain received field strength.

When a different transmitting power is used, a correction factor needs to be taken into account in the calculations.

To better understand the calculation of the range achieved, let us consider an example of voice communication in the 2 MHz frequency band with the following parameters:

- The transmitting power is 10 watts (i.e., 20 dB below the 1 kW reference transmitter).
- The fading margin is 3 dB.
- The noise level in the environment of the receiver is 32 dBuV/m.

The voice communication requires typically an SNR of 9 dB, which means that the signal needs to be 9 dB above the noise level, that is, $32 + 9 = 41$ dBuV/m. Assuming a 3 dB of fading margin, this value becomes 44 dBuV/m. Finally, since the transmitting power is 20 dB below the reference one considered by the curves and entering with the correction factor, the level becomes $44 + 20 = 64$ dBuV/m. Entering with such level into the 2 MHz curve, we obtain an approximate range of 140 kilometers. An alternative way to calculate the range or signal strength, entering with the same input parameters, can be performed using the GRWAVE simulator.

In the case that the path between a transmitter and a receiver is composed of different sections with different terrains, the calculation can be performed as a combination of different paths with different terrains. This method is known as the Millington method [Burrows 1949].

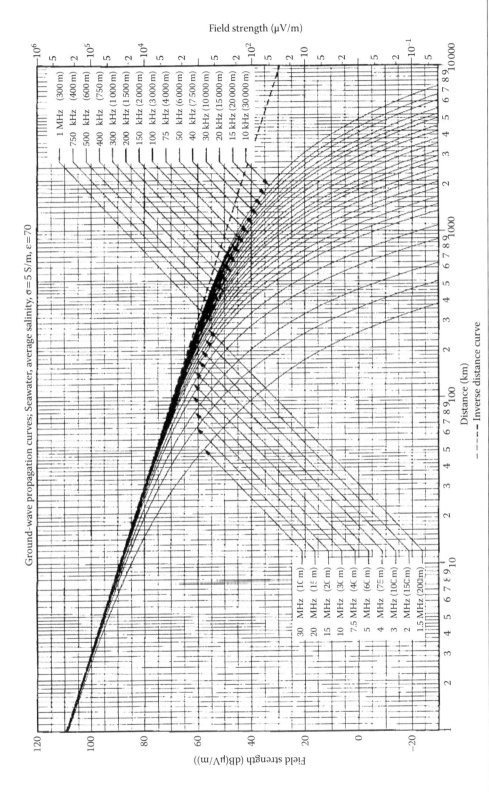

Figure 5.18 Field-strength curves as a function of distance with frequency as a parameter (for seawater with σ = 5 S/m and ε = 70) (From ITU-R Recommendation P.368–7. 1992. *Ground-Wave Propagation Curves for Frequencies between 10 kHz and 30 MHz.* With permission.).

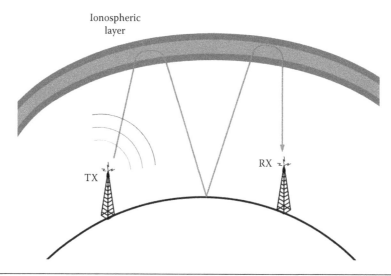

Figure 5.19 Refraction of electromagnetic waves in the ionospheric layers.

5.1.5 Ionospheric Propagation

Long-range radio communications can be achieved by different means: the modern type of long-range communication is normally achieved with satellite communication. Nevertheless, this can also be achieved using the so called short wave or high frequency (HF) communications, whose waves propagate at long range using the ionosphere.* In fact the ionospheric propagation can be the mode to support a long range communication link using frequencies from few hundred of kHz up to few dozens of MHz.

Ionospheric propagation consists of successive refraction in the ionosphere layers and successive reflection in the earth's surface. This can be seen from Figure 5.19. In fact, as per Snell's law, the gradual and successive refraction in the ionosphere can also be viewed as a reflection phenomenon.

With such propagation, and by choosing the correct carrier frequency, time of the day, and angle of incidence, a communication link can be established between any two points in the earth. The price to pay is the reduced bandwidth,† which typically characterizes the sky wave (as well as the groundwave).

The ionosphere is normally viewed as plasma with low level of ionization, composed of free electrons in a medium where they can collide with heavier particles. This plasma is normally characterized by two physical parameters: the number of electrons per volume unity and the number of collisions that electrons suffer per time unit. Moreover, the number of electrons per volume unity (N_e/m^3) shows a high variability over the day, seasons, and solar cycle. Note that the solar cycle corresponds to 11 years.

A complete reflection of an electromagnetic wave is experienced in the ionosphere if its frequency is equal to the critical frequency, as long as the propagation direction is

---

* This propagation type is also referred to as sky wave.
† Which translates in a reduced data rate.

perpendicular to the ionosphere. This characteristic is normally used by an ionosonde. Note that the critical frequency is a function of the level of ionization experienced in such part of the ionosphere.

Carrier frequencies higher than the critical frequency will cross the ionosphere. Furthermore, due to Martyn's law, the refraction occurs when the incident angle is not perpendicular to the ionosphere, even with a carrier frequency higher than the critical frequency.

Martyn's law is defined by

$$\omega = \omega_C \sec \theta_0 \qquad (5.48)$$

where θ_0 is the incident angle (measured from the perpendicular to the ionosphere layer) and where ω is the angular carrier frequency ($\omega = 2\pi f$ and f is the carrier frequency). Moreover, ω_C is the critical angular frequency. Note that Equation 5.48 shows that the ionosphere can reflect much higher frequencies with oblique propagation than with vertical propagation. This is the ideal condition to achieve long distances.

As described in Chapter 4, the refraction phenomenon is modeled by Snell's law as $n_1 \sin \theta_1 = n_2 \sin \theta_2$, where n stands for the refraction index of the medium and θ stands for the angle from the vertical. A wave that crosses a border between a medium with refraction index n_1 into a medium with refraction index n_2 suffers a refraction, that is, a deviation or bending in the wave's direction corresponding to $\theta_2 - \theta_1$.

The ionosphere is structured in layers, with different characteristics, namely D, E, and F layers (by ascending order of altitude). Some layers may present sublayers and some may be absent in certain period of the day. Note that the stronger ionization occurs normally at altitudes between 200 and 400 km. Figures 5.20 and 5.21 depict the typical ionospheric layers, respectively, for the day and night.

During the day, the typical layers are D layer (between 50 and 90 km), E layer (between 90 and 140 km), F1 layer (between 140 and 200 km), and F2 layer (above 230 km). Note that the F2 layer is the most important, as it is present even in the absence of sun (during the night). Furthermore, since it is placed at higher altitudes, it also allows establishing communication links at longer ranges. Moreover, it refracts higher frequencies.

Figure 5.21 depicts the common layers present at night. It is seen that, during the night, F1 and F2 layers merge, creating a single F layer. Moreover, while the E layer tends to be present during the night, the D layer is normally absent.

In addition to the aforementioned layers present during the day and night, E sporadic may also be present under certain conditions of the ionosphere (during the day or night). This layer may refract the same frequencies as the F layer.

As can be seen from Figure 5.17, for a certain carrier frequency, very low or very high angles of incidence results in an absence of reflection* by the ionospheric layer.

---

* In fact, the ionospheric layer does not reflect electromagnetic waves. Successive refraction is experienced in accordance with Snell's law. Macroscopically, this can be viewed as a reflection.

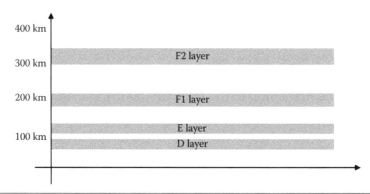

Figure 5.20 Representation of the ionospheric layers (day).

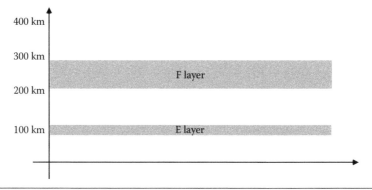

Figure 5.21 Representation of the ionospheric layers (night).

A high angle (from the normal to the layer) originates the wave absorption by the ionosphere, whereas a low angle originates that the wave crosses the layer without having been reflected.

Carrier frequencies too high or too low results that the rays are not reflected. In the example of Figure 5.17, a higher frequency is used to achieve F layer. This frequency is not reflected by the E layer, except at very high angles of incidence.

Note that the frequency and angle necessary to reach a certain destination differ from those parameters necessary to reach a different destination.

The optimum angles and frequencies present several levels of variation, namely the following:

- It varies from year to year depending on the sun spot number (SSN) as a function of the sun intensity. A higher sun activity results in higher electronic density, which results in lower frequencies for the same destination range (and same angle of incidence).
- It varies with the season: because of higher sun intensity, the ionospheric layers are more intense in the summer. Consequently, the summer frequencies are typically higher than those in the winter.

- It varies with the time of the day: as seen from Figures 5.20 and 5.21, due to the absence of the sun, the level of ionization during the night is lower. This results in a lower frequency for the same distance and angle of incidence.
- It varies with the latitude: since the solar incidence angle is lower at high latitudes, the electronic density of layers decreases at high latitudes.

5.2 Satellite Communication Systems

One of the most important advantages of satellites relies on its wide coverage, which translates in service availability in remote areas. Satellites can be used for many different purposes. They can be used for broadcast of radio or television channels, for point-to-point or point-to-multipoint communications, for capture of images, for meteorological purposes, and so on.

The first satellite used for communication was the moon in 1958. An electromagnetic beam was sent toward a specific position in the moon, which reflected it backward to the earth.

Afterwards, Echo I and Telstar, in 1962, incorporated an active repeater onboard it. First applications consisted of intercontinental transmissions of television and communications with ships at sea. Later on, satellites started being used for intercontinental exchange of voice and finally for data and positioning systems.

The basic principles of satellite communications were not deeply modified over time. A satellite has one or several transponders, each one operating in a different frequency band. This consists of a receiver, followed by a frequency translator, an amplifier, and a transmitter. The translator is necessary as the uplink and downlink frequencies are different. The main functionality consists of receiving a signal, amplifying it and sending it back to the earth, that is, acting as a repeater.

Depending on the orbit altitude and attitude, there are different types of orbits: geostationary earth orbit (GEO), medium earth orbit (MEO), low earth orbit (LEO), and highly elliptical orbit (HEO). The LEO altitude varies between 300 and 2000 km, whereas the MEO orbit corresponds to an altitude between 5,000 and 15,000 km. The GEO altitude is typically 35,782 km. Finally, the HEO presents an elliptical orbit with perigee at very low altitudes (typically 1000 km) and with the apogee at high altitudes (between 39,000 and 53,600 km). The different orbits are plotted in Figure 5.22.

5.2.1 Physical Analysis of Satellite Orbits

To understand how satellites stand in space, it is worth introducing some physical concepts. This analysis allows us deducting the altitude, speed, and period for each different orbit.

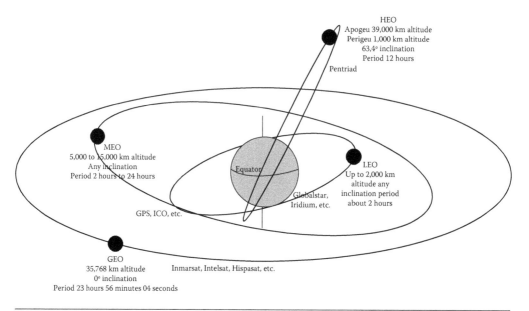

Figure 5.22 Plot of different orbits.

The widely known Newton's law of universal gravitation establishes the attraction force between two objects due to gravity. Particularly, considering that the two objects are the earth and a satellite, the law of universal gravitation becomes

$$f_a = G \frac{m_E \cdot m_{SAT}}{d^2} \tag{5.49}$$

where, in Equation 5.49, we have assumed m_E as the mass of the earth and m_{SAT} as the mass of the satellite. G stands for the gravitational constant ($6.674 \times 10^{-11} \, N \cdot \le m^2 \cdot kg^{-2}$). Moreover, d stands for the distance between the mass center of the earth and the mass center of the satellite.

The magnitude of the centripetal force of an object of mass m_{SAT} moving at a speed v_{SAT} along a path with a curvature radius r_{SAT} becomes

$$f_C = m_{SAT} \cdot \frac{v_{SAT}^2}{r_{SAT}} \tag{5.50}$$

Using the equality $v_{SAT} = \omega_{SAT} \cdot r_{SAT}$, Equation 5.50 becomes

$$f_C = m_{SAT} \cdot \omega_{SAT}^2 \cdot r_{SAT} \tag{5.51}$$

where ω_{SAT} stands for the angular speed of the satellite.

For the satellite to stand in the sky, the attraction force defined by Equation 5.49 needs to equal the centripetal force defined by Equation 5.51. Equaling the two forces and making $d = r_{SAT}$, we obtain

Table 5.1 Orbit Period As a Function of Orbit Altitudes and Radius

ALTITUDE (KM)	ORBIT RADIUS (KM)	ORBIT PERIOD (HOUR)
0 (earth's surface)	6373	1.4068
300	6673	1.5073
5000	11373	3.35
35782	42155	23.9327

$$G \frac{m_{\mathrm{E}} \cdot m_{\mathrm{SAT}}}{r_{\mathrm{SAT}}^2} = m_{\mathrm{SAT}} \cdot \omega_{\mathrm{SAT}}^2 \cdot r_{\mathrm{SAT}} \tag{5.52}$$

which leads us to

$$\omega_{\mathrm{SAT}} = \sqrt{\frac{G \cdot m_{\mathrm{E}}}{r_{\mathrm{SAT}}^3}} \tag{5.53}$$

Since $\omega_{\mathrm{SAT}} = 2\pi / T_{\mathrm{SAT}}$, we arrive at the third Kepler's law:

$$T_{\mathrm{SAT}}^2 = \frac{4\pi^2}{G \cdot m_{\mathrm{E}}} r_{\mathrm{SAT}}^3 \tag{5.54}$$

where T_{SAT} is the orbit's period of the satellite.

Entering with the earth's mass $m_{\mathrm{E}} = 5.9737 \times 10^{24}$ kg and with the gravitational constant G into Equation 5.54, we can finally enunciate the relationship between the radius of the satellite's orbit and its period as

$$T_{\mathrm{SAT}}^2 = 9.9022 \times 10^{12} \cdot r_{\mathrm{SAT}}^3 \tag{5.55}$$

Expressing the distance in kilometers and the period in hours, and isolating the satellite radius, Equation 5.55 becomes [Kadish 2000]

$$r_{\mathrm{SAT(km)}} = 5076 \cdot T_{\mathrm{SAT\_(hour)}}^{2/3} \tag{5.56}$$

Noting that r_{SAT} stands for the radius of the satellite's orbit, we know that

$$r_{\mathrm{SAT}} = r_{\mathrm{E}} + h_{\mathrm{SAT}} \tag{5.57}$$

where r_{E} corresponds to the earth's radius and h_{SAT} stands for the orbit's altitude.*

Entering with the earth radius $r_{\mathrm{E}} = 6373$ km, Equation 5.56 can now be expressed as a function of the orbit's altitude h_{SAT}, in kilometers, as

$$\begin{aligned} h_{\mathrm{SAT(km)}} &= r_{\mathrm{SAT(km)}} - r_{\mathrm{E}} \\ &= 5076 \cdot T_{\mathrm{SAT\_(hour)}}^{2/3} - 6373 \end{aligned} \tag{5.58}$$

Table 5.1 shows different orbit periods, expressed in hours, for several orbit's altitudes and radius. Note that the LEO altitude is between 300 and 2000 km, whose period is around 2 hours. With regard to the MEO, its orbit altitude is between

---

* In the case of the GEO, the orbit's altitude is typically 35,782 km.

5,000 and 15,000 km, whose orbit period spans from around 3 hours up to almost 9 hours. Finally, the GEO orbit is at the altitude of 35,782 km, whose orbit period equals the day period (23.9327 hours). At this altitude, GEO satellites go around the earth in a west to east direction at the same angular speed as the earth's rotation.

5.2.2 Characteristics of Different Orbits

Satellite communications can be viewed as a type of cellular communications, whose coverage is much higher (due to higher altitude of the satellite). This can be seen from Figure 5.23.

Depending on the distance to the earth, there are different types of satellite's orbits. Nevertheless, there are two layers around the earth where locating satellites should be avoided, due to high electromagnetic radiation, which may deteriorate the satellite equipments. These two layers are entitled Van Hallen layers, whose altitudes are 2000–5000 km and 15,000–20,000 km.

5.2.2.1 Geostationary Earth Orbit   The most well-known type of orbit is the GEO. This orbit is called geostationary because the position relative to any point in the earth is kept stationary. The first satellites being used for mobile communications were launched in 1970 and were of GEO type. The GEO altitude is typically 35,782 km above the equator and, since it is geostationary, its period equals the earth rotation period. Since they have a geostationary orbit, they are relatively easy to control. Moreover, because of the high altitude, the coverage is maximized, corresponding

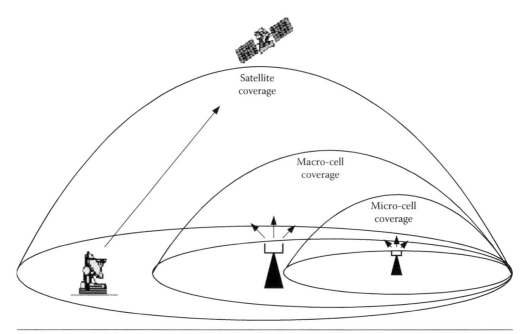

Figure 5.23   Indicative difference between satellite and cellular coverage areas.

to approximately one third of the earth's surface, which makes the communication service available to a wide number of potential users.

GEO-type satellites were unable to provide services to small mobile terminals, such as the existing cellular telephones, due to the following main reasons:

- High path loss, which results from the enormous distance from the earth.
- Low antenna gain of the satellite transponder, to allow covering a wide area of the earth surface (typically one third of the earth's surface). An indicative throughput available by a transponder is typically limited to 72 Mbps. Covering a wider area means that the throughput per user* is very much reduced
- Low power spectral density as a result of the low power available onboard the transponder (typical 10 dBW) and the enormous distance from the earth

Therefore, high power and high antenna gain were basic requirements of the earth stations to allow the connection establishment with a GEO satellite, which translates to low mobility and high dimensions.

In addition to the limitations of GEO satellites, since GEO orbits are located at an altitude of around 36,000 km, the round trip distance is approximately 72,000 km. This distance corresponds, at a speed of light, to a delay of 240 ms, which is much higher than in the case of MEO or LEO. This represents a high latency introduced in signals. In case the two terminals are not served by the same satellite, a double hop may be necessary. In this case, this latency increases to approximately one half of a second, which is a value that may bring problems for voice or for data communications.[†]

With the enormous growth of the telecommunications industry and the development of the new services such as the Internet and multimedia applications, the satellite operators viewed MEO and LEO as great potential business.

Although the LEO and MEO present lower footprints than GEO, requiring several satellites to allow an adequate coverage, other technological achievements facilitated the implementation of this satellites, such as direct connection between different satellites, on board switching and routing, advanced antenna systems, and so on.

5.2.2.2 Medium and Low Earth Orbit   With altitudes below that of the GEO, the MEO and LEO orbits overcame many of the limitations experienced with GEO satellite communications. As can easily be concluded from the orbit's name, the LEO is placed at a low altitude (between 300 and 2000 km) and the LMEO is located at a medium altitude (between 5000 and 5,000 km). These orbits can be seen from Figure 5.22.

---

* The throughput per user is the total throughput divided by the number of potential users within the coverage area.

[†] With such delay, the stop and wait data link layer protocol results into a very inefficient use of the channel.

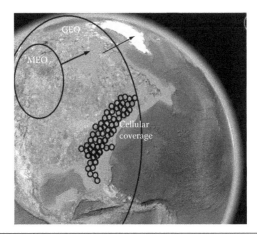

Figure 5.24   Example of geostationary earth orbit and medium earth orbit footprints against cellular coverage.

Contrarily to the GEO, these orbits are not geostationary and, as expressed by Equation 5.58, their period increases with the decrease of the orbit's altitude. Consequently, as these satellites are permanently moving around the earth, the coverage of a certain region needs to make use of several different satellites. Note that their orbits can be any, around the earth, namely above the equator, above a meridian, or with any inclination. In any case, the center of the orbit is always the center of the earth. Figure 5.24 depicts indicative footprints for GEO and MEO satellites, as well as the coverage made available by a cellular base station in a part of the east coast of the United States. The arrow connected to the MEO footprint circle represents the direction of the satellite moving, which corresponds to the direction of the footprint movement.

It is not possible to keep satellites below an altitude of 200 km, due to the enormous heating and deterioration that satellites are subject to, as well as the great tendency to change their orbits. This would translate to an enormous use of the engines and fuel to correct the orbits. Since the amount of fuel onboard satellite is limited, this is not a viable solution.

From the telecommunications point of view, lower altitudes of satellites translate in lower path losses. In order to establish a link with a GEO, it is normally required to make use of a parabolic antenna (high gain). Nevertheless, a link can normally be established with a MEO or LEO making use of an omnidirectional antenna.

As the satellites and their footprints are permanently moving (they are not stationary), a connection may be initiated with a satellite and, after a certain period of time, the connection may be handed over to another satellite. Therefore, the level of complexity necessary to manage such handover is increased, relatively to the GEO. Furthermore, the moving orbits require a higher level of adjustments from the control station.*

---

* A control station is a station in the earth that communicates with the satellite to send orders to adjust the orbit, the speed, the coarse, altitude, and so on. These orders also relate to adjustments in transmitting power, frequencies, antenna direction, and so on.

It is worth noting that the latest developments already allow the GEO-type satellites to work with higher power spectral densities, providing, however, low data rate services for terminals with omnidirectional antennas (e.g., the Fleetphone used by Inmarsat constellation). This is mainly achieved with the implementation of advanced antenna systems, which enables satellite antenna gains higher than 40 dBi.

5.2.2.3 Highly Elliptical Orbit Another type of orbit is the HEO. It presents an elliptical orbit with perigee at very low altitudes (typically 1000 km) and with the apogee at high altitudes (between 39,000 and 53,600 km). These orbits can be utilized for military observation or for meteorological purposes, with the perigee above the region to observe. Moreover, since GEO satellites do not cover the poles (they are located above the equator), HEO satellites are useful to provide communication services to regions with high latitudes. Nevertheless, since they are not stationary, the service is only available when the satellites pass over the region of interest, which occurs close to the perigee.

Table 5.2 presents a comparison among different satellite constellations.

5.2.3 Satellite's Link Budget Analysis

As can be seen from Figures 5.25 and 5.26, the satellites can be used for telecommunications in two basic modes: point-to-point and point-to-multipoint modes. In the point-to-point mode, the satellite acts as a repeater between two terminals. In this case, the exchange of data is normally performed in both directions (bidirectional). In the point-to-multipoint mode, the satellite acts as a repeater between a transmitting station and many receiving stations. This is normally used for broadcast, such as television or radio broadcast (unidirectional).

Table 5.2 Advantages and Disadvantages of Different Orbits

	ADVANTAGES	DISADVANTAGES
LEO	1. Can operate with low power levels and reduced antenna gains	1. Complex control of satellites
	2. Reduced delays	2. Frequent handovers
		3. High Doppler effect
		4. High number of satellites
MEO	1. Acceptable propagation delay and link budget, but worse than in the LEO case	
GEO	1. Reduced number of satellites and, consequently, simplest solution	1. Requires high antenna gains and powers to overcome increased path loss
		2. Difficult to operate with handheld terminals
	2. No need for handover	3. High delays (240 ms)
		4. Reduced minimum elevation angles for high latitudes which translates in high fading effects
HEO	1. High minimum elevation angles even for high latitudes	1. Requires high antenna gains and powers to overcome increased path loss
	2. Enables coverage of very specific regions	2. Extremely high delays, except in the perigee

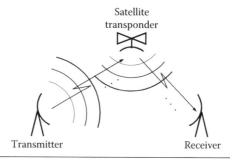

Figure 5.25 Generic diagram of a point-to-point satellite communication system.

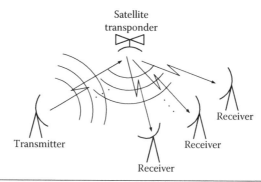

Figure 5.26 Generic diagram of a point-to-multipoint (broadcast) satellite communication system.

The selection of the frequencies for use by satellite communications are a function of several factors. The frequency must be sufficiently high such that the desired directivity and bandwidth is achieved. As can be seen from Equation 5.19, the directivity increases with the decrease of the wavelength, that is, it increases with the frequency. On the other hand, increasing the frequency also increases the path loss (see Equation 5.4). Figure 5.27 depicts the attenuation as a function of the frequency for a GEO satellite (distance corresponding to 36,000 km).

As expected, higher frequencies correspond to higher attenuation levels, which may bring link budget limitations.

In any case, the purpose is to maximize the received signal power, as defined by Equation 5.8, such that it is above the receiver's sensitivity threshold and such that the carrier-to-noise ratio C/N defined by Equation 5.14 is maximized. As per the Shannon capacity equation (see Chapter 3), a higher SNR allows transmitting at higher data rates. Note that there is a different correspondence between SNR and carrier-to-noise ratio for each different modulation schemes [Carlson 1986].

The frequency bands normally assigned to satellite communications are the L band (1–2 MHz), the C band (4–6 MHz), the X band (7–8 MHz), the Ku band (12–14 MHz), and the Ka band (18–22 MHz). The mostly used band is the C, using the 6 GHz band in the uplink,* whereas the 4 GHz band is normally adopted for the

---

* Uplink is defined as the link between a station in the earth and the satellite transponder.

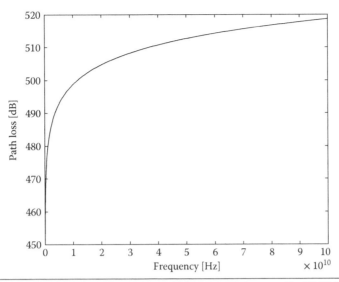

Figure 5.27 Path loss [dB] as a function of the frequency [Hz] for a geostationary earth orbit (36,000 km altitude).

downlink.* Since this pair of frequency bands is currently saturated, the second mostly used spectrum for satellite communications is currently the Ku band: 14 GHz in the uplink and 12 GHz in the downlink. Moreover, since higher carrier frequencies present typically higher bandwidths, the Ku band tends to be the preferable. However, it is more subject to rain attenuation.

Note that, to avoid interference, uplink and downlink frequencies should be different. Furthermore, the uplink frequency band is typically higher than the downlink frequency band. Since the earth station has typically more power available than the satellite transponder (typically 10 dBW), the higher attenuations originated by the higher frequency in the uplink is overcome by this additional transmitting power. In addition, since the satellite's beam is directed towards the earth (high noise temperatures), the satellite's receiver experiences higher noise power than earth stations. The use of higher uplink frequencies leads to higher satellite receiver's antenna gain, which allows maximizing the SNR.

The signal from the satellite's earth station is received with the power defined by Equation 5.11 (see Figures 5.28 and 5.29). Figure 5.28 depicts the propagation path between the satellite and the two end stations that are subject to the propagation adversities as described in Section 5.1.1 for the direct wave propagation. The propagation channel considered in a satellite link, at the earth station side, is typically modeled by a Ricean distribution. As previously described, it consists of a line-of-sight component to which a Rayleigh distribution is added. The Rayleigh distribution models the several reflected, diffracted, and scattered rays in buildings, street, trees, and so on. At high latitudes, as the satellite's inclination decreases, the strength of reflected waves become more predominant (and the line-of-sight components become weaker)

---

* Downlink is defined as the link between the satellite transponder and a station in the earth.

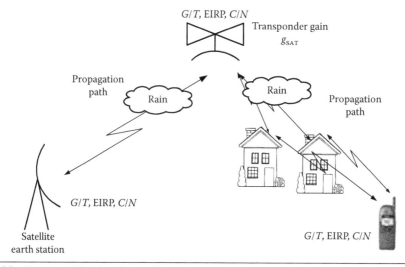

Figure 5.28   Typical satellite link with a satellite earth station and a mobile station.

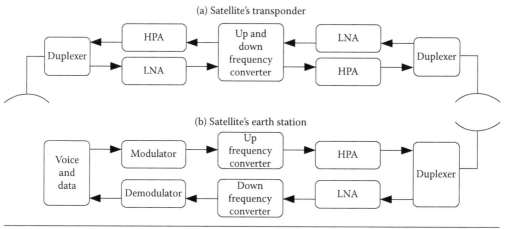

Figure 5.29   Scheme of (a) satellite's transponder and (b) satellite's earth station.

and the resulting signal is more subject to fading. Moreover, shadowing may also be important, especially in urban scenarios.

Note that additional attenuations such as antenna misalignments, rainfall, fog, diffraction caused by buildings, scattering originated by trees, or reflections caused by buildings and streets may also be quantified and taken into account in the computation of the receiving signal power, as previously described. The received carrier-to-noise ratio C/N can be computed using Equation 5.17, which already takes into account the receiver's merit factor G_R/T_n specified for the satellite transponder, namely the thermal noise captured by the receiving antenna and the noise introduced by the low noise amplifier (LNA) and introduced by the high power amplifier (HPA).

Since the uplink frequency is higher than the downlink frequency, the up-down frequency converter depicted in Figure 5.29 is responsible for the down frequency conversion.

As a satellite consists basically of a repeater placed at high altitude, the satellite's downlink transmitting power P_{E_D} consists of the uplink (satellite) received signal power P_{R_U} multiplied by the transponder's amplification gain g_{SAT}, becoming

$$P_{E\_D} = P_{R\_U} \cdot g_{SAT} \tag{5.59}$$

Usually, the most critical link is in the downlink direction due to the limited transmitting power P_E, which is available on board the satellite and due to the low antenna gain g_E (limited by its size). Therefore, the most critical receiver is the earth station. The downlink received signal power P_{R_D} becomes

$$
\begin{aligned}
P_{R\_D} &= P_{R\_U} \cdot g_{SAT} \cdot A_{tt\_D} \\
&= P_{E\_D} \cdot A_{tt\_D}
\end{aligned}
\tag{5.60}
$$

where A_{tt_D} stands for the downlink attenuation.

As described in Section 3.3.5, the total received noise power N_{TOTAL} in the downlink becomes

$$N_{TOTAL} = N_U \cdot f_{SAT} \cdot g_{SAT} \cdot A_{tt\_D} + N_D \tag{5.61}$$

where N_U stands for the uplink noise power, N_D stands for the downlink noise power, and f_{SAT} stands for the satellite noise factor. Note that N_{TOTAL} includes the contribution of the noise in the uplink and downlink paths.

Consequently, the received C/N becomes

$$(C/N)_{TOTAL} = \frac{P_{R\_D}}{N_{TOTAL}} \tag{5.62}$$

Alternatively, we can also compute the $(C/N)_{TOTAL}^{-1} = 1/(C/N)_{TOTAL}$ as

$$
\begin{aligned}
(C/N)_{TOTAL}^{-1} &= \frac{N_{TOTAL}}{P_{R\_D}} \\
&= \frac{N_U \cdot f_{SAT} \cdot g_{SAT} \cdot A_{tt\_D} + N_D}{P_{R\_U} \cdot g_{SAT} \cdot A_{tt\_D}} \\
&= \frac{N_U \cdot f_{SAT}}{P_{R\_U}} + \frac{N_D}{P_{R\_D}} \\
&= \frac{f_{SAT}}{(C/N)_U} + \frac{1}{(C/N)_D} \\
&= \frac{f_{SAT}}{(C/N)_U} + \frac{f_{SAT}}{f_{SAT} \cdot (C/N)_D}
\end{aligned}
\tag{5.63}
$$

and therefore, the received C/N can also be computed as

$$(C/N)_{\text{TOTAL}} = \frac{(C/N)_{\text{U}} + f_{\text{SAT}} \cdot (C/N)_{\text{D}}}{f_{\text{SAT}}} \qquad (5.64)$$

In the case of a double hop satellite link, the computation of the resulting C/N can be obtained from reusing Equation 5.64 iteratively, as follows:

- Compute the C/N at the input of the earth station, as previously described.
- Compute the C/N at the input of the second satellite's transponder (using the earth station's noise factor f_{EARTH}).
- Compute the C/N at the input of the second earth station (using the second satellite's noise factor $f_{\text{SAT\_2}}$).

Alternatively, as defined in Section 3.3.5, one could compute the $(C/N)_{\text{OUT}} = \dfrac{(C/N)_{\text{IN}}}{f_{\text{OUT}}}$, where the overall noise factor is $f_{\text{OUT}} = f_1 + \dfrac{f_2 - 1}{g_1} + \dfrac{f_3 - 1}{g_1 g_2} + \dfrac{f_4 - 1}{g_1 g_2 g_3} + \ldots + \dfrac{f_N - 1}{\prod\limits_{i=1}^{N-1} g_i}$. Note that we may view the path loss attenuation as a device gain and the thermal noise as the noise generated in an electronic equipment. Consequently, we may process jointly different propagation paths and electronic components (e.g., satellite's transponder, satellite earth station, etc.), using the same principle, and computing the resulting noise factor f_{OUT}, which is then used to compute the resulting $(C/N)_{\text{OUT}}$.

5.3 Terrestrial Microwave Systems

A terrestrial microwave system consists of a bidirectional radio link between two sites that use directional antennas (typically parabolic shape). Since it consists of a radio link, all the link budget and carrier-to-noise ratio calculations defined in Section 5.1.1 are also applicable. Moreover, in case the path between two interconnecting sites is not in line of sight, a repeater may be incorporated. This can be due to the earth curvature or due to the existence of obstacles. In this case, as a satellite may also be viewed as a repeater, the same principles as those deducted for the satellite link are also applicable in this case.

Typical parameters used in microwave systems are the following:

- Carrier frequency band: 10 GHz or below
- Transmitting power: 1 W
- Antennas gain: 35 dBi
- Link distance: up to around 30 km (may be extended using repeaters)

Microwave systems have been widely used to interconnect different sites, such as cellular base stations, local area networks, for exchange of television or radio channels between broadcast stations, and so on. Therefore, it is used for any type of media, such as voice, data, television, and so on. Microwave systems can be viewed as an alternative to fiber optic or coaxial cables, due to its higher implementation simplicity.

Taking into account the earth's curvature, the radio horizon of a microwave link is limited to

$$d_{\rm h} = \left[(r_{\rm E} + h)^2 - r_{\rm E}^2 \right]^{1/2} \simeq \sqrt{2 r_{\rm E} h} \qquad (5.65)$$

where $r_{\rm E} = 6373$ km, which stands for the earth's radius and h stands for the antenna height (it is assumed that both antennas are placed at the same heights).

To identify whether a terrestrial microwave link is clear of obstacles, one needs to analyze the Fresnel ellipsoids (see Figure 5.30). Assuming that $r_1, r_2 \gg D$, we have

$$D_{\rm n} = \sqrt{\frac{4 n \lambda r_1 r_2}{r_1 + r_2}} \qquad (5.66)$$

where $D_{\rm n}$ stands for the nth order ($n = 1, 2, \ldots$) diameter of the Fresnel ellipsoid.

Note that Equation 5.66 was deducted making the difference of path distance between the direct wave and reflected wave as $\Delta d = d_{\rm D} - d_{\rm R} = n\lambda/2$ (see Figure 5.31). Even values of n correspond to destructive interference between direct and reflected waves, whereas odd values of n correspond to constructive interference.

To assure that the received signal level is not 1 dB below the signal received in free space, the first Fresnel ellipsoid should be clear of obstacles. Note that the distance where $D_{\rm n}/2$ is counted refers to the position between r_1 and r_2 (see Figure 5.30). This

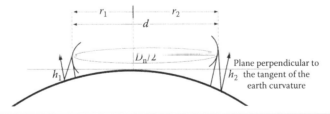

Figure 5.30 Terrestrial microwave link.

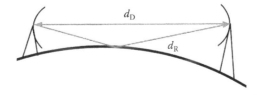

Figure 5.31 Direct and reflected waves of a microwave link.

refers to any position in the path, where an obstacle may exists and where one intends to identify whether it interferes with the terrestrial microwave link. Naturally, in case the link is clear of obstacles, the only limitation is the earth curvature. In this case, assuming that both antennas height are the same, the bottleneck occurs typically at the midway of the link, that is, for $r_1 = r_2$.

End of Chapter Questions

1. Which types of fading do you know? Characterize each one.
2. Which types of satellite orbits do you know?
3. From the known orbits, which one has the lower orbit?
4. What are the advantages and disadvantages of the LEO relating to the GEO orbit?
5. What is the difference between carrier-to-noise ratio and signal-to-noise ratio?
6. What is the typical performance measure used in digital communications?
7. What are the differences between reflection, diffraction, and scattering?
8. Which type of communication can be viewed as an alternative to satellite communication, in order to achieve a long range?
9. What is the difference between ground wave and surface wave?
10. What is the difference between surface wave and ionospheric wave?
11. What are the common ionospheric layers present during the day? And during the night?
12. What is the relationship between the altitude of a ionospheric layer and range?
13. What is the relationship between the altitude of a ionospheric layer and frequency?
14. What is the relationship between E_b/N_0 and the C/N?
15. What is the effect of a reflected wave, as compared to free space? Is it constructive or destructive?
16. Describe the model used to quantify the diffraction effect.
17. What are the parameters that improve the received signal strength of a receiver subject to diffraction?
18. Which measures can be used to mitigate the negative effects of a reflected wave?
19. For both free space propagation and a propagation model with a reflected wave, what is the relationship between the received power strength and the distance?
20. According to the Friis formula, what is the received power, for a 1 kW transmit power, a 10 kilometer distance and using both isotropic antennas?
21. What is the free space path loss equation?
22. What is the relationship between bit energy, bit period, and received power?

23. What are the statistical distributions that characterize the fast fading? What are the differences among them?
24. What is the statistical distribution that characterizes the slow fading?
25. For the surface propagation, what is the received signal strength, assuming a transmit power of 5 kW, and a range of 100 kilometers?
26. What is the silence zone?

6

SOURCE CODING AND TRANSMISSION TECHNIQUES

As previously described, one of the most important advantages of digital transmission relies on the ability to perform regeneration, although at the cost of a higher signal bandwidth, as compared to analog signals. Regeneration of signals allows partially removing the effects of channel impairments, such as noise, interference, attenuation, or distortion.

Chapter 1 stated that, depending on whether signals are transmitted in analog or in digital form and assuming that the source data is digital, such signals require passing through a process of modulation* or line coding,† respectively (see Figure 6.1). This is performed using a specific modulation scheme or digital coding that aims to optimize and adapt the source signal to the transmission medium.

The transmission of signals using line-coding techniques is performed in baseband, whereas the transmission associated with modulated signals makes use of a carrier to modulate a signal around a certain carrier frequency (bandpass). This may represent an advantage, as it is possible to select a transmitting frequency band whose channel frequency response presents less channel impairments.

The output of a line encoder is a digital signal, as it consists of discrete voltages that encode the source logic states. Consequently, it can be stated that the line coding is used when the transmission medium is digital. On the other hand, the output of a modulator is an analog signal, as it modulates a carrier, which is an analog signal. Therefore, a modulator is a device that performs the conversion from digital signals into analog, whereas a demodulator is a device that performs the conversion from analog signals into digital.

Regardless of whether the generated bits are transmitted in digital or analog, the bits are subject to a conversion into symbols (by either a line encoder or a modem) before the transmission process takes place. At the transmitter side, each bit, or group of bits, is encoded into one or more symbols. Conversely, at the receiver side, symbols are converted into bits. A symbol with M different signal levels allows $\log_2 M$ bits being encoded (by each symbol‡). The bit rate R_B is a function of the source encoder.

---

* The modulation and demodulation is performed using a MODEM, which implements the digital to analog conversion (modulation) at the transmitter side and the opposite at the receiver side (demodulation).

† Line coding is also referred to as digital coding.

‡ As described in Chapter 5, the bit rate R_B and symbol rate R_S are related by $R_B = R_S \cdot \log_2 M$.

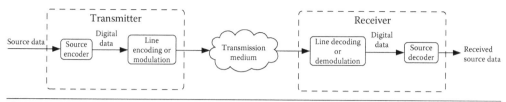

Figure 6.1 Generic block diagram of a digital communication chain.

A more efficient source encoder allows transmitting more information source using a lower bit rate (for the same quality).

The Nyquist theorem can be expressed by rewriting Equation 3.20 for the minimum baseband bandwidth B_{min} (roll-off factor $\alpha = 0$) as

$$R_B = 2B_{min} \times \log_2 M \tag{6.1}$$

where B_{min} stands for the bandwidth in Hertz. Moreover, the symbol rate (baud rate) R_S, expressed in symbols per second (symb/s) or baud, relates to the bit rate R_B and to M through

$$R_S = R_B / \log_2 M \tag{6.2}$$

We may rewrite Equation 6.1, taking into account the equivalence Equation 6.2 as

$$B_{min} = R_S/2 \tag{6.3}$$

From Equation 6.3, we know that a signal with a symbol rate of 1 Msymb/s requires a minimum baseband bandwidth of $B_{min} = 10^6/2 = 500$ kHz. Naturally, assuming a raised cosine pulse,* the required bandwidth is maximized with a roll-off factor $\alpha = 1$, which corresponds to twice the minimum bandwidth (see Chapter 3, Equations 3.19 and 3.20). From Equation (3.19), the maximum baseband bandwidth becomes

$$B_{max} = R_S \tag{6.4}$$

This means that the baseband bandwidth is bounded by $R_S/2 < B < R_S$.

As can be seen from Figure 6.2, in the case of bandpass (carrier modulated) signals, the negative baseband part of the spectrum is also transmitted. In this case, from Equation (3.19), the minimum bandwidth is doubled and Equation 6.3 becomes

$$B_{min} = R_S \tag{6.5}$$

Equation 6.3 allows obtaining the bandwidth occupied by a digital signal with a certain symbol rate. In case the signal is analog and we are interested in quantifying

---

* Naturally, the bandwidth can be higher for other pulses (e.g., rectangular).

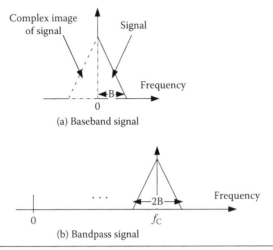

Figure 6.2   Spectrum of (a) baseband and (b) bandpass signals.

the sample rate f_a necessary to perform its analog to digital conversion (digitization), Equation 6.3 can be used by replacing R_S into f_a, resulting in

$$f_a \geq 2B \tag{6.6}$$

where B is the highest frequency component present in the analog signal. Equation 6.6 translates the Nyquist sampling theorem. This assures that the recovered analog signal from its samples is not subject to distortion.

6.1 Source Coding

The source coding is the process of transforming information media (source) into data. The information source can be voice, audio, text, video, images, and so on. Different source coding techniques exist, with different efficiencies and performances. Some more efficient source coding techniques allow encoding the same information media at the cost of a lower bit rate R_B. According to Equation 6.1, for the same modulation or line encoding technique, a lower bit rate results in a reduced bandwidth.

6.1.1 Voice

The voice is an analog information source, consisting of the most important information source used by humans to communicate. Its frequency domain spreads from around 80 to 12,000 Hz,* with the most important frequency components in the band 150–8000 Hz. However, it has been proven that its quality is acceptable and understandable if we restrict the signal to the band 300–3400 Hz. This consists of a trade-off between quality/intelligibility and transmission bandwidth.

---

* The ear response is typically 20 Hz to 20 kHz for a young person, degrading with the age.

Moreover, because of its irregular structure, the voice presents many pauses. This is because the voice signal is only present approximately 40% of the time. In channels presenting scarce bandwidth, these pauses can be used to interpolate other voice signals, using a mechanism known as time assignment speech interpolation (TASI).

6.1.1.1 Analog Audio   Analog data take continuous values of intensity within certain intervals. Legacy communication systems such as analog telephony or analog radio broadcast use analog voice. In this case, the analog signal is not subject to a digitization process. Nevertheless, to reduce its bandwidth and decrease the level of channel impairments (e.g., attenuation, distortion, etc.), a bandpass filtering* is normally applied to the signal.

A processing normally used in time division multiplexing (TDM) consists of simply transmitting a sample of an analog signal at fixed intervals corresponding to the sampling period. This source encoding technique is known as pulse amplitude modulation (PAM). Each PAM sample has, at sampled instants, the amplitude as the sampled analog signal. The sampling period T_a is obtained from the sampling rate f_a defined by Equation 6.6, as $f_a = 1 / T_a$. Since a sample signal only occupies a part of the time of the channel, the remaining time can be used to send samples of other analog signals. The generation of a PAM sample from an analog signal can be seen from Figure 6.3.

Knowing that a bandpass filtered voice signal has a spectrum in the range 300 Hz–3.4 kHz, according to Equation 6.6, the sampling rate is $f_a \geq 2 \times 3.4 = 6.8$ ksamples/s. This results in a sampling period of $T_a \leq \dfrac{1}{6.8 \times 10^3} \approx 147.06\ \mu s$.

Other analog source encoding techniques exist, such as the pulse position modulation (PPM) or the pulse duration modulation (PDM). While in the PAM, each sample presents the amplitude of the sampled analog signal, the PPM samples present a constant amplitude, but the amplitude of the analog signal is encoded by varying the position of the PPM sample. In the case of the PDM, the amplitude of the sample is also kept fixed, while the width of the pulse is used as a variable to encode the amplitude of the analog signal.

6.1.1.2 Digital Audio   Contrary to analog signals, digital signals take discrete values within certain intervals. Examples of digital signals are numbers, text, or binary signals. A binary signal is a special case of a digital signal, where the discrete values are only two (logic state 0 or 1, encoded with certain voltages). As previously described, an advantage of binary signals relies on their improved resistance against the channel impairments and regeneration.

6.1.1.2.1 Pulse Code Modulation   It is a source coding technique that comprises the digitization of analog voice, as well as the encoding of each sample of voice signal using a total

---

* For analog telephony, the bandpass filtering is applied within the range 300 Hz–3.4 kHz.

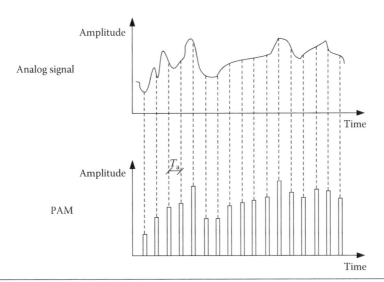

Figure 6.3   Analog signal and corresponding pulse amplitude modulation signal.

of eight binary digits (bits). The source encoding and decoding is performed in a device named codec, being responsible for performing the following individual operations:

- Sampling
- Quantization
- Coding

The sampling of an analog signal must follow the steps previously described for the sampling Nyquist theorem. Using a sampling rate equal or higher than the double of the maximum frequency component present in the analog signal assures that the recovered analog signal from its samples is not subject to distortion. The minimum sampling rate of a bandpass filtered voice signal is $f_a = 2 \times 3.4 = 6.8$ ksamples/s. Nevertheless, the pulse code modulation (PCM) uses a sampling rate of 8 ksamples/s.[*] The signal resulting from the sampling operation corresponds to the PAM signal depicted in Figure 6.3.

The quantization process considers the approximation of the sampled analog signal[†] to a certain finite value.[‡] The quantization process results in a certain level of quantization noise, which results from the difference between the input signal and its quantization level. The level of quantization noise can be minimized by increasing the number of finite values L (quantization levels). However, since the quantization levels are encoded by a certain number of bits, increasing the number of quantization levels results in the increase of the number of bits, which translates in an increase of the data rate. The PCM technique considers a total of 256 quantization levels.

---

[*] Some authors express sampling rate in Hertz, instead of samples/s (samples per second).
[†] Which may take an infinite range of values.
[‡] The closer finite value to the analog value.

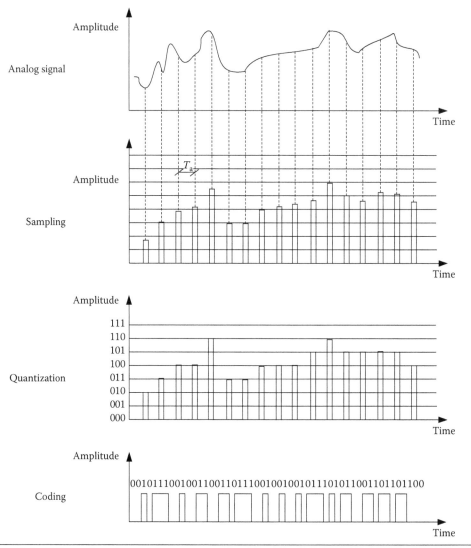

Figure 6.4 Uniform pulse code modulation.

Finally, coding is the process of transforming a quantization level (finite value) into a certain number of bits. The number of bits N_b is given by $N_b = \log_2 L$, where L corresponds to the number of quantization levels. The PCM technique considers eight bits to encode a total of $2^8 = 256$ levels.* Knowing that the bit rate is given by $R_b = f_a \times N_b$ and that the sample rate of PCM is 8 ksamples/s, the resulting bit rate of the PCM technique becomes $R_b = 64$ kbps. Figure 6.4 depicts the principles used in each individual PCM operation.

The music is another signal with interest. The sample rate typically used in music signals is 44 ksamples/s, whereas its uniform coding comprises 16 bits per sample. This results in a bit rate $R_b = 700$ kbps.

---

* Since $N_b = \log_2 L \Leftrightarrow L = 2^{N_b}$.

The above description of the PCM corresponds to the uniform PCM. The uniform PCM means that the quantization levels are uniformly distributed in amplitude. Nevertheless, it is known that lower amplitudes present a higher probability of occurrence than higher amplitudes. To minimize the quantization noise without increasing the number of quantization levels, a solution is to consider narrower quantization intervals at lower amplitudes and wider quantization intervals at higher amplitudes. This adapts the quantization levels to the statistic of the voice signal, and therefore, it allows reducing the quantization noise without having to increase the number of quantization levels. Alternatively, this allows a reduction in the source data rate for the same level of quantization noise. This technique is known as nonuniform PCM. Such nonuniform quantization is obtained using a logarithmic compression characteristic, as applied in the real PCM systems normalized by the ITU-T Recommendation G.711. Two main techniques exist [ITU-T G.711]: the μ law used in the United States and Japan, and the A law, adopted by Europe (among other countries).

The μ law is defined by

$$y = \text{sgn}(x)\frac{\ln(1+\mu|x|)}{\ln(1+\mu)} \tag{6.7}$$

where x stands for the signal amplitude of the input signal $(-1 \le x \le 1)$, $\text{sgn}(x)$ is the sign function of x, and μ is a parameter used to define the compression degree. Its typical value is $\mu = 255$. Finally, y is the quantized level (output).

The A law is defined by

$$y = \begin{cases} \text{sgn}(x)\dfrac{A|x|}{1+\ln(A)} & 0 \le |x| \le \dfrac{1}{A} \\[2ex] \text{sgn}(x)\dfrac{1+\ln|Ax|}{1+\ln(A)} & \dfrac{1}{A} \le |x| \le 1 \end{cases} \tag{6.8}$$

where A has a typical value of $A - 87.6$. From Equation 6.8 and from Figure 6.5, we can conclude that there are two regions in the A law:

- Low amplitude region $(x \le 1/A)$, with a linear variation
- High amplitude region $(1/A \le x \le 1)$, with a logarithmic variation (see Figure 6.5)

It is worth noting that the μ law leads to a slightly less amount of the quantization noise* than the A law, for lower signal levels, whereas the A law tends to achieve better performance at higher amplitudes.

---

* Which translates in a higher SNR for lower signal levels.

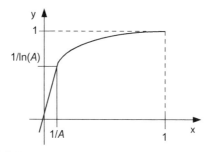

Figure 6.5   A law compression characteristic.

The A law uses 13 segments, with seven positives and seven negatives. Those two segments that cross the origin are linear, and therefore, they are accounted for as a single segment. Each 8-bit PCM word is encoded as follows:*

- The first bit represents the polarity of the input analog sample.
- The following three bits represent the segment (000 and 001 are used by the central segment).
- The following (and last) four bits encode the linear quantization within each segment.

Comparing logarithmic A law or μ law against linear algorithm, a reduction in the dynamic range from 12 bits to 8 bits is achieved. This translates in a throughput reduction from 96 to 64 kbps.

6.1.1.2.2 Differential Source Coding Techniques   An alternative solution to encode a source analog signal relies on encoding the difference between the actual signal sample and the previous signal sample (or, alternatively, the predicted signal sample†). This is especially useful for signals such as voice or video, as those signals tend to suffer a small variation from sample to sample. Since the variation range of the difference signal is lower than that of the samples, a lower number of bits are necessary to encode the signal, to achieve the same performance.

A widely known differential voice coding technique is the delta modulation. This transmits a single bit to encode a sample. Its logic state is 1 if the signal increases its level (comparing to the previous sample) or 0 otherwise. To achieve an acceptable performance, the sampling frequency needs to be higher than that of the Nyquist sampling theorem. An important advantage of the delta modulation relies on the simplicity of the hardware used to implement it.

Contrarily to the delta modulation, which considers the transmission of the difference signal between the actual and the previous signal sample, the differential PCM transmits the difference signal between the actual and the predicted signal sample. The prediction is obtained by extrapolation from the previous samples. A common

---

* The μ law uses 15 segments, instead of 13.
† The predicted signal sample is obtained statistically by making use of the previous samples.

implementation of the extrapolation is making use of a transversal filter. The length of the transversal filter corresponds to the number of previous samples that are accounted for in the prediction of the actual sample. A technique that normally improves the differential PCM technique is using a transversal filter whose coefficients adapts to the variation of the signal. The resulting technique is known as adaptive differential PCM (ADPCM) [ITU-T G.726] and allows a signal quality equivalent to that obtained with the PCM, but using data rates from 16 to 40 kbps.*

6.1.1.2.3 Linear Predictive Coding   The previously described codec involve the transmission of the signal itself or, alternatively, of the difference signal. There is another group of codec that are specific of voice signals (vocoder), which simply involve the transmission of a certain number of parameters characteristic of the voice signal. Such parameters allow obtaining a replica of the original signal, by synthesis, at the receiver side. The analysis and generation of the key parameters at the transmitter side is normally employed making use of a transversal filter. A vocoder includes two individual steps: (1) the voice analysis at the transmitter side, and (2) the synthesis of the voice from the received parameters at the receiver side.

The linear predictive coding (LPC) includes a transversal filter that models the transfer function of the voice, as well as the type of excitation associated to the voice signals. The LPC is used in GSM for the transmission of voice at 8.0 kbps. A specific type of LPC, entitled LPC-10, allows the transmission of voice, in digital form, at a rate of 2.4 kbps over high frequency (HF) radio channels.

Another vocoder technique used in radio channels is the mixed-excitation linear predictive (MELP), using rates of 2.4 kbps [MIL-STD-3005 1999], 1.2 kbps [Wang 2000], and 600 bps [Chamberlain 2001]. Its advantage relies on the high quality of the voice obtained at a reduced data rate,† as well as the improved performance obtained even under noisy conditions.

The moving picture experts group (MPEG) layer III (commonly referred to as MP3) is widely used to compress music, at a low data rate of 112–128 kbps, while enabling CD-quality music wideband audio (10–22 kHz). This is a very economic way of storing and transferring music. The layer III of the MPEG protocol (described below for video) defines the audio compression technique that is sent or stored in a synchronized manner with the MPEG video.

6.1.2 Video

The video is widely used as an information source to communicate. Examples of systems that make use of video are analog and digital television, videoteleconference, interactive television, and high definition television, among others.

---

* In this case, the sample rate corresponds to the Nyquist sampling rate.
† The voice quality obtained with the MELP is much better than that obtained with the LPC-10.

The video information source consists of successive images. To give the idea of continuity, a minimum of 20 images per second are required. In television, 25–30 images per second are transmitted. Moreover, to give an idea of continuity, each image must be composed of a sufficient number of lines. A low-resolution image has 300–400 lines, whereas a high resolution has more than 1000 lines. Finally, a line is composed of a number of image elements, known as pixels.* Each pixel is characterized by its relative position, luminance, and color.

6.1.2.1 Analog Video   The European video system is entitled phase alternation line (PAL). It considers a total of 625 transmitted lines, from which only 575 lines are visible. Moreover, the PAL system considers 25 images per second. The number of transmitted image elements are $M = ABC$, where A corresponds to the number of visible lines (575), B corresponds to the number of pixels per line (572), and C stands for the number of images per second (25). This leads to $M = 575 \times 572 \times 25 = 8,222,500$ image elements. Since each image element can be viewed as a signal sample, according to the Nyquist sampling theorem of Equation 6.6, we can conclude that the PAL system requires a minimum bandwidth of $B = M/2 = 4.1$ MHz. Taking into account the return of the beam, the required bandwidth becomes 5.5 MHz[†] for luminance. Adding to this value the bandwidth required for audio and color, the required bandwidth of a video system using the PAL standard becomes approximately 8 MHz.

6.1.2.2 Digital Video   As previously described, the transmission of data in the digital format allows a higher level of data integrity, which translates in a better reproduction of the original information at the receiver side. In addition, the Internet world demands for the total digitization of the different information sources. The video signal is analog by nature, and the digital video is obtained from the analog video by digitization.

We have seen that an analog video signal (PAL system) requires a bandwidth of approximately 8 MHz. According to the sampling theorem of Equation 6.6, this corresponds to a total of $M = 16$ Msamples/s. Assuming that each image element is quantized and encoded using 10 bits/sample, we obtain the data rate of 160 Mbps. Once again, using the Nyquist sampling theorem of Equation 6.6, we conclude that the minimum bandwidth required to accommodate a digital video signal corresponds to 80 MHz. Such a value is ten times higher than the required bandwidth necessary to accommodate an analog video signal. This is the price to pay for the additional video quality, which is inherent to digital signals. Nevertheless, there are algorithms that allow the digitization of video at a lower data rate, as described in the following.

---

* In fact, an image is composed of a number of pixels, which correspond to the number of pixels in a line multiplied by the number of lines per image.
† This value is much higher than the 3.4 kHz bandwidth required for telephony.

6.1.2.2.1 Moving Picture Experts Group   Similar to the ITU-T G.711 codec used for voice, there are several codec used for video. MPEG is the mostly known video algorithm, presenting different versions. The MPEG [Watkinson 2001] comprises a suite of standards that define the way video is digitized and compressed. These standards encode the video using a differential procedure.* Different MPEG standards present different bandwidth requirements, compression procedures, and resolutions. On the other hand, the digital video broadcast (DVB) defines the physical layer and data link layer used in the distribution of digital video. Since this section refers to source coding, the DVB is not included here, being described in Chapter 10.

The MPEG-2 is a video compression algorithm using data rates from 3 to 100 Mbps, comprising different resolutions from 352×240 to 1920×1080. This standard is used in DVB and high definition television (HDTV).

The MPEG-4 is another ISO standard for video and audio encoding and compression. It also includes other types of media such as images, text, graphics, and so on. The compression supports the transmission of MPEG-4 video and audio in 64 kbps channels (such as in an ISDN B channel).

Since the MPEG encoder is much more complex than the decoder, the MPEG is referred to as an asymmetric coding mechanism. This characteristic is important, as the MPEG is widely used for video broadcast, and where the receiver is intended to be kept with low complexity and cost. Moreover, only the decoder is standardized, whereas the encoder is kept as a proprietary algorithm.

6.2 Line Coding

Line coding is used for the transmission of signals in baseband. Line coding is also referred to as digital baseband modulation or simply as digital coding. A block diagram composed of a line encoder/decoder is plotted in Figure 6.6. As can be seen, the line encoder/decoder normally has embedded the error control capability. This is implemented using an error correction code or an error detection code associated to retransmission of data. Both these techniques are covered in Chapter 10.

The selection of the digital encoding technique depends on different criteria, such as

- Required bandwidth: according to the Nyquist theorem, for a certain fixed throughput, if a symbol encodes a higher number of bits (i.e., for higher values of M), the corresponding signal bandwidth is reduced.
- Level of immunity to the channel impairments: if the level of the transmitted symbol that encodes the logic state one is sufficiently different from the level that encodes the logic state zero, then the level of noise or interference necessary to originate a corrupted bit needs also to be high. Nevertheless, the

---

* Differential algorithms translate in lower data rates. Since the variation range of the difference signal is lower than that of the samples, a lower number of bits are necessary to encode the signal.

Figure 6.6   A communication chain including a line encoder/decoder.

required transmit power level increases with the increase of the difference between the levels of the two logic states.

- Self-synchronization of the receiver from the received signal: using a digital encoding technique that presents zero crossing in all symbols allows the receiver to extract a clock signal for synchronism* purposes. This avoids spending bandwidth for the transmission of a synchronism signal using a different channel or multiplexed with the data. Nevertheless, this type of zero crossing symbols are typically associated to a higher bandwidth.

- Error detection capability: even without the use of an error detection code, some sequence of levels are not allowed by certain digital encoding techniques. As an example, a Manchester symbol considers always a transition from $+V$ to $-V$ (logic state 0) or from $-V$ to $+V$ (logic state 1). The absence of such transition signalizes the receiver that the symbol is subject to an error.

- Zero value average level: the use of symbols with zero value average level facilitates the amplification process of transmitters.

The symbols used by line codes may be grouped into different classifications:

- Unipolar: a logic state is represented by a positive or negative voltage, whereas the other logic state corresponds to an absence of voltage. An advantage of this technique relies on its simplicity and energy saving, which result from the absence of signal corresponding to one of the logic states.

- Polar: a logic state is represented by a positive voltage, whereas the other logic state corresponds to a negative voltage. According to the central limit theorem, the probability of occurrences of the two logic states is equal. This results in a signal with zero mean value.

- Bipolar: a certain logic state can be represented by a positive or a negative voltage V, whereas the other is represented by an absence of voltage. Moreover, the pole of the voltage V varies sequentially. This represents an advantage as an absence of the expected pole variation allows the receiver to detect an error.

- Biphase: each logic state is represented by a transition from a positive to a negative voltage or the reverse. This assures zero crossing in all symbols, which is an important characteristic for synchronism purposes.

---

* This corresponds to physical layer synchronism, that is, the ability of a receiver to know the exact instant of start and end of a symbol.

6.2.1 Return to Zero

The return to zero line coding belongs to the group of unipolar techniques. It is characterized by representing the logic state one with a certain voltage ($+V$) for half of the bit duration and an absence of voltage during the remaining half of the bit duration. Moreover, the logic state zero is represented by an absence of voltage. This can be seen from Figure 6.7. Alternatively, the representation of the two logic states can be reversed, or the voltage V can be a negative, instead of a positive value.

6.2.2 Non-Return to Zero

The non-return to zero (NRZ) line coding belongs to the group of unipolar techniques. It is characterized by representing the logic state one with a certain voltage ($+V$) and the logic state zero with an absence of voltage. Alternatively, the logic state zero can be represented by a certain voltage ($+V$), while the logic state one is represented by an absence of voltage. Both cases can be seen from Figure 6.8 (options (a) and (b)).

6.2.3 Non-Return to Zero Inverted

The NRZ inverted (NRZ-I) technique belongs to the group of unipolar line coding techniques. It is characterized by representing the logic state one with a transition (from 0 to V or from V to 0) and the logic state zero by an absence of transition (see Figure 6.9). Since the encoded signal is a function of the difference between the

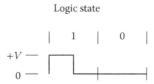

Figure 6.7   Unipolar return to zero.

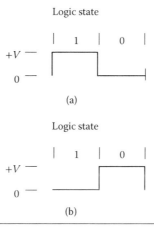

Figure 6.8   Unipolar nonreturn to zero (both options (a) and (b) are possible).

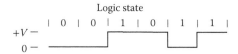

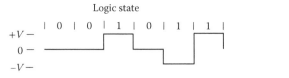

Figure 6.9 Nonreturn to zero inverted.

Figure 6.10 Bipolar alternate mark inversion.

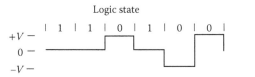

Figure 6.11 Pseudoternary.

previous and following logic state, this technique is also referred to as differential. Alternatively, the representation of the two logic states can be reversed. Note that the voltage V can be a positive or a negative value.

6.2.4 Bipolar Alternate Mark Inversion

The bipolar alternate mark inversion (Bipolar AMI) line coding belongs to the group of bipolar line coding techniques. As can be seen from Figure 6.10, it is characterized by representing the logic state zero with an absence of voltage and the logic state one by an alternating positive and negative voltage ($+V$ or $-V$). Two important characteristics of this line coding techniques rely on the inherent ability to detect errors (two consecutive $+V$ or $-V$ are impossible conditions) and on its zero mean value.

6.2.5 Pseudoternary

The pseudoternary line coding technique corresponds to the biphase AMI with the difference that the logic state one is represented by an absence of voltage, whereas the logic state zero is alternately represented by a positive and a negative voltage ($+V$ or $-V$). This is depicted in Figure 6.11.

6.2.6 Manchester

The Manchester line coding belongs to the group of biphase line coding technique. It is characterized by representing the logic state one with a positive transition at the half bit duration (from 0 to $+V$) and the logic state zero with a negative transition at

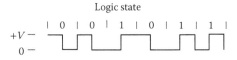

Figure 6.12 Manchester.

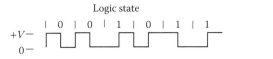

Figure 6.13 Differential Manchester.

the half bit duration (from +V to 0). This can be seen from Figure 6.12. An important advantage of the Manchester technique relies on the inherent synchronism capability, which results from the fact that at least one transition exists per bit duration. Alternatively, the transitions can be reversed for the two logic states, or the considered voltage can be negative, instead of positive.

6.2.7 Differential Manchester

Similarly to the Manchester technique, the differential Manchester belongs to the group of biphase line coding technique. It is characterized by the existence of a transition at half bit duration. However, this transition is not used to encode the bits. It is only used for synchronism purposes. The transition at the beginning of the bit duration is the one that encodes the bits as follows: the existence of a transition corresponds to the logic state zero and the absence of a transition represents the logic state one. This can be seen from Figure 6.13.

6.2.8 Two Binary One Quaternary

Two binary one quaternary (2B1Q) consists of a line coding technique that encodes groups of two bits into a single voltage. The increased code efficiency of the 2B1Q results an important advantage, as compared to previous line coding techniques. To allow this, a total of four voltages are required (two positive and two negative), as described in Table 6.1.

The 2B1Q line coding technique is plotted in Figure 6.14.

Table 6.1 2B1Q Line Coding

LOGIC STATE OF TWO BITS GROUP	SIGNAL LEVEL
00	$-V_2$
10	$+V_2$
11	$+V_1$
01	$-V_1$

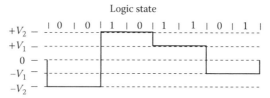

Figure 6.14 2B1Q line coding.

6.3 Modulation Schemes

A modem implements the modulation at the transmitter side, whereas the demodulation process is carried out at the receiver side. The modulation involves the process of encoding one or more source bits into a modulated carrier wave. An important advantage of using a modem, instead of line coding, relies on the ability to select a frequency band where the channel impairments are less destructive to the transported signal. Contrarily, line-encoding techniques always transmit the signals in the baseband, even though if the level of distortion or attenuation is high. Let us consider a voice graded twisted pair as an example, whose bandwidth is limited to the frequency range 300 Hz–3.4 kHz. Above this upper frequency, the attenuation level becomes too high, which translates in a high level of distortion. If one intends to transmit a symbol rate whose bandwidth is higher than that required for voice, the solution is to select a carrier frequency to place the signal (bandpass signal) where the channel impairments are less intense than those experienced in baseband. This is the operation typically implemented by a dial-up or digital subscriber line (DSL) modem. A block diagram of a communication chain including a modem is plotted in Figure 6.15. As can be seen, a modem typically has embedded the error control capability, which consists of error correction or error detection (normally associated to retransmission). Error control techniques are covered in Chapter 10.

The three elementary modulation schemes involve amplitude, frequency, or phase of a carrier, respectively, as parameters to encode source bits [Proakis 1995], as defined in the following subsections.

6.3.1 Amplitude Shift Keying

The amplitude shift keying (ASK) is an elementary modulation scheme that uses the amplitude of a carrier as the parameter to encode bits. As can be seen from Figure 6.16, the logic state one is represented by the transmission of a carrier wave and the logic state zero is represented by the absence of carrier transmission. Alternatively, the encoded logic states can be reversed.

A predefined parameter is the frequency of the carrier wave. Such a parameter is known as the transmitter and receiver. Moreover, the amplitude is another predefined

Figure 6.15   A communication chain including a modem.

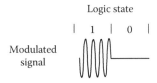

Figure 6.16   Amplitude shift keying.

parameter. A high amplitude level is more resistant to a corruption of bit,* but requires a higher power from the transmitter.

The ASK can easily be extended to $M = 4$ by adopting four amplitude levels (one of this may correspond to the absence of carrier). This results in the 4-ASK modulation, where each symbol encodes $\log_2 4 = 2$ bits.

6.3.2 Frequency Shift Keying

The frequency shift keying (FSK) is an elementary modulation scheme that uses frequency of a carrier as the parameter to encode bits. As can be seen from Figure 6.17, the logic state one is represented by the transmission of a carrier wave with a frequency f_1, while the logic state zero is represented by the transmission of a carrier wave with a frequency f_2. To minimize the occurrence of corrupted bits, the frequencies f_1 and f_2 should be sufficiently far apart to avoid being affected by frequency oscillations or Doppler effects.

6.3.3 Phase Shift Keying

The phase shift keying (PSK) is an elementary modulation scheme that uses phase of a carrier as the parameter to encode bits. As can be seen from Figure 6.18, the logic state one is represented by the transmission of a carrier wave with the phase 0 and the logic state zero is represented by the transmission of a carrier wave with the phase π. Alternatively, the encoded logic states can be reversed.

The described modulation scheme corresponds to binary PSK (BPSK[†]), as there are only two phases (each symbol represents a single bit). However, we may use a symbol to encode more than one bit. Figure 6.19 plots the constellation and the

---

* That is, receive a logic state one when the transmitted logic state was a zero or the reverse.
† The BPSK modulation corresponds to 2-PSK ($M = 2$).

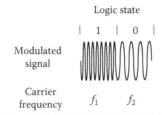

Figure 6.17   Frequency shift keying.

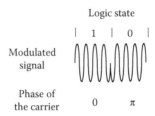

Figure 6.18   (Binary) phase shift keying.

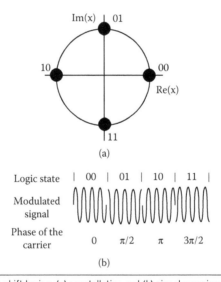

Figure 6.19   Quadrature phase shift keying: (a) constellation and (b) signal mapping.

signal mapping of the quadrature phase shift keying (QPSK). A QPSK symbol presents four discrete values ($M = 4$), which allows encoding two bits ($\log_2 4 = 2$). While the BPSK only uses the real representation (phases 0 and π), the QPSK uses both real and the imaginary parts. This results in a higher spectral efficiency (bit/s/Hz), that is, a higher bit rate transported in the same signal bandwidth, as defined by Equations 6.2 and 6.3. The price to pay for this additional bit rate is the worse bit error probability that results from the decreased minimum Euclidian distance, as compared to BPSK modulation. The minimum Euclidian distance is the minimum distance between two constellation points. For a lower Euclidian distance, a lower level of noise is enough to originate a corrupted bit. This is the reason why higher order modulations require higher levels of SNR. While in the case of the BPSK, the minimum Euclidian distance is 2, in the case of the QPSK, such value is reduced to $\sqrt{2}$.

The bit error probability for M-PSK modulations is presented in Chapter 5, as a function of both E_b/N_0 and C/N.

6.3.4 Multilevel Quadrature Amplitude Modulation Constellations

Multilevel quadrature amplitude modulation (M-QAM) is considered an attractive technique to achieve high throughput within a limited spectrum, due to its high spectral efficiency. Therefore, it has been proposed for wireless systems by several authors [Webb 1994] [Webb 1995] [Goldsmith 1997]. In fact, 16-QAM modulation has already been standardized for the high speed downlink packet access (HSDPA) mode of the UMTS by 3GPP [Marques da Silva, Correia, and Dinis 2009].

M-QAM constellations can be viewed as a mix between PAM and M-PSK. While the M-PSK considers all different constellation points with the same amplitude but with different phases, different constellation points of M-QAM modulation present different amplitudes but with different phases, as well as different phases.

Figure 6.20 shows the constellation points of the 16-QAM modulation. The $M = 16$ levels allows each symbol to encode a total of $\log_2 16 = 4$ bits. As can be seen, moving from one constellation point into an adjacent only changes one bit. A mapping of bits into symbols that follows this rule is known as gray mapping. Most frequent channel impairments (noise, interference, etc.) only originate the movement from one constellation point into its adjacent, which results in a single corrupted bit. As an example, the constellation point 1101 is characterized by having the phase $\pi/4$ and the amplitude $\sqrt{2}a/2$ (a is the Euclidian distance, i.e., the minimum distance between two adjacent constellation points). On the other hand, the constellation point 0000 is characterized by having the phase $3\pi/4$ and the amplitude $3\sqrt{2}a/2$.

The bit error probability for the M-QAM modulation is presented in Chapter 5.

It is worth noting that the modulation scheme may adapt dynamically to the channel conditions. A noisy channel,* or a channel that is more subject to any type of interference, should use a lower order modulation scheme (e.g., QPSK), in order to achieve the same performance. Contrarily, a channel whose level of impairments is lower allows using a higher order modulation scheme (e.g., 16-QAM). As described in Chapter 10, the adaptive modulation and coding (AMC) technique considers the change of the modulation scheme/order and/or code rate dynamically as a function of the SNR.

---

* Which corresponds to a lower SNR level.

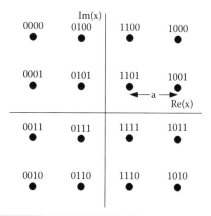

Figure 6.20   Constellation of 16-quadrature amplitude modulation.

6.4 Coding Efficiency of a Symbol

The coding efficiency of a symbol used in a line coding or a modem consists of the quotient between the source bit rate R_B and the modulation frequency f_M used to encode the source bit rate. The modulation frequency corresponds to the number of different discrete levels per second (amplitude, frequency, phase, or a combination of these elements), which is used to encode the source bit rate. We may express the coding efficiency as

$$\eta = R_B / f_M \tag{6.9}$$

Using a code whose symbol encodes more than one bit (e.g., QPSK*) results in a coding efficiency higher than one. Contrarily, a code that needs more than one level to represent one bit (e.g., Manchester[†]) presents a coding efficiency lower than one.

6.5 Scrambling of Signals

Scrambling consists of an operation that aims to improve the signal quality by changing a sequence and logic state of bits. This is achieved by splitting a long sequence of bits into groups and applying the scrambling operation to each group of bits individually. The scrambling operation presents the following advantages:

- Synchronism: since it breaks long sequence of bits with the same logic state, it increases the number of logic state transitions. This results in an improved capability of the receiver to extract the clock signal from the received signal.
- Error control: after the scrambling of signals, some sequence of bits become impossible. Detecting an impossible sequence of bits gives the receiver the ability to detect an error.

---

* The coding efficiency of the QPSK modulation scheme is 2.
[†] The coding efficiency of the Manchester code is 0.5.

- Security: a third party that intercepts a message cannot decode the data without having knowledge about the generator polynomial. Therefore, the scrambling operation can be viewed as a type of encryption.

Figure 6.21 shows an example of a scrambling encoder (scrambler). A scrambling encoder is implemented using feedback shift registers. The output of the scrambler depicted in Figure 6.21 is given by

$$b_n = a_n \oplus b_{-2} \oplus b_{-5} \tag{6.10}$$

Note that the symbol $\oplus$ in Equation 6.10 stands for modulo 2 adder (XOR). The corresponding generator polynomial is

$$P(x) = 1 + x^{-2} + x^{-5} \tag{6.11}$$

The encoder is normally initialized by filling all registers of the shift register with zero value bits. Then, for each data bit that is fed, the shift register shifts once to the right. An important parameter of a scrambler is the constraint length. It corresponds to the number of previous input bits that a certain output bit depends on. This corresponds to the number of memory registers.

Figure 6.22 shows an example of a scrambling decoder (descrambler). It is implemented using feedforward shift registers. The output of the descrambler depicted in Figure 6.22 is given by

$$c_n = b_n \oplus b_{-2} \oplus b_{-5} \tag{6.12}$$

This corresponds to the decoder of the scrambling encoder depicted in Figure 6.21.

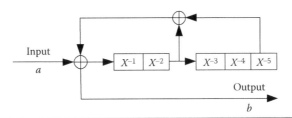

Figure 6.21 Example of a scrambling encoder.

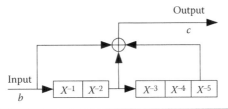

Figure 6.22 Example of a scrambling decoder.

6.6 Multiplexing

Multiplexing consists of a mechanism that allows the sharing of communication resources among different channels. It includes a multiplexer (MUX) at the transmitter and a demultiplexing (DEMUX) at the receiver side. The multiplexing is the operation of encapsulating different channels into a single structured signal for transmission into a common transmission medium. Figure 6.23 shows a generic block diagram of a link using multiplexing.

Depending on whether different channels are transported in different frequency subcarriers or different time slots, the multiplexing technique is referred to as frequency division multiplexing (FDM) or as TDM. Figure 6.24 depicts the frequency and time characteristics of both FDM and TDM signals. Independent channels transmitted in different frequency bands or different time slots are uncorrelated. The uncorrelation is assured by making use of guard bands or guard times between adjacent subcarriers or time slots (see Figure 6.24).

In case the resources are directly shared by different users,* the multiplexing is referred to using equivalent designation followed by the word access, that is, frequency division multiple access or time division multiple access. When different users transmit

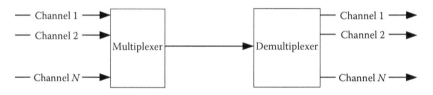

Figure 6.23   Generic block diagram of a link using multiplexing.

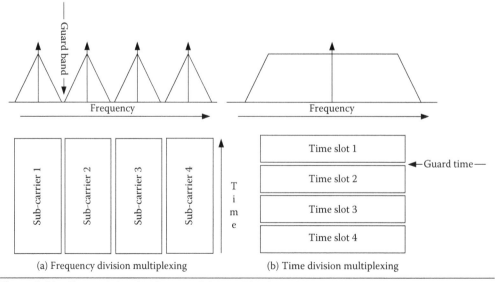

Figure 6.24   Characteristics of (a) frequency division multiplexing and (b) time division multiplexing TDM signals.

---

* Where each user has a different signal.

simultaneously, using the same bandwidth, but using the spread spectrum transmission technique with different spreading codes, the multiple access technique is referred to as code division multiple access (CDMA). This topic is covered in Chapter 7. Another multiple access technique widely used in LAN and MAN is the carrier sense multiple access with collision detection (CSMA-CD) or the carrier sense multiple access with collision avoidance (CSMA-CA). These multiple access techniques are used to perform a statistical multiplexing of the resources, instead of rigidly allocating resources to users (that, at a certain moment, may not need to use). CSMA-CD and CSMA-CA are described in Chapter 10.

6.6.1 Frequency Division Multiplexing

Frequency division multiplexing considers the simultaneous transmission of different channels in different frequency bands. The uncorrelation of signals (channels) among different sub-bands is assured using guard bands. A guard band consists of a frequency band that does not include the transmission of any signal. This is especially important because the limits of frequency bands used for the transmission of signals are not abrupt. This depends on the cut-off response of the bandpass filters, amplifiers, and so on.

FDM signals are currently used by cable television operators to distribute analog or digital television channels. In the past, FDM was widely implemented in coaxial cables, which was the transmission medium used to interconnect different telephonic switching nodes.

Figure 6.25 shows the processing of a FDM transmitter, assuming double side band (DSB) signals. As can be seen, it consists of modulating different channels with different subcarriers $(f_1, f_2,..., f_N)$, followed by an adder module (i.e., sum at signal level). Afterwards, the resulting signal is carrier modulated around certain carrier frequency f_C and bandpass filtered in order to remove the (negative) frequencies below the carrier frequency f_C.

In case the channel consists of an analog telephony channel, the bandwidth B corresponds to 3.4 kHz. In case of digital signals, the minimum bandpass signal bandwidth B_{min} can be deducted from Equation 6.5 as*

$$B_{min} = \frac{R_B}{\log_2 M} \tag{6.13}$$

The total transmitting signal has a bandwidth corresponding to the sum of the elementary signal bandwidths plus the sum of the guard bands existing between adjacent sub-bands.

---

* Note that the maximum bandpass signal bandwidth is calculated for roll-off factor $\alpha = 1$, resulting in $B_{max} = \frac{2R_B}{\log_2 M}$.

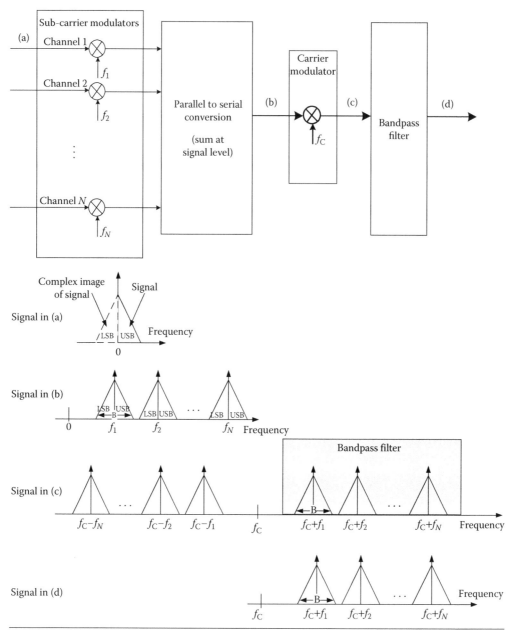

Figure 6.25 Processing of a frequency division multiplexing transmitter with the corresponding signals (double side band signals are assumed).

As can be seen from the plotted figure, the carrier modulation operation translates the baseband signal from the frequency 0 to the frequency f_C, generating frequencies above and below the carrier frequency. This results from the fact that the carrier modulation operation corresponds to a multiplication of a signal in the time domain by a sinusoide $\cos(j\omega)$. Since $\cos(j\omega) = (1/2)e^{j\omega} + (1/2)e^{-j\omega}$, and since $F[v(t)^* e^{j\omega}] = V(f - fc)$, we conclude that the Fourier transform (spectrum) of a time domain signal $v(t)$ multiplied by a complex carrier results in the spectrum of the

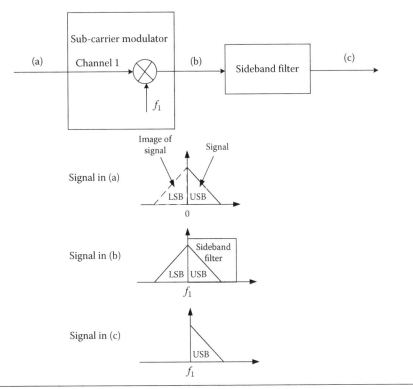

Figure 6.26 Generation of a single side band signal (upper side band).

signal $V(f)$ translated in the spectrum as $V(f-fc)$. Taking into account the frequency translation with both $e^{j\omega}$ and $e^{-j\omega}$, we obtain the plotted signal in Figure 6.25c. To avoid duplication of signals, the bandpass filtering operation assures that only frequency components above the carrier frequency are transmitted. This translates in power and spectrum saving. The transmitted signal corresponds to the plotted signal in Figure 6.25d.

Note that the signals of elementary channels can be either double side band or single side band (SSB). In the latter case, after the subcarrier modulation process, one of the sidebands needs to be filtered (using a sideband filter[*]). SSB signals only consider the transmission of the lower side band (LSB) or upper side band (USB) resulting in a power and spectrum saving [Carlson 1986]. The generation of SSB signals is depicted in Figure 6.26, for the special case of the USB.

Figure 6.27 shows the generic block diagram of a FDM receiver along with the corresponding signals. The first operation consists of bandpass filtering the received signal with a bandwidth corresponding to approximately the bandwidth occupied by all transported channels. This filter is also referred to as the receiving filter, and its main objective consists of removing all the noise present outside the band of the signal of interest. After the filtering operation, the carrier demodulation

---

[*] A sideband filter can be viewed as a bandpass filter.

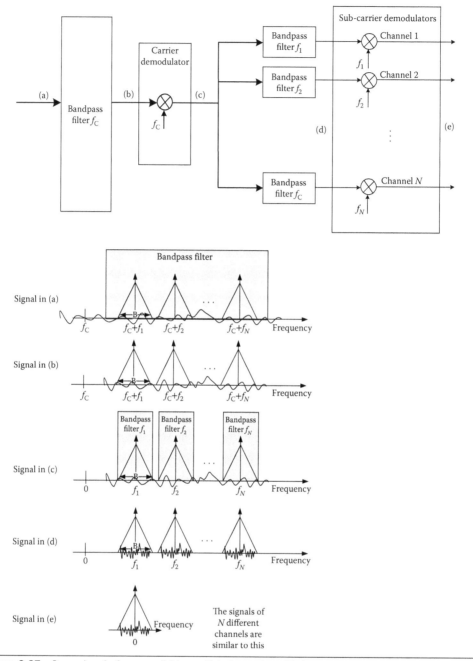

Figure 6.27 Processing of a frequency division multiplexing receiver with the corresponding signals.

operation is performed. This consists of performing a translation of the bandpass signal from the carrier frequency f_C into the baseband (i.e., frequency zero). This is followed by an operation of filtering centered in the subcarrier frequency. This isolates each of the signals, after which independent subcarrier demodulation is performed to recover the replicas of the transmitted signals (corresponding to different channels).

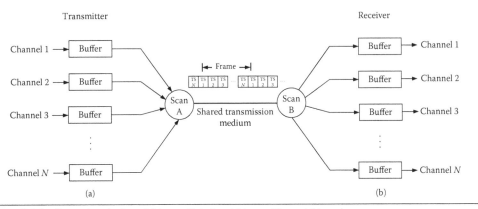

Figure 6.28 Processing of a time division multiplexing: (a) transmitter and (b) receiver.

6.6.2 Time Division Multiplexing

TDM considers the transmission of different channels in different uncorrelated time slots, but using a single carrier frequency. Since a synchronism among different channels in the transmitter and the receiver is required, this multiplexing technique is also referred to as synchronous TDM (STDM).

There are two different interposition methods: bit interposition and word interposition. The former method includes the transmission of a single bit in each time slot. On the other hand, the word interposition comprises the transmission of a group of bits* in each time slot.

Figure 6.28 shows the block diagram of a TDM transmitter and receiver.

The channels 1 to N can be either analog or digital. In case of analog, each channel is typically encoded using the PAM, as previously described. In this case, one sample of a channel (analog signal) is transmitted in each time slot. Since the scan of the receiver (scan B) is synchronized with the scan of the transmitter (scan A), one sample of channel n is transmitted in time slot n, being extracted at the receiver side. In case the signal is digital, one or more bits are transmitted in each time slot. A common source coding technique is the PCM. Both, for analog or digital signals, the sampling period $T_a = 1/f_a$ needs to follow the Nyquist sampling theorem as defined by Equation 6.6, which corresponds to $T_a \leq 1/2B$. Assuming a sampling frequency of 8 ksamples/s, the sampling period becomes $T_a = 125$ μs.

The total throughput of the shared transmission medium is equal or higher than the sum of elementary channel's throughputs. As can be seen from Figure 6.24, the uncorrelation between adjacent time slots is assured by using guard times. This corresponds to time periods between adjacent time slots without data being transmitted. As can be seen from Figure 6.28, the group of all time slots (1, 2, ..., N) is referred to as a frame, and its transmission is repeated every frame interval. This frame consists of a transport frame, not a data link layer frame.† This frame does not include any kind

---

* Using the PCM coding, a word consists of a group composed of eight bits (one signal sample).
† The data link layer is described in Chapter 10.

of error control (detection or correction). Error control capability is performed for the data link layer frame.* Nevertheless, the transport frame needs to include some additional bits for signalization of begin and end of a frame (using, e.g., a flag). This allows the scan A and scan B devices to keep synchronization.

End of Chapter Questions

1. What is the difference between a line coder and a modem?
2. Which kind of line encoding techniques do you know?
3. What is the difference between FDM and TDM?
4. Which kind of multiplexing techniques do you know?
5. What is a scrambler used for?
6. What does coding efficiency stands for?
7. What is the minimum baseband bandwidth required to accommodate a baseband signal which comprises a symbol rate of 2 Msymb/s?
8. What is the minimum passband bandwidth required to accommodate a carrier modulated signal which comprises a symbol rate of 2 Msymb/s?
9. Describe the block diagram of a FDM transmitter.
10. Describe the block diagram of a FDM receiver.
11. Describe the block diagram of a TDM transmitter and receiver.
12. How can we generate an SSB signal?
13. What can be the advantages of transmitting signals using a modem, instead of a line encoder?
14. What is the bandpass filter used for in a FDM receiver?
15. What does Euclidian distance stands for?
16. Why should the Euclidian distance be maximized?
17. Define the Manchester line encoding technique.
18. Define the bipolar AMI line encoding technique.
19. Define the non-return to zero inverted line encoding technique.
20. What is the difference between amplitude, frequency, and phase shift keying?
21. Give an example of a modulation scheme that comprises both amplitude and phase shift keying.
22. What does a differential codec stands for? What is its advantage relating to a nondifferential codec?
23. What is the advantage of digital source data, relating to analog data?
24. Which types of voice codec do you know?
25. What are the differences between the ITU-T G.711 standard versions used in the US and in Europe?

---

* For example, the point-to-point protocol and the HDLC data link layer protocols use cyclic redundancy codes as error detection codes.

26. What is the advantage of a logarithmic voice codec, relating to a uniform voice codec?
27. How does the ITU-T G.711 standard use the eight bits to encode a sample signal of analog voice?
28. What is the difference between a MPEG and DVB?
29. Describe the PAL video system. What is the bandwidth of a PAL video signal?
30. Considering an analog signal is intended to be digitized, what is the minimum sampling rate that can be used?
31. What is the sampling rate used in PCM?

7

ADVANCED TRANSMISSION TECHNIQUES TO SUPPORT CURRENT AND EMERGENT MULTIMEDIA SERVICES

7.1 Advances in Wireless Systems and Their Technical Demands

Today, the challenge facing the mobile telecommunications industry is how to continually improve the end-user experience and to offer appealing services through a delivery mechanism that offers improved speed, service attractiveness, and service interaction. Furthermore, to deliver the required services to the users with the minimum cost, the technology should allow better and better performances, higher throughputs, improved capacities, and higher spectral efficiencies. The following sections describe several measures to reach such desiderates.

The bandwidth requirements for future wireless systems present a considerable challenge since multipath propagation leads to severe time dispersion effects. In this case, conventional time-domain equalization schemes are not practical. Block transmission techniques, with appropriate cyclic prefixes (CP) and employing frequency-domain equalization (FDE) techniques, have been shown to be suitable for high data rate transmission over severely time-dispersive channels [Falconer et al. 2002], and therefore presenting advantages for use with emergent wireless systems. Orthogonal frequency-division multiplexing (OFDM) technique is the most popular modulation based on this technique. OFDM technique has been selected for long-term evolution (LTE), as opposed to wideband code division multiple access (WCDMA) which is the air interface technique that has been selected by European Telecommunications Standard Institute (ETSI) for Universal Mobile Telecommunication System (UMTS). Single-carrier modulation using FDE is an alternative approach based on this principle [Sari et al. 1994; Falconer et al. 2002]. Due to the lower envelope fluctuations of the transmitted signals (and implicitly a lower peak-to-mean envelope power ratio [PMEPR]), single carrier–frequency domain equalization (SC-FDE) schemes (also entitled as single carrier–frequency division multiple access [SC-FDMA]) are especially interesting for the uplink transmission [Falconer et al. 2002]. Furthermore, multiple input multiple output (MIMO) schemes enhanced with state-of-the-art receivers is also normally associated to multimedia broadcast and multicast services (MBMS) in order to improve the overall system performance in terms of capacity, spectral efficiency, and coverage.

7.2 Spread Spectrum Communications

Digital communications employing spread spectrum signals are characterized by using a bandwidth, B_{wd}, much greater than the information bit rate. This means that for a spread spectrum signal, we have $SF = B_{wd}/R_b >> 1$, where R_b is the information bit rate and SF is the bandwidth expansion factor (called spreading factor or processing gain). This bandwidth expansion can be accomplished through different types of spread spectrum techniques [Glisic and Vucetic 1997; Proakis 2001]:

- Direct sequence (DS): Figure 7.1 shows the general scheme for a DS spread spectrum system. The information symbols are encoded and then modulated in combination with a pseudorandom sequence. In the receiver, the demodulator removes the pseudorandom impression from the received signal. DS techniques can be performed in two distinct ways. In the most common techniques, the transmitted bits are encoded using a code with a given coding rate (say 1/2 or 1/3) and then the multiplication by the pseudorandom sequence increases the transmitted signal bit rate. This can be seen as applying a repetition code to the encoded bits and then adding the pseudorandom signature to the signal [Proakis 2001]. An alternative way of implementing DS techniques corresponds to performing all the spreading of the signal using a low-rate code and then adding the pseudorandom signature to the signal. This approach is usually referred to as code spread (CS).
- Frequency hopping (FH): In this case, the available bandwidth is divided into several contiguous sub-band frequency slots. A pseudorandom sequence is used for selecting the frequency slot for transmission in each signaling interval.
- Time hopping (TH): In this method, a time interval is divided into several time slots and the coded information symbols are transmitted in a time slot selected according to a pseudorandom sequence. The coded symbols are transmitted in the selected time slot as blocks of one or more codewords.
- Hybrid techniques: DS, FH, and TH can be combined to obtain other types of spread spectrum signals. For example, a system can use a combination of DS and FH, where the transmitted signal is associated with two code sequences. One of the sequences is multiplied by the signal to be transmitted, whereas the second is used for selecting the frequency slot for transmission in each signaling interval.

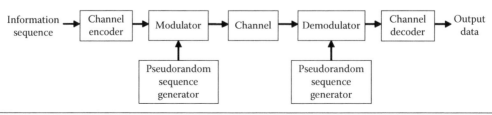

Figure 7.1   General scheme of a direct sequence spread spectrum system.

In the following section, we will only consider DS spreading techniques. One common application of DS spread spectrum signals is code division multiple access (CDMA) communications, where several users share the same channel bandwidth for transmitting information simultaneously. If the spreading of each user is the joint effect of the channel encoder and the spreading code, then the transmission is designated as DS-CDMA. If the bandwidth expansion is performed by the channel encoder alone, then the transmission is referred to as CS-CDMA. In [Viterbi 1990], it is shown that this technique can achieve maximum theoretical performance. In fact, DS-CDMA can be regarded as a special case in CS-CDMA. All the users can transmit in the same frequency band and may be distinguished from each other by using a different pseudorandom sequence. If all the signals are transmitted with the symbols synchronized between them and we have a flat fading channel, then for each symbol of a user, there will be only one interfering symbol from each of the other users. This is called synchronous CDMA transmission and is usually employed in the downlink connection between a beam station (BS) and the users in a cell. If each transmitted symbol interferes with two symbols of any other user, then the transmission is asynchronous. This is the common method in the uplink connection.

7.3 Code Division Multiple Access

Code division multiple access system consists of different spread spectrum transmissions [Ojanperä and Prasad 1998] [Holma and Toskala 2000], each one associated to a different user's transmission using a different (ideally orthogonal) spreading sequence. Narrowband code division multiple access system was adopted in the nineties by IS-95 standard in the United States. Afterwards, UMTS proceed with the utilization of CDMA, in this particular case wideband CDMA.

Figure 7.2 depicts a simplified block diagram of a spread spectrum transmitter, where the first block consists of a symbol modulator (mapper), responsible for the conversion of source bits into symbols. The resulting signal is then sampled for spreading purposes. The spreader consists of a block that performs the multiplication of the sampled versions of the symbols with the samples of the spreading sequence. Finally, the signal is passed through a band limited pulse shaping filter and carrier modulator. At the receiver, the signal is convoluted again with exactly the same spreading sequence.

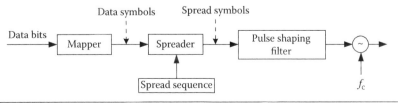

Figure 7.2 Generic block diagram of a spread spectrum transmitter.

The operations performed by a CDMA transmitter can be, as seen in Figure 7.3, applied to a binary phase shift keying (BPSK) modulated signal. In this figure T_S and T_C stand for the symbol and chip period and S_F stands for the spreading factor.

The main characteristics of spread spectrum systems are the additional resistance to interference and the possibility to take advantage of the multipath channel in order to exploit multipath diversity. This leads to an improved performance and spectral efficiency, as compared to narrowband signals. The relationship between the power spectral densities of a spread spectrum signal and that of the original signal corresponds to the spreading factor. Figure 7.4 depicts the exposed concept.

Assuming that different signals use orthogonal spreading sequences, the corresponding signals are also orthogonal and the same spectrum can be shared among different signals. Moreover, WCDMA consists of a CDMA system whose spread bandwidth

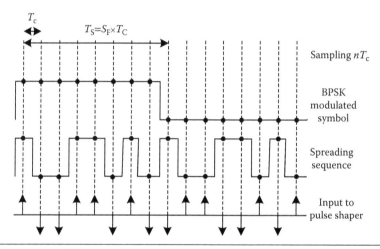

Figure 7.3   Generation of a BPSK spread spectrum signal (spreading factor = 8).

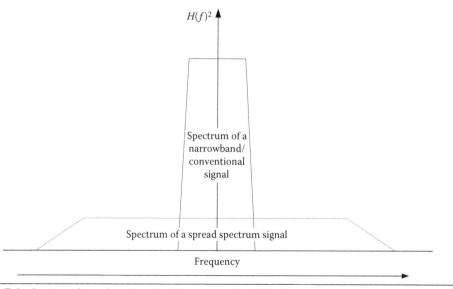

Figure 7.4   Spectrum of spread spectrum signal versus narrowband signal.

is typically higher than the coherence bandwidth of the channel. This allows a better exploitation of multipath diversity but requires higher spectrum availability.

The typical WCDMA receiver is a RAKE receiver [Glisic and Vucetic 1997], which has several fingers to detect different multipaths of the channel. In multipath environment, since each finger of the RAKE receiver discriminates a different multipath, the combination of the signals from different fingers with a maximum ratio combining tends to achieve a performance improvement compared to a single decorrelator. For this reason, it is normally stated that a WCDMA system jointly with RAKE receiver is able to exploit multipath diversity. As the spreading factor increases, the resolution of the RAKE receiver also increases, allowing better discrimination of the several propagation paths, increasing the diversity order, and potentially, improving the performance.

Assuming correct synchronization, the received signal at the RAKE receiver output is the original signal plus higher frequency components, which are not part of the original signal and are to be filtered. If there is any undesired interfering signal at the receiver, the spread spectrum signal will affect it just as it did to the original signal at the transmitter, spreading it to the bandwidth of the spread spectrum signal. Thus, the neglecting effect of the interfering signal is less powerful than in conventional narrowband signals. Furthermore, if the spreading sequences of the different signals that share the same spectrum are not perfectly orthogonal at the receiver side (due to lack of orthogonality itself, due to lack of synchronization, due to the multipath effect, etc.), then the resulting signal is composed of the desired signal plus the noise and a component called multiple access interference (MAI), which can be mitigated by use of a multiuser detector (MUD) for WCDMA signals [Marques da Silva 2003a] [Marques da Silva 2005a] [Marques da Silva 2005c].

7.3.1 General Model

To model a DS-CDMA system, we will first consider single-user transmission. Figure 7.5 represents the basic communication model.

According to this model, a sequence of symbols, s_i, representing the information to be transmitted, enters the modulator that outputs a sequence of ideal pulses $\delta(t - jT_c)$ modulated by the spreaded symbols $\sqrt{E_c} \cdot s_{[j/SF]} \cdot c_j$ (T_{chip} is the chip duration, c_j is the jth chip of the spreading sequence, and E_c is the average chip energy). This sequence passes through a shaping filter with impulse response $h_T(t)$ and frequency response

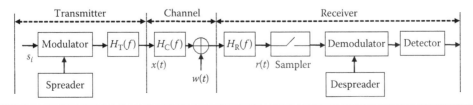

Figure 7.5   Basic DS-CDMA communication link.

$H_T(f)$, resulting in a transmitted signal that can be written in the following equivalent low-pass form

$$x(t) = \left(\sqrt{E_c} \sum_{j=0}^{\infty} s_{\lfloor j/SF \rfloor} \cdot c_j \cdot \delta(t - jT_{chip}) \right) * h_T(t) = \sqrt{E_c} \sum_{j=0}^{\infty} s_{\lfloor j/SF \rfloor} \cdot c_j \cdot h_T(t - jT_{chip}). \quad (7.1)$$

The signal is then transmitted through the channel, which is modeled by a time invariant linear system with impulse response $h_C(t)$ (frequency response $H_C(f)$), using, for example, the model defined in Appendix A. At the receiver, the signal passes through a filter with impulse response $h_R(t)$ (frequency response $H_R(f)$). The resulting signal is then obtained as the convolution of the transmitted signal with $h_C(t)$ and $h_R(t)$, that is,

$$r(t) = x(t) * h_C(t) * h_R(t) + n(t) \quad (7.2)$$

where $n(t)$ is the noise component at the output of the receiver filter, that is, it is given by

$$n(t) = w(t) * h_R(t) \quad (7.3)$$

where $w(t)$ denotes the white noise at the receiver input. To design the shaping filter, $H_T(f)$, and the reception filter, $H_R(f)$, we will consider an ideal channel, that is, $h_C(t) = \delta(t)$ ($H_C(f) = 1$). Therefore, the received signal can be written as

$$r(t) = \sqrt{E_c} \sum_{j=0}^{\infty} s_{\lfloor j/SF \rfloor} \cdot c_j \cdot p(t - jT_{chip}) + n(t) \quad (7.4)$$

where $p(t)$ corresponds to the time response of the cascade of the transmitter and receiver filters

$$p(t) = h_T(t) * h_R(t). \quad (7.5)$$

After the filter, the signal is sampled at rate f_{chip} ($f_{chip} = 1/T_{chip}$). The sequence of samples r_k can be represented as (considering no delay in the transmission)

$$r_k \equiv r(t = kT_{chip}) = \sqrt{E_c} \sum_{j=0}^{\infty} s_{\lfloor j/SF \rfloor} \cdot c_j \cdot p(kT_{chip} - jT_{chip}) + n(kT_{chip}) \quad (7.6)$$

or

$$r_k = \sqrt{E_c} \sum_{j=0}^{\infty} s_{\lfloor j/SF \rfloor} \cdot c_j \cdot p_{k-j} + n_k$$
$$= \sqrt{E_c} s_{\lfloor k/SF \rfloor} \cdot c_k \cdot p_0 + \sum_{\substack{j=0 \\ j \neq k}}^{\infty} s_{\lfloor j/SF \rfloor} \cdot c_j \cdot p_{k-j} + n_k \quad (7.7)$$

In the last passage, we admitted that $p_0 = 1$. The second term of the last equation represents the intersymbol interference (ISI), which is not desired. The condition for no ISI is

$$p(kT_{\text{chip}}) = p_k = \begin{cases} 1, & k = 0 \\ 0, & k \neq 0 \end{cases} \qquad (7.8)$$

According to *Nyquist pulse shaping criterion*, the necessary and sufficient condition for $p(t)$ to obey the above condition is its Fourier transform $P(f)$ that satisfies [Proakis 2001]

$$P_{\text{eq}}(f) = \sum_{l=-\infty}^{\infty} P\left(f + \frac{l}{T_{\text{chip}}}\right) = T_{\text{chip}} \qquad (7.9)$$

where

$$P(f) = H_{\text{T}}(f)H_{\text{R}}(f) \qquad (7.10)$$

There are several functions satisfying the Nyquist criterion. One of the most common functions used for $P(f)$ is the family of raised cosine functions that can be expressed as

$$P(f) = \begin{cases} T_{\text{chip}}, & |f| \leq \dfrac{1-\beta}{2T_{\text{chip}}} \\[2mm] \dfrac{T_{\text{chip}}}{2}\left\{1 + \cos\left[\dfrac{\pi T_{\text{chip}}}{\beta}\left(|f| - \dfrac{1-\beta}{2T_{\text{chip}}}\right)\right]\right\}, & \dfrac{1-\beta}{2T_{\text{chip}}} < |f| \leq \dfrac{1+\beta}{2T_{\text{chip}}} \\[2mm] 0, & |f| > \dfrac{1+\beta}{2T_{\text{chip}}} \end{cases} \qquad (7.11)$$

where β is called the roll-off factor with $0 \leq \beta \leq 1$. This factor represents the fraction of excess bandwidth beyond the Nyquist frequency $1/(2T_{\text{chip}})$, and therefore $P(f)$ has a bandwidth of

$$B_{\text{wd}} = \frac{\beta + 1}{2T_{\text{chip}}}. \qquad (7.12)$$

The corresponding impulse response in the time domain of the raised cosine function $P(f)$ is

$$p(t) = \text{sinc}\left(\frac{\pi t}{T_{\text{chip}}}\right)\frac{\cos\left(\dfrac{\pi \beta t}{T_{\text{chip}}}\right)}{1 - \left(\dfrac{2\pi t}{T_{\text{chip}}}\right)^2} \qquad (7.13)$$

The receiver filter can be matched to the transmitter filter, that is, $h_R(t) = h_T^*(-t)$ and thus $H_T(f)$ and $H_R(f)$ will both have root raised cosine responses, that is,

$$\left|H_T(f)\right| = \left|H_R(f)\right| = \sqrt{\left|P(f)\right|} \tag{7.14}$$

As an example, in UMTS, the shaping filter $h_T(t)$ is a root-raised cosine with a roll-off factor $\beta = 0.22$ [3GPP 25.101-v6.6.0].

Therefore, if $p(t)$ satisfies the *Nyquist pulse shaping criterion*, then Equation (7.7) reduces to

$$r_k = \sqrt{E_c} s_{\lfloor k/SF \rfloor} \cdot c_k + n_k \tag{7.15}$$

The sequence of samples r_k is then despreaded by multiplying it by the conjugate of the spreading sequence and averaging over sets of SF chips. The resulting decision variables z_i for each information symbol, i, can be expressed as

$$z_i = \frac{1}{SF} \sum_{k=i \cdot SF}^{(i+1) \cdot SF - 1} c_k^* \cdot r_{0,k} \tag{7.16}$$

Although the channel effect was considered ideal for obtaining (7.14), in a real system the receiver has to compensate its effect. This can be accomplished either by including $h_C(t)$ in the function $p(t)$ (resulting in $p(t) = h_T(t) * h_R(t) * h_C(t)$), and then computing the receiver filter as $H_R(f) = P(f)/[H_T(f)H_C(f)]$ (this corresponds to channel equalization) where $P(f)$ satisfies the Nyquist criterion (using (7.11) for example), or using some other channel compensation processing technique, as will be shown further ahead.

7.3.2 Narrowband CDMA

CDMA systems can be considered narrowband or wideband depending on the mobile propagation conditions according to the definitions presented above. An example of a cellular system that employs narrowband CDMA technique is the IS-95 standard. If the transmitted signal bandwidth is lower than the coherence bandwidth of the channel for the environments for which the system was projected, then there will be only one distinguishable received replica of the signal. In this case, the system is narrowband CDMA.

Let us consider now a multiuser environment where $N_u + 1$ users are transmitting simultaneously as shown in Figure 7.6.

In this case, the signal transmitted by each user u can be expressed in a form similar to Equation 7.1 as

$$x_u(t) = \sqrt{E_c} \sum_{j=0}^{\infty} s_{u,\lfloor j/SF \rfloor} \cdot c_{u,j} \cdot h_T(t - jT_{chip}) \tag{7.17}$$

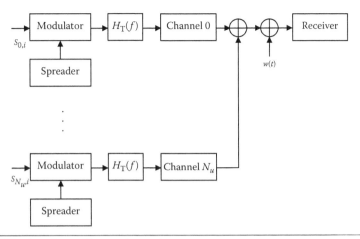

Figure 7.6 CDMA scheme in a multiuser environment.

where $s_{u,\lfloor j/\text{SF} \rfloor}$ is the information symbol and $c_{u,j}$ is the symbol of the spreading sequence for user u.

First we will admit that the CDMA system is narrowband, and thus, the channel impulse response (CIR) of each user can be modeled as

$$h_{C,u}(t) = \alpha_u \delta(t - \tau_u) \qquad (7.18)$$

where α_u is a complex attenuation (α_u is a random process dependent of the time but since we are admitting that $f_D T_{\text{chip}} \ll 1$, with f_D denoting the Doppler frequency ($f_D = f_c v/c$, with f_c denoting the carrier frequency, v the terminal speed, and c the speed of light) we can assume it is approximately constant during a symbol period). In Figure 7.7, a basic receiver for user 0 is shown. First the received signal passes through the matched filter $h_R(t)$. From Equation 7.2, this signal can be written as

$$r(t) = \sqrt{E_c} \sum_{u=0}^{N_u} \sum_{j=0}^{\infty} s_{u,\lfloor j/\text{SF} \rfloor} \cdot c_{u,j} \cdot \alpha_u \cdot p(t - jT_{\text{chip}} - \tau_u) + n(t) \qquad (7.19)$$

The filter output is then sampled at times $t = kT_{\text{chip}} + \tau_0$ for extracting user 0 according to

$$
\begin{aligned}
r_{0,k} &\equiv r(t = kT_{\text{chip}} + \tau_0) \\
&= \sqrt{E_c} \sum_{u=0}^{N_u} \sum_{j=0}^{\infty} s_{u,\lfloor j/\text{SF} \rfloor} \cdot c_{u,j} \cdot \alpha_u \cdot p(kT_{\text{chip}} + \tau_0 - jT_{\text{chip}} - \tau_u) + n(kT_{\text{chip}} + \tau_0) \\
&= \sqrt{E_c} \cdot s_{0,\lfloor k/\text{SF} \rfloor} \cdot c_{0,k} \cdot \alpha_0 + \sqrt{E_c} \sum_{u=1}^{N_u} \sum_{j=0}^{\infty} s_{u,\lfloor j/\text{SF} \rfloor} \cdot c_{u,j} \cdot \alpha_u \cdot p(kT_{\text{chip}} + \tau_0 - jT_{\text{chip}} - \tau_u) + n_{0,k}
\end{aligned}
\qquad (7.20)
$$

Note that it was taken into account that $p(t)$ satisfies the *Nyquist pulse shaping criterion* and thus Equation 7.8 is valid. In (7.20), the second term represents the interference component, from the other users (MAI) which, for an asynchronous transmission, has contributions from all the transmitted chips of those users since generally $\tau_0 - \tau_u \neq aT_{\text{chip}} \quad (\forall a \in \mathbb{Z})$.

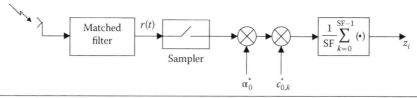

Figure 7.7   DS-CDMA receiver scheme for user 0.

The channel is compensated by multiplying the sampled sequence by the complex conjugate of α_u (considering perfect channel knowledge). The resulting sequence is then despreaded by multiplying it by the complex conjugate of the respective spreading sequence, $c_{0,j}$, and each set of SF samples belonging to the same information symbol is summed. The decision variable $z_{0,i}$ for the ith information symbol, can be expressed as

$$
\begin{aligned}
z_{0,i} &= \frac{1}{SF} \sum_{k=i\cdot SF}^{(i+1)\cdot SF-1} \alpha_0^* \cdot c_{0,k}^* \cdot r_{0,k} \\
&= \frac{\sqrt{E_c}\cdot|\alpha_0|^2}{SF} \sum_{k=i\cdot SF}^{(i+1)\cdot SF-1} s_{0,\lfloor k/SF\rfloor} \cdot |c_{0,k}|^2 \\
&+ \frac{\sqrt{E_c}}{SF} \sum_{k=i\,SF}^{(i+1)SF-1} \sum_{u=1}^{N_u} \sum_{j=0}^{\infty} \alpha_0^* \cdot c_{0,k}^* \cdot s_{u,\lfloor j/SF\rfloor} \cdot c_{u,j} \cdot \alpha_u \cdot p(kT_{chip}+\tau_0-jT_{chip}-\tau_u) \\
&+ \frac{1}{SF} \sum_{k=i\cdot SF}^{(i+1)\cdot SF-1} \alpha_0^* \cdot c_{0,k}^* \cdot n_{0,k} \\
&= \sqrt{E_c}\cdot|\alpha_0|^2 \cdot s_{0,i} + \frac{\sqrt{E_c}}{SF}\sum_{u=1}^{N_u}\sum_{j=0}^{\infty} \alpha_0^* \cdot s_{u,\lfloor j/SF\rfloor} \cdot \alpha_u \cdot \sum_{k=i\cdot SF}^{(i+1)\cdot SF-1} c_{0,k}^* \cdot c_{u,j} \cdot p(kT_{chip}+\tau_0-jT_{chip}-\tau_u) \\
&+ \frac{1}{SF} \sum_{k=i\cdot SF}^{(i+1)\cdot SF-1} \alpha_0^* \cdot c_{0,k}^* \cdot n_{0,k}
\end{aligned}
\tag{7.21}
$$

In this equation, it was assumed that $|c_{0,k}|^2 = 1$. In Equation 7.21, the second term represents the interference from the other users (MAI), which depends on the cross-correlation between the desired spreading code, and the spreading code of interferer u weighted by the pulse shaping function $p(t)$, which depends on the relative delays. In a downlink synchronous transmission, the fading coefficients and the delays are all the same, $\alpha_0 = \alpha_u$ and $\tau_0 = \tau_u\ (u = 1...N_u - 1)$, and Equation 7.21 simplifies to

$$
z_{0,i} = \sqrt{E_c}\cdot|\alpha|^2\cdot s_{0,i} + \frac{\sqrt{E_c}}{SF}\cdot|\alpha|^2\cdot\sum_{u=1}^{N_u} s_{u,i} \cdot \sum_{k=i\cdot SF}^{(i+1)\cdot SF-1} c_{0,k}^* \cdot c_{u,k} + \frac{\alpha^*}{SF} \sum_{k=i\cdot SF}^{(i+1)\cdot SF-1} c_{0,k}^* \cdot n_{0,k} \tag{7.22}
$$

In this case, to minimize the multiuser interference, it is only necessary to employ spreading sequences with low cross-correlation values. For some values of SF (e.g., if SF is a power of 2) it is possible to design sequences where

$$
\sum_{k=0}^{SF-1} c_{0,k}^* \cdot c_{u,k} = 0 \tag{7.23}
$$

that is, orthogonal sequences in which case there will be no multiuser interference (in a synchronous and for the single path propagation channel). This is what is done in the downlink connection of UMTS [3GPP 25.213-v6.1.0].

7.3.3 Wideband CDMA

If the transmitted signal bandwidth is greater than the coherence bandwidth of the channel then it will be possible to resolve several multipath components and it will correspond to a wideband CDMA system. The UMTS system, which employs DS-CDMA transmission techniques, is a wideband system. The CIR for WCDMA systems can be described using a tapped delay line model [Silva 2003], where the L multipaths are considered discrete. Considering that the channel is time invariant, the CIR for each user can be written as

$$h_{c,u}(t) = \sum_{l=1}^{L} \alpha_{u,l} \cdot \delta(t - \tau_{u,l}) \tag{7.24}$$

where $\alpha_{u,l}$ is the fading coefficient affecting the transmitted signal of user u for the lth propagation path and $\tau_{u,l}$ is the respective time delay with $\tau_{u,l} = \tau_{u,1} + m_l T_{chip}$, $m_l \in \mathbb{Z}$ (the relative path delays for the same user are integer multiples of T_{chip}).

Since in wideband CDMA systems the multipath replicas arriving with relative delays higher than the chip duration carry information about the transmitted signal, then there is the possibility that when a replica is severely attenuated due to the fading the others may be received in more favorable conditions. Due to the good autocorrelation properties of the spreading codes usually used in CDMA systems, it is possible to distinguish and extract the strongest replicas (with relative delays higher than the chip duration) present in the received signal and thus obtain diversity. These replicas can be extracted and combined using a RAKE receiver [Price 1958] to help recovering the transmitted signal, as shown in Figure 7.8. To maximize the resulting signal-to-noise ratio (SNR) at the receiver, the extracted replicas are weighted using the complex conjugate of the respective fading coefficient and then added. This technique is designated as maximal ratio combining (MRC) [Proakis 2001].

According to the scheme shown in Figure 7.8, first, the received signal goes through a matched filter similar to what was done in the narrowband CDMA receiver. From Equations 7.2, 7.17, and 7.24, the filtered signal $r(t)$ can be expressed as follows:

$$r(t) = \left[\sum_{u=0}^{N_u-1} x_u(t) * h_{C,u}(t) \right] * h_{\text{R}}(t) + n(t)$$

$$= \sqrt{E_c} \sum_{u=0}^{N_u} \sum_{l=1}^{L} \sum_{j=0}^{\infty} \alpha_{u,l} \cdot s_{u,\lfloor j/SF \rfloor} \cdot c_{u,j} \cdot p(t - jT_{\text{chip}} - \tau_{u,l}) + n(t) \tag{7.25}$$

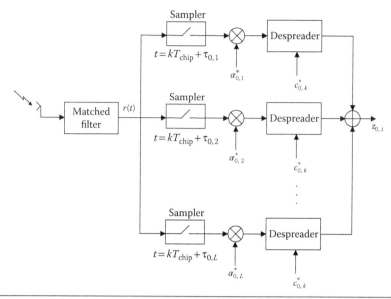

Figure 7.8   RAKE receiver scheme for user 0.

The rest of the RAKE is composed of up to L parallel branches denoted as fingers. Each finger extracts one of the received replicas for the target user. The processing steps inside each finger are similar to the ones performed in the receiver for narrowband CDMA system. Therefore, first the filter output is sampled at times $t = kT_{\text{chip}} + \tau_{0,f}$ for extracting user 0 in finger f. This can be expressed as follows:

$$
\begin{aligned}
r_{0,k,f} &\equiv r(t = kT_{\text{chip}} + \tau_{0,f}) = \sqrt{E_c} \sum_{u=0}^{N_u} \sum_{l=1}^{L} \sum_{j=0}^{\infty} \alpha_{u,l} \cdot s_{u,\lfloor j/\text{SF} \rfloor} \cdot c_{u,j} \cdot p(kT_{\text{chip}} + \tau_{0,f} - jT_{\text{chip}} - \tau_{u,l}) \\
&\quad + n(kT_{\text{chip}} + \tau_{0,f}) \\
&= \sqrt{E_c} \cdot s_{0,\lfloor k/\text{SF} \rfloor} \cdot c_{0,k} \cdot \alpha_{0,f} + \sqrt{E_c} \sum_{\substack{l=1 \\ l \neq f}}^{L} \sum_{j=0}^{\infty} \alpha_{0,l} \cdot s_{0,\lfloor j/\text{SF} \rfloor} \cdot c_{0,j} \cdot p(kT_{\text{chip}} + \tau_{0,f} - jT_{\text{chip}} - \tau_{0,l}) \\
&\quad + \sqrt{E_c} \sum_{u=1}^{N_u} \sum_{l=1}^{L} \sum_{j=0}^{\infty} \alpha_{u,l} \cdot s_{u,\lfloor j/\text{SF} \rfloor} \cdot c_{u,j} \cdot p(kT_{\text{chip}} + \tau_{0,f} - jT_{\text{chip}} - \tau_{u,l}) + n_{0,k,f} \\
&= \sqrt{E_c} \cdot s_{0,\lfloor k/\text{SF} \rfloor} \cdot c_{0,k} \cdot \alpha_{0,f} + \sqrt{E_c} \sum_{\substack{l=1 \\ l \neq f}}^{L} \alpha_{0,l} \cdot s_{0,\lfloor (k+(\tau_{0,f}-\tau_{0,l})/T_{\text{chip}})/\text{SF} \rfloor} \cdot c_{0,k+(\tau_{0,f}-\tau_{0,l})/T_{\text{chip}}} \\
&\quad + \sqrt{E_c} \sum_{u=1}^{N_u} \sum_{l=1}^{L} \sum_{j=0}^{\infty} \alpha_{u,l} \cdot s_{u,\lfloor j/\text{SF} \rfloor} \cdot c_{u,j} \cdot p(kT_{\text{chip}} + \tau_{0,f} - jT_{\text{chip}} - \tau_{u,l}) + n_{0,k,f}
\end{aligned}
\tag{7.26}
$$

To obtain this equation, it was assumed that $p(t)$ satisfies the *Nyquist pulse-shaping criterion* and thus Equation 7.8 is valid.

The channel is compensated by multiplying the sampled sequence by the complex conjugate of $\alpha_{u,f}$ (assuming perfect channel knowledge). The resulting sequence is then despreaded and the result is summed with the outputs of the other fingers. The decision variable $z_{0,i}$ for the ith information symbol can be expressed as follows:

$$z_{0,i} = \frac{1}{\text{SF}} \sum_{f=1}^{L} \sum_{k=i\cdot\text{SF}}^{(i+1)\cdot\text{SF}-1} \alpha_{0,f}^{*} \cdot c_{0,k}^{*} \cdot r_{0,k,f}$$

$$= \sqrt{E_c} \cdot s_{0,i} \cdot \sum_{f=1}^{L} |\alpha_{0,f}|^2 + \frac{\sqrt{E_c}}{\text{SF}} \sum_{f=1}^{L} \sum_{k=i\cdot\text{SF}}^{(i+1)\cdot\text{SF}-1} \sum_{\substack{l=1 \\ l \neq f}}^{L} \alpha_{0,f}^{*} \cdot c_{0,k}^{*} \cdot \alpha_{0,l} \cdot s_{0,\lfloor (k+(\tau_{0,f}-\tau_{0,l})/T_{\text{chip}})/\text{SF} \rfloor} \cdot c_{0,k+(\tau_{0,f}-\tau_{0,l})/T_{\text{chip}}} \quad (7.27)$$

$$+ \frac{\sqrt{E_c}}{\text{SF}} \sum_{u=1}^{N_u} \sum_{f=1}^{L} \sum_{k=i\cdot\text{SF}}^{(i+1)\cdot\text{SF}-1} \sum_{l=1}^{L} \sum_{j=0}^{\infty} \alpha_{0,f}^{*} \cdot c_{0,k}^{*} \cdot \alpha_{u,l} \cdot s_{u,\lfloor j/\text{SF} \rfloor} \cdot c_{u,j} \cdot p(kT_{\text{chip}} + \tau_{0,f} - jT_{\text{chip}} - \tau_{u,l})$$

$$+ \frac{1}{\text{SF}} \sum_{f=1}^{L} \sum_{k=i\cdot\text{SF}}^{(i+1)\cdot\text{SF}-1} \alpha_{0,f}^{*} \cdot c_{0,k}^{*} \cdot n_{0,k,f}$$

In this equation, the second term represents interference caused by the user's own signal due to the delayed multipath replicas and the third term represents the interference component from all the multipath replicas of the other users. These interference components can be reduced through the use of MUD schemes.

Much research has been undertaken in the area of the MUD [Marques da Silva 2003a] [Marques da Silva 2000] for WCDMA technology. Optimal MUD, usually known as maximum likelihood sequence detector, are too complex for practical application. Its complexity increases exponentially with the increase of the number of users. On the other hand, suboptimal MUD has a complexity that increases linearly with the increase of the number of users. The latter can take two different forms: linear MUD (e.g., decorrelating [Marques da Silva 2005a], minimum mean square error [MMSE] [Glisic 1997]), and subtractive MUD (e.g., successive interference cancellation and parallel interference cancellation [Marques da Silva 2003a] [Glisic 1997]). Linear suboptimal detectors apply a linear transformation to the bank output of the conventional detectors, in order to decrease the level of MAI seen by each user.

There are two main types of implementations for subtractive MUDs: the parallel interference cancellation scheme (PIC) [Varanasi 1990] and the successive interference canceller [Patel 1994] [Johansson 1999]. The main difference between both schemes relies on how the interference subtraction is performed on the received signal. The PIC removes all the interfering signals simultaneously after they are detected, whereas the SIC detector relies on removing the interfering signals from the received signal, one at a time as they are detected.

Nevertheless, a MUD is normally employed at the BS (uplink) where there is enough power processing capability and where it is easier to know/estimate the uplink CIR and the spreading and scrambling sequences of interfering users. Such power processing capability is normally not available at the mobile station (MS) side, being for this reason important to employ alternative schemes to improve the performance.

If the transmission is synchronous and comes from a BS (downlink), the channel coefficients and delays do not depend on the user, $\alpha_{0,l} = \alpha_{u,l}$ and $\tau_{0,l} = \tau_{u,l}$ ($u = 1 \ldots N_u - 1$, $l = 1 \ldots L$), and this equation simplifies to

$$
z_{0,i} = \sqrt{E_c} \cdot s_{0,i} \cdot \sum_{f=1}^{L} |\alpha_f|^2 + \frac{\sqrt{E_c}}{\text{SF}} \sum_{u=1}^{N_u} s_{u,i} \cdot \sum_{f=1}^{L} |\alpha_f|^2 \cdot \sum_{k=i \cdot \text{SF}}^{(i+1) \cdot \text{SF}-1} c_{0,k}^* \cdot c_{u,k}
$$

$$
+ \frac{\sqrt{E_c}}{\text{SF}} \sum_{u=0}^{N_u} \sum_{f=1}^{L} \sum_{k=i \cdot \text{SF}}^{(i+1) \cdot \text{SF}-1} \sum_{\substack{l=1 \\ l \neq f}}^{L} \alpha_f^* \cdot \alpha_l \cdot c_{0,k}^* \cdot c_{u,k+(\tau_f - \tau_l)/T_{\text{chip}}} \cdot s_{u,\lfloor (k+(\tau_f - \tau_l)/T_{\text{chip}})/\text{SF} \rfloor} \quad (7.28)
$$

$$
+ \frac{1}{\text{SF}} \sum_{f=1}^{L} \sum_{k=i \cdot \text{SF}}^{(i+1) \cdot \text{SF}-1} \alpha_f^* \cdot c_{0,k}^* \cdot n_{0,k,f}
$$

In this equation, the interference is grouped in a different form. The second term represents interference from the signals of the other users time aligned with the desired signal, whereas the third term represents the interference caused by the multipath replicas of all signals. These two terms are considered as MAI (see Chapter 3). The second term can be cancelled if orthogonal spreading sequences are employed. In this case, Equation 7.23 is valid and Equation 7.28 is simplified to

$$
z_{0,i} = \sqrt{E_c} \cdot s_{0,i} \cdot \sum_{f=1}^{L} |\alpha_f|^2 + \frac{\sqrt{E_c}}{\text{SF}} \sum_{u=0}^{N_u} \sum_{f=1}^{L} \sum_{k=i \cdot \text{SF}}^{(i+1) \cdot \text{SF}-1} \sum_{\substack{l=1 \\ l \neq f}}^{L} \alpha_f^* \cdot \alpha_l \cdot c_{0,k}^* \cdot c_{u,k+(\tau_f - \tau_l)/T_{\text{chip}}} \cdot s_{u,\lfloor (k+(\tau_f - \tau_l)/T_{\text{chip}})/\text{SF} \rfloor}
$$

$$
+ \frac{1}{\text{SF}} \sum_{f=1}^{L} \sum_{k=i \cdot \text{SF}}^{(i+1) \cdot \text{SF}-1} \alpha_f^* \cdot c_{0,k}^* \cdot n_{0,k,f} \quad (7.29)
$$

Nevertheless, the type of spreading sequences to use must be carefully selected since orthogonal spreading sequences may originate high level of MAI caused by the multipath environment, even in a synchronous environment (downlink).

7.4 Orthogonal Frequency-Division Multiplexing

OFDM is a transmission technique adopted by many high data rate communication systems such as the standards IEEE 802.11n (Institute of Electrical and Electronics Engineers) and IEEE 802.16e (2005), being suitable for frequency-selective fading channels [Cimini 1985] [Liu 2005]. In opposition to the conventional single-carrier transmission techniques where the information symbols are transmitted in a single stream, OFDM technique splits the symbols in several lower rate streams, which are then transmitted in parallel subcarriers. As a consequence, the symbol period is increased, making the signal less sensitive to ISI. To avoid interference between subcarriers, the several streams should be transmitted in orthogonal subcarriers.

It is known that sinusoids with frequencies spaced by $1/T$ form an orthogonal basis set in a T-duration interval and a periodic signal with period T can be represented as a linear combination of the orthogonal sinusoids. This means that the orthogonality between subcarriers is assured by using the discrete Fourier transform (DFT) and inverse DFT (IDFT). In practice, OFDM is normally implemented through an efficient technique called fast Fourier transform (FFT) and inverse FFT (IFFT). Therefore, OFDM signals are commonly generated by computing the N-point IFFT, where the input of the IFFT is the frequency-domain representation of OFDM signal. The output of the IFFT is the time-domain representation of the OFDM signal, the N-point IFFT out is defined as a useful OFDM symbol. This "time domain" OFDM signal is composed of N subcarriers, as depicted in Figure 7.9b, as opposed to a single-carrier signal (also depicted in Figure 7.9a) [Marques da Silva 2010].

Although with OFDM signals the symbol stream is split into several parallel substreams with lower rate (each one associated to a different subcarrier), ISI can still occurs within each substream. To mitigate the effects of ISI caused by channel delay spread, each block of N IDFT coefficients is typically preceded by a cyclic prefix or a guard interval consisting of Ng samples (Ng stands for the number of samples at the CP), such that the length of the CP is, at least, equal to the time span of the channel (channel length). The CP is simply a repetition of the last Ng time-domain symbols. The prefix insertion operation is illustrated in Figure 7.10.

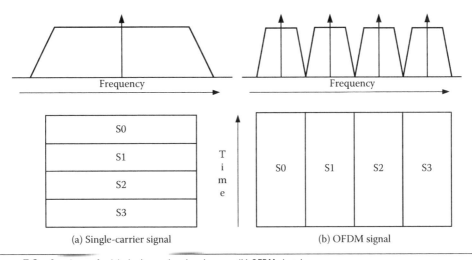

Figure 7.9   Spectrum of a (a) single-carrier signal versus (b) OFDM signal.

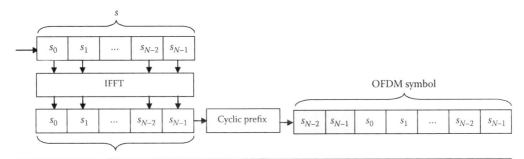

Figure 7.10   Cyclic prefix insertion.

Although most of the processing of an OFDM transmitter is implicitly performed using IDFT function (as depicted in Figure 7.11a), the elementary processing of an OFDM transmitter is similar to that of a frequency-division multiplexing (FDM) transmitter. Figure 7.12 shows the implicit processing of an OFDM transmitter. As can be seen, such processing includes the FDM processing that consists of modulating different channels with different subcarriers, followed by an adder module (i.e., sum at signal level). Contrarily, the OFDM processing is that the source is a single channel (instead of multiple channels, as considered by a FDM transmitter), which is split into lower data rate channels.

Therefore, the basic idea of the OFDM transmission technique consists of splitting a higher rate into a group of parallel lower rate streams and modulating each lower rate stream with a subcarrier in such a way that the resulting parallel signals are ideally uncorrelated. Another difference between FDM and OFDM signals relies on

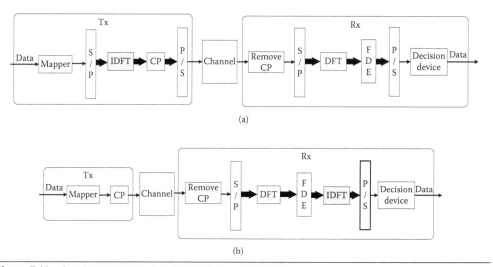

(a)

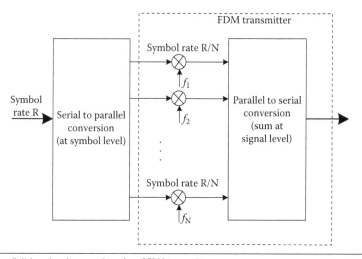

(b)

Figure 7.11 Generic transmission chain for (a) OFDM and (b) SC-FDE.

Figure 7.12 Implicit baseband processing of an OFDM transmitter.

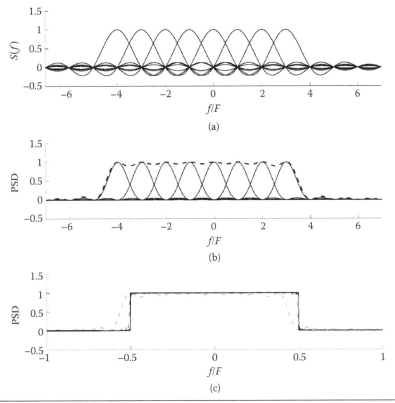

Figure 7.13 Transmitted OFDM signal: (a) individual subcarriers, (b) PSD of subcarriers and envelope, and (c) PSD of envelope signal.

how these subcarriers are uncorrelated. While FDM signals use guard bands to assure uncorrelation between adjacent subcarriers,* in OFDM, the uncorrelation is implemented using the IDFT (at the transmitter) and DFT (at the receiver) processing. This results in a much more efficient manner, which typically translates in much higher channel efficiency. As can be seen from Figure 7.13, OFDM signals from adjacent subcarriers present some level of overlapping in the frequency domain. Nevertheless, the mathematical OFDM implementation using the IDFT/DFT assures that the signals are uncorrelated at the receiver. Although these uncorrelated and parallel subcarrier signals are frequency separated, they are summed in time (second box in Figure 7.12), before being transmitted together. It is worth noting that a typical limitation in OFDM signals results from the fact that the spurious free dynamic range of an OFDM amplifier is very demanding, in order to respond to this amplitude variation that results from the instantaneous sum of N subcarrier signals (parallel to serial conversion). This is more visible for higher number of subcarriers.

Since the symbol rate of each subcarrier signal has a symbol rate N times lower than that of the original signal, the level of ISI that results from the multipath channel is much lower. Another great advantage of OFDM signals results from the fact that,

---

* The FDM spectrum in plotted in Chapter 6

even though if a subcarrier experiences a deep fading or other type of interference, since the symbols are interleaved (serial to parallel conversion), these errors can easily be recovered using error correction techniques.

The subcarrier frequencies are known by the receiver, allowing the recovery of the original symbol stream from the received signal. Nevertheless, there are a number of nonideal effects that degrade the performance of OFDM systems, such as the local oscillator offset,[*] the window location offset,[†] carrier interference, and so on. These impairments result in the inability of the receiver to track the subcarrier frequencies, which results in some level of degradation. Moreover, as mentioned earlier, the use of low dynamic range devices in the transmission chain heavily degrades the performance of OFDM signals.

The subcarrier spacing is determined by the IDFT size N and input sampling rate of the IDFT. A subset of active subcarriers is mapped with the data modulation symbols and pilot symbols. The remaining subcarriers are left inactive prior to the IDFT to simplify the transmitter implementation.

If the CP length is greater than the length of the channel, the linear convolution of the transmitted sequence of IDFT coefficients with the discrete-time channel is converted into a circular convolution. As a result, the effects of the ISI and inter-carrier interference (ICI) are completely and easily removed. After removal of the guard interval, each block of N received samples is converted back to the frequency domain using a DFT. Each of the N frequency-domain samples are processed with a simple one-tap frequency-domain equalizer and applied to a decision device of a metric computer.

In conventional time-domain signals, the equalization process consists of a series of convolution operations, whose length is proportional to the time span of the channel. This means that for severely time-dispersive channels, the receiver can be very complex. In OFDM systems, the equalization is performed as a simple multiplication of the OFDM signal spectrum with the frequency response of the channel. This represents a great advantage in terms of processing requirements and effectiveness, as compared to the equalization process normally employed in time-domain signals.

A variant of conventional OFDM schemes is orthogonal frequency-division multiple access (OFDMA), where the multiple access is achieved by assigning subsets of OFDM subcarriers to individual users, allowing simultaneous low data rate transmission from several users.

From the mathematical point of view, the OFDM signal associated to the lth transmitted block has the following form

$$s_l(t) = \sum_{n=-N_G}^{N-1} s_{n,l} h_T(t - nT_S) \tag{7.30}$$

---

[*] Local oscillator offset (also known as local oscillator frequency offset of the receiver), results in loss of orthogonality between adjacent subcarriers.

[†] The window location offset may place some subcarriers out of the receiver's windows.

with T_S denoting the symbol duration, N_G denoting the number of samples at the CP, and $h_T(t)$ is the adopted pulse shaping filter. The block $\{s_{n,l}; n = 0,1,...,N-1\}$ is the IDFT of the block $\{S_{k,l}; n = 0,1,...,N-1\}$, where $S_{k,l}$ denotes the transmitted symbols associated to the kth subcarrier, that is, the kth symbol of the lth transmitted block. The signal $s_l(t)$ is transmitted over a time-dispersive channel, the received signal is sampled and the CP is removed. The resulting the time-domain block is as follows:

$$\{y_{n,l}; n = 0,1,...,N-1\}. \tag{7.31}$$

If the length of the CIR is smaller than $N_G T_S$, then the DFT of the block $\{y_{n,l}; n = 0,1,...,N-1\}$ is $\{Y_{k,l}; k = 0,1,...,N-1\}$, with $Y_{k,l} = S_{k,l}H_k + N_{k,l}$, where H_k denotes the channel frequency response associated to the kth subcarrier and $N_{k,l}$ denotes the channel noise. This means that a frequency-selective channel behaves as a flat fading channel at the subcarrier level. Therefore, we can easily invert the channel effects

$$\tilde{S}_{k,l} = \frac{Y_{k,l}}{H_{k,l}} = \frac{Y_{k,l}H^*_{k,l}}{|H_{k,l}|^2} \tag{7.32}$$

For phase shift keying (PSK) constellations, the information is on the phase and this equalization process is simply accomplished through $\tilde{S}_{k,l} = Y_{k,l}H^*_{k,l}$.

It is worth noting that the asynchronous digital subscriber line (ADSL) implements a type of OFDM transmission technique entitled discrete multitone (DMT). In order to better optimize the signal to the transmission channel, the DMT presents additional features, namely the ability to remove certain subcarriers and the ability to adjust the modulation order and type, independently for each subcarrier.

7.5 Single Carrier–Frequency Domain Equalization

Although OFDM is the most popular block transmission technique, the same concept can be used with single-carrier modulations. In fact, single-carrier modulation using FDE is an alternative approach based on this principle. As with OFDM, with SC-FDE the data blocks are preceded by a CP, long enough to cope with the overall channel length. Due to the lower envelope fluctuations of the transmitted signals (and implicitly a lower PMEPR), SC-FDE schemes are especially interesting when a low-complexity and efficient power amplification is required [Falconer 2002]. As can be seen from Figure 7.11, whereas the OFDM transmitter includes the computation of the IDFT, the SC-FDE transmitter is a regular time domain one with the exception that the CP is added. Nevertheless, after reception (and after removing the CP), the SC-FDE receiver computes the DFT, before performing the FDE, followed by the IDFT computation. As the signal is transmitted in blocks (to which the CP is added and to which the equalization is performed), SC-FDE transmission is also considered as a block transmission technique.

From the mathematical point of view, SC-FDE signals are similar to OFDM signals, that is, the lth-transmitted block has the following form

$$s_l(t) = \sum_{n=-N_G}^{N-1} s_{n,l} h_T(t - nT_S) \qquad (7.33)$$

Once again with T_S denoting the symbol duration, N_G denoting the number of samples at the CP, and $h_T(t)$ denoting the adopted pulse-shaping filter. However, the lth time-domain block to be transmitted $\{s_{n,l}; n = 0,1,..., N-1\}$ is directly obtained from the data signal, without employing IDFT operation (i.e., the symbols are transmitted in the time domain, not in the frequency domain).

Assuming that the CP is longer than the overall CIR of each channel, the lth frequency-domain block before the FDE block (i.e., the DFT of the lth received time-domain block, after removing the CP) is $\{y_{n,l}; n = 0,1,..., N-1\}$ and the corresponding frequency-domain block (i.e., the corresponding DFT) is $\{Y_{k,l}; k = 0,1,..., N-1\}$, with

$$Y_{k,l} = S_{k,l} H_{k,l} + N_{k,l} \qquad (7.34)$$

Once again, $H_{k,l}$ denotes the channel frequency response for the kth subcarrier and lth time-domain block (the channel is assumed invariant in the frame). $N_{k,l}$ is the frequency-domain block channel noise for that subcarrier and the lth block. The block $\{S_{k,l}; n = 0,1,..., N-1\}$ is the DFT of the block $\{s_{n,l}; n = 0,1,..., N-1\}$.

Although the situation is similar to the OFDM case, it is not desirable to invert perfectly the channel in the SC-FDE case because this might lead to noise enhancement effects that spread to all data symbols (in the OFDM case, these noise enhancement effects are restricted to the symbols associate to subcarriers in deep fading and does not spread to other data symbols because the data is transmitted in the frequency domain). This means the FDE should be optimized under the MMSE criterion and the samples at the FDE output are given by [Marques da Silva 2010]

$$\tilde{S}_{k,l} = \frac{Y_{k,l} H_{k,l}^*}{\left(\alpha + \left|H_{k,l}\right|^2\right)} \qquad (7.35)$$

with

$$\alpha = \frac{\mathrm{E}\left\{\left|N_{k,l}\right|^2\right\}}{\mathrm{E}\left\{\left|S_{k,2l-j}\right|^2\right\}} \qquad (7.36)$$

7.5.1 *Iterative Block–Decision Feedback Equalizer Receivers*

It is well known that decision feedback equalizers (DFEs) [Proakis 2001] can significantly outperform linear equalizers (in fact, DFEs include as special case linear equalizers). Time-domain DFEs have good performance/complexity trade-offs, provided that the CIR is not too long. However, if the CIRs expand over a large number of symbols (such as in the case of severely time-dispersive channels), conventional time-domain DFEs are too complex. For this reason, a hybrid time-frequency SC-DFE was proposed in [Benvenuto 2002], employing a frequency-domain feedforward filter and a time-domain feedback filter. This hybrid time-frequency-domain DFE has a better performance than a linear FDE. However, as with conventional, time-domain DFEs, it can suffer from error propagation, especially when the feedback filters have a large number of taps. A promising iterative block-DFE (IB-DFE) approach for SC transmission was proposed in [Benvenuto 2002] and extended to transmit/diversity scenarios in [Dinis 2003]. Within these IB-DFE schemes, both the feedforward and the feedback parts are implemented in the frequency domain, as depicted in Figure 7.14.

For a given ith iteration, the IB-DFE output samples are given by

$$\tilde{S}_k^{(i)} = F_k^{(i)} Y_k - B_k^{(i)} \hat{S}_k^{(i-1)}, \tag{7.37}$$

where $\{F_k^{(i)}; k = 0,1,...,N-1\}$ and $\{B_k^{(i)}; k = 0,1,...,N-1\}$ denote the feedforward and feedback equalizer coefficients, respectively, and $\{\hat{S}_k^{(i-1)}; k = 0,1,...,N-1\}$ is the DFT of the hard-decision block $\{\hat{s}_n^{(i-1)}; n = 0,1,...,N-1\}$, of the $(i-1)^{\text{th}}$ iteration, associated to the transmitted time-domain block $\{s_n; n = 0,1,...,N-1\}$.

The forward and backward IB-DFE coefficients $\{F_k^{(i)}; k = 0,1,...,N-1\}$ and $\{B_k^{(i)}; k = 0,1,...,N-1\}$, respectively, are chosen so as to maximize the "Signal-to-Interference plus Noise Ratio" (SINR). In [Dinis 2003], it is shown that the optimum feedforward and feedback coefficients are given by

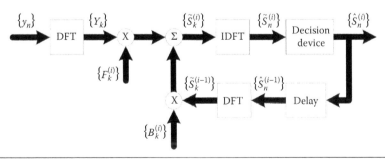

Figure 7.14   IB-DFE receiver structure.

$$F_k^{(i)} = \frac{\kappa_F^{(i)} H_k^*}{\alpha + \left[1 - (\rho^{(i-1)})^2\right] |H_k|^2} \tag{7.38}$$

and

$$B_k^{(i)} = \rho^{(i-1)}(F_k^{(i)} H_k - 1), \tag{7.39}$$

respectively, where $\kappa_F^{(i)}$ is selected to ensure that $\gamma^{(i)} = 1$ and

$$\rho^{(i)} = \frac{E[s_n^* \hat{s}_n^{(i)}]}{E[|s_n|^2]} \tag{7.40}$$

is a measure of the reliability of the decisions used in the feedback loop. Since the IB-DFE coefficients take into account the overall block reliability, the error propagation problem is significantly reduced. Consequently, the IB-DFE techniques offer much better performances than the noniterative methods. In fact, the IB-DFE schemes can be regarded as low-complexity turbo equalizers [Tuchler 2002], since the feedback loop uses the equalizer outputs instead of the channel decoder outputs. For the first iteration, we do not have any information about s_n, which means that $\rho = 0$, $B_k^{(0)} = 0$, and $F_k^{(0)} = \frac{\kappa_F^{(0)} H_k^*}{\beta + |H_k|^2}$. Therefore, the IB-DFE reduces to a linear FDE. Clearly, for the first iteration ($i = 0$), no information exists about S_k and the correlation coefficient in (7.40) is zero.

After that first iteration, and if the residual bit error rate (BER) is not too high, we can use the feedback coefficients to eliminate a significant part of the residual interference. When $\rho \approx 1$ (after several iterations and/or moderate-to-high SNR), we have an almost full cancellation of the residual ISI through these coefficients, whereas the feedforward coefficients perform an approximate matched filtering. Clearly, (7.37) could be written as

$$\tilde{S}_k^{(i)} = F_k^{(i)} Y_k - B_k^{(i)} \overline{S}_k^{(i-1)}, \tag{7.41}$$

with

$$\overline{S}_k^{(i-1)} = \rho^{(i-1)} \hat{S}_k^{(i-1)}. \tag{7.42}$$

Since $\rho^{(i-1)}$ can be regarded as the blockwise reliability of the estimate $\hat{S}_k^{(i-1)}$, $\overline{S}_k^{(i-1)}$ is the overall block average of $S_k^{(i-1)}$ at the FDE output. To improve the performances, we could replace the "block wise averages" by "symbol averages," which can be done as described in the following.

If we assume that the transmitted symbols are selected from a quadrature phase shift keying (QPSK) constellation under a Gray mapping rule (the generalization to other cases is straightforward), that is, $s_n = \pm 1 \pm j = s_n^I + j s_n^Q$, with $s_n^I = \text{Re}\{s_n\}$ and $s_n^Q = \text{Im}\{s_n\}$ (and similar definitions for $\tilde{s}_n$, $\overline{s}_n$ and $\hat{s}_n$), then it can be shown that the log-likelihood ratios (LLRs) of the "in-phase bit" and the "quadrature bit," associated to s_n^I and s_n^Q, respectively, are given by

$$L_n^I = 2\tilde{s}_n^I / \sigma_p^2 \tag{7.43}$$

and

$$L_n^Q = 2\tilde{s}_n^Q / \sigma_p^2 \tag{7.44}$$

where

$$\sigma_p^2 = \frac{1}{2} E[|s_n - \tilde{s}_n|^2] \approx \frac{1}{2N} \sum_{n=0}^{N-1} E[|\hat{s}_n - \tilde{s}_n|^2]. \tag{7.45}$$

Under a Gaussian assumption, it can be shown that the mean value of s_n conditioned to the FDE output $\tilde{s}_n$ is

$$\begin{aligned} \bar{s}_n &= \tanh\left(\frac{L_n^I}{2}\right) + j\tanh\left(\frac{L_n^Q}{2}\right), \\ &= \rho_n^I \hat{s}_n^I + j\rho_n^Q \hat{s}_n^Q \end{aligned} \tag{7.46}$$

where the hard decisions $\hat{s}_n^I = \pm 1$ and $\hat{s}_n^Q = \pm 1$ are defined according to the signs of L_n^I and L_n^Q, respectively, and ρ_n^I and ρ_n^Q can be regarded as the reliabilities associated to the "in-phase" and "quadrature" bits of the nth symbol, given by

$$\rho_n^I = \frac{E[s_n^{I*}\hat{s}_n^I]}{E[|s_n^I|^2]} = \tanh\left(\frac{|L_n^I|}{2}\right) \tag{7.47}$$

and

$$\rho_n^Q = \frac{E[s_n^{Q*}\hat{s}_n^Q]}{E[|s_n^Q|^2]} = \tanh\left(\frac{|L_n^Q|}{2}\right) \tag{7.48}$$

(for the first iteration, $\rho_n^I = \rho_n^Q = 0$ and $\bar{s}_n = 0$).

The feedforward coefficients are still obtained from (7.38), with the block wise reliability given by

$$\rho^{(i)} = \frac{1}{2N} \sum_{n=0}^{N-1} (\rho_n^{I(i)} + \rho_n^{Q(i)}), \tag{7.49}$$

Therefore, the receiver with "blockwise reliabilities," denoted in the following as IB-DFE with hard decisions, and the receiver with "symbol reliabilities," denoted in the following as IB-DFE with soft decisions, employ the same feedforward coefficients; however, in the first, the feedback loop uses the "hard-decisions" on each data block, weighted by a common reliability factor, whereas in the second the reliability factor changes from symbol to symbol (in fact, the reliability factor is different in the real and imaginary component of each symbol).

It is also possible to define turbo FDE receivers based on IB-DFE receivers that, as conventional turbo equalizers, employ the channel decoder outputs instead of the uncoded "soft decisions" in the feedback loop.

7.6 Diversity Combining Algorithms

Before MIMO systems are introduced, it is worth describing diversity combining algorithms, as MIMO systems can be viewed as transmit and receiving diversity. Therefore, those signals need to be properly combined to allow the exploitation of diversity.

Multipath fading is an important problem experienced in wireless communications. Fading results in fluctuations in the signal amplitude, which degrades the BER performance at the receiver. Diversity schemes attempt to mitigate this problem by finding independently faded paths in the mobile radio channel. Usually this involves providing replicas of the transmitted signal over time, frequency, space, or multipath. Diversity is the most important contributor to reliable wireless communications. It can be processed at the transmitter or at the receiver side (or both).

Note that both WCDMA and OFDM are efficient schemes because they explore diversity. WCDMA allows the exploitation of multipath diversity (with the RAKE receiver), whereas OFDM transmission technique allows the exploitation of frequency diversity.

The diversity system must optimally combine the receiver-diversified waveforms so as to maximize the resulting signal quality. There are several ways of combining the signals received from the different types of diversity, namely selection combining, MRC, equal ratio combining (EGC), the mean square error (MSE) based combining, and so on. In addition, this combining may be either performed at the received signal level or at the symbol level. These combining schemes are explained in the following paragraphs.

7.6.1 Selection Combining

The combiner selects the strongest signal from the total signals at its input. This criteria is used by the selective transmit diversity (TD) scheme, where the receiver selects one of the several transmitting antennas. Similarly, selection combining can also be employed to select one out of several signals received from different receiving antennas.

7.6.2 Maximal Ratio Combining

The combiner performs a weighted sum of all signals at its input. This criterion is optimum only in the presence of noise, since it maximizes the SNR, while it minimizes the noise. The resulting SNR level of the signal at the combiner's output corresponds to the sum of the elementary SNRs (combiner's input SNRs).

In the case of wideband DS-CDMA signals, the MRC is optimum only when interference does not exist. It may correspond to a synchronized network with flat fading and with orthogonal spreading sequences. Figure 7.15 shows a two-branch Maximum Ratio Combiner performed at the receiver's side, where $\odot$ stands for the convolution operation. Since this scheme is generic, the spreading and pulse shaping were not considered. As can be seen, at each branch of the receiver, the signal

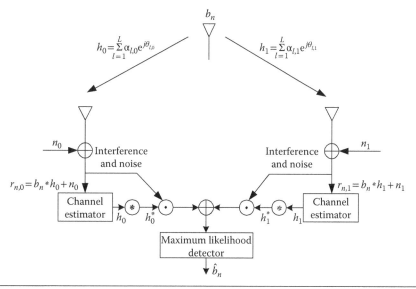

Figure 7.15 Two-branch maximum ratio combiner (at the receiver).

is convolved by the complex conjugate of the channel, to allow removing the phase. Furthermore, since the signals from different branches are in phase, and weighted by the square of the absolute value of the (sum of the) channel coefficients (after the channel complex conjugation convolution), they are summed. This corresponds to the maximum ratio combining criteria, as the signals are properly weighted by the square of the absolute values of the channel gains.

7.6.3 Equal Gain Combining

The combiner performs a coherent sum of all signals at its input. While providing diversity, this criterion tends, in most scenarios, to achieve a worse performance than the MRC. If the signal from one branch has a deep fade, this effect is felt in the resulting signal since the combining weights are equal in the several branches. This scheme presents a great advantage as knowledge about the channel state information (CSI) is not required, and therefore, estimation circuits are not needed, leading to a very simple combiner.

7.6.4 MSE-Based Combining

This criterion tends to lead to better performances in the presence of interference, namely, multipath interference or MAI. The combining weights are calculated based on information provided by some way. A traditional way consists of using a pilot or a training sequence that allows the receiver to periodically evaluate a coefficient that corresponds to some difference (MSE) between the transmitted signal and the received one. Let us consider that we have available N receiving branches that are intended to be combined to provide diversity (e.g., output from the N fingers from L multipaths or output from the N receiving antennas).

Let us denote $\hat{b}_n$ as the output data symbol from the nth receiving branch and $\hat{\mathbf{c}}_n$ as the output training symbol vector from the nth receiving branch. Similarly, $\tilde{\mathbf{c}}$ stands for the known training symbol vector $\mathbf{c}$, needed to implement the MSE-based combining algorithm.

Before it proceeds with the MSE-based combining criteria, it is important to normalize the signals involved. Therefore, the normalization quantity for each of the nth (receiving) branch becomes

$$a_n = E\left[\hat{\mathbf{c}}_n ./ \tilde{\mathbf{c}}\right] \tag{7.50}$$

where the operand ./ stands for the point-by-point division.

Since it is intended to calculate the expected value of $E[\hat{\mathbf{c}}_n ./ \tilde{\mathbf{c}}]$, $\tilde{\mathbf{c}}$ and $\hat{\mathbf{c}}_n$ should be considered as vectors (group of symbols), instead of a single value (symbol), such that an occurrence of a symbol error in the detection within the training sequences vector is averaged. Therefore, the length of the vectors $\hat{\mathbf{c}}_n$ and $\tilde{\mathbf{c}}$ shall be as maximum as possible so as to lead to the expected value. The normalized training vector becomes

$$\hat{\mathbf{c}}_n^{\text{normalized}} = \frac{\hat{\mathbf{c}}_n}{a_n}. \tag{7.51}$$

We are now able to apply the MSE-based combining criteria to each branch of the N independent and identical distributed (i.i.d.) branches, during training sequence as follows:

$$\begin{aligned}
\text{MSE}(\mathbf{c}_n) &= E\left[\left|\hat{\mathbf{c}}_n^{\text{normalized}} - \tilde{\mathbf{c}}\right|^2\right] \\
&= E\left[\left|\frac{\hat{\mathbf{c}}_n}{a_n - \tilde{\mathbf{c}}}\right|^2\right] \\
&= E\left[\left|\frac{\hat{\mathbf{c}}_n}{E\left[\frac{\hat{\mathbf{c}}_n \cdot}{\tilde{\mathbf{c}}}\right]} - \tilde{\mathbf{c}}\right|^2\right]
\end{aligned} \tag{7.52}$$

After all of the described procedure has been implemented to each one of the N receiving branches during the training sequence to allow the calculation of $\text{MSE}(\mathbf{c}_n), n = 1, ..., N$, we are now able to weight the received symbols from different branches, before combining them. This makes

$$\begin{aligned}
\hat{b}_n^{\text{weighted}} &= \frac{\hat{b}_n^{\text{detected}}}{\text{MSE}(\mathbf{c}_n)} \\
&= \frac{\hat{b}_n^{\text{detected}}}{E\left[\left|\frac{\hat{\mathbf{c}}_n}{E\left[\hat{\mathbf{c}}_n ./ \tilde{\mathbf{c}}\right] - \tilde{\mathbf{c}}}\right|^2\right]}
\end{aligned} \tag{7.53}$$

where $\hat{b}_n^{\text{weighted}}$ stands for the new normalized estimated value of the symbol b (data transmitted symbol) obtained from the nth receiving branch and $1/\text{MSE}(\mathbf{c_n})$ means the combining weight corresponding to the nth receiving branch. From Equation, it is viewed that a received signal closer to the transmitted one results in a more powerful weight. The opposite also applies. The combined data symbol $\hat{b}$, resulting from the N (receiving) branches becomes

$$
\begin{aligned}
\hat{b} &= \sum_{n=1}^{N} \hat{b}_n^{\text{weighted}} \\
&= \sum_{n=1}^{N} \frac{\hat{b}_n^{\text{detected}}}{\text{MSE}(\mathbf{c_n})}, \\
&= \sum_{n=1}^{N} \frac{\hat{b}_n^{\text{detected}}}{E\left[\left|\dfrac{\hat{\mathbf{c}}_n}{E\left[\dfrac{\hat{\mathbf{c}}_n \cdot}{\tilde{\mathbf{c}}}\right] - \tilde{\mathbf{c}}}\right|^2\right]}
\end{aligned}
\tag{7.54}
$$

which corresponds to the estimate of the transmitted symbol b, obtained using diversity from the N receiving branches.

7.7 RAKE Receiver

In a multipath channel, the original transmitted signal is reflected from obstacles such as buildings and mountains and the receiver sees several copies of the signal with different delays. If signals arrive shifted in time more than one chip apart from each other, the RAKE receiver can resolve them. Actually, from each multipath signal's point of view, other multipath signals can be regarded as interference and is partially suppressed by the processing gain. However, a further benefit is obtained if the resolved multipath signals are combined using a RAKE receiver. Hence, the signal waveform of CDMA signals facilitates the utilization of multipath diversity. Expressing the same phenomenon in the frequency domain means that the bandwidth of the transmitted signal is larger than the coherence bandwidth of the channel, and the channel presents frequency selectivity. In the case of single-path, multipath diversity provided by the RAKE receiver would not bring any added value.

A RAKE receiver consists of a bank of decorrelators, each receiving a different multipath signal. After despreading by the decorrelators, the signals are combined using, usually, the MRC algorithm, as described in Section 7.6.2. Since the received multipath signals present uncorrelated fading coefficients, diversity order, and thus, performance is improved. Figure 7.16 illustrates the principle of the RAKE receiver. After spreading

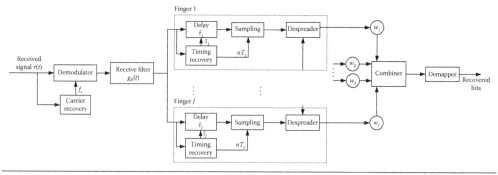

Figure 7.16   Scheme of a RAKE receiver.

and modulation in the transmitter, the signal is passed through a multipath channel, which can be modeled by a tapped delay line. In the multipath channel, there are L multipath components with different delays $(\tau_1, \tau_2, ..., \tau_L)$, attenuation factor $(\alpha_1, \alpha_2, ..., \alpha_L)$, and phase components $(\theta_1, \theta_2, ..., \theta_L)$, each corresponding to a different propagation path. The RAKE receiver has, ideally, a different finger for each multipath component. In each finger, the received signal is correlated by the spreading sequence, which is time aligned earlier with the delay of the corresponding multipath signal. After despreading, each captured multipath signal (in each jth finger of the RAKE) is weighted and they are then combined. Moreover, when the MRC algorithm is considered to combine the several multipath signals, each finger aligns the phase of the signal corresponding to that multipath and weights it by the channel attenuation, in order to allow a coherent sum. Therefore, when the RAKE considers a MRC algorithm, the signal in each finger is weighted by the complex conjugate of the corresponding multipath coefficient $\left(w_j = \left(\alpha_j\, e^{j\theta} \right)^* = \alpha_j\, e^{-j\theta} \right)$ to allow a coherent sum. Each finger of the RAKE receiver corresponds to a decorrelator chain depicted inside each box of Figure 7.16. Due to the mobile movement, the scattering environment will change, and thus, the delays and attenuation factors will change as well. Therefore, it is necessary to measure the tapped delay line profile and to reallocate RAKE fingers whenever the delays have changed by a significant amount. Small-scale changes, less than one chip, are taken care of by a code tracking loop, which tracks the time delay of each multipath signal.

The full exploitation of DS-CDMA advantages is only achieved by the use of the RAKE receiver because this is the tool to take advantage of multipath diversity. In CDMA, the utilization of the spread spectrum with a RAKE receiver allows using the frequency selectivity of the channel as a kind of diversity. The resolution of the RAKE receiver is the chip period. This is the reason why the resolution is improved when the spreading factor is increased. With a lower chip period, it better discriminates different multipaths, allowing higher order of multipath diversity. On the other hand, when the chip period is too long, the several multipaths are received within the same chip period, providing chip interference and not allowing the corresponding discrimination and diversity.

The implicit diversity provided by the RAKE receiver is similar to the explicit diversity provided by the antenna diversity, as well as the corresponding diversity equations.

Although the MRC is usually employed in the RAKE receiver, some other algorithms defined in Section 7.6 may, as well, be employed for multipath combining. The RAKE receiver, depicted in Figure 7.16, consists of a bank of J fingers, each correlating to a different delay of the received signal.

This scheme is similar to a single decorrelator, but the RAKE considers a different finger (correlator) for each multipath that is intended to be captured, in order to provide diversity. The finger outputs are then combined to form a decision statistic. The structure is equivalent to more practical forms in which the received signal is first filtered with a chip pulse shaping matched filter and despreading is performed using received chip sample and the spreading sequence.

The jth finger is used to despread the received spread spectrum signal, resulting in the set of despread values:

$$\hat{b}_{j,k,n} = \sum_{m=[(n-1)\cdot S_F]+1}^{n\cdot S_F} y_{j,m} \cdot s_{k,m}^*, n = 1..P, j = 1..J \tag{7.55}$$

where $y_{j,m}$ stands for the mth sample (sampled at a chip rate) of the output of the receiving filter (chip shaping), with a delay corresponding to the jth finger (i.e., to the corresponding multipath) of the RAKE receiver. Furthermore, $\hat{b}_{j,k,n}$ corresponds to the nth estimated symbol of the kth user, whose estimate is performed using the signal of the jth captured finger (whose value is based on the corresponding delayed version of the received signal $r(t - \tau_j)$). These despread values are combined using combining coefficients $w_1, w_2, ..., w_J$ to produce a decision statistic as described in section 7.6.

$$\hat{b}_{k,n} = \sum_{j=1}^{J} w_j \cdot \hat{b}_{j,k,n} \tag{7.56}$$

where $\hat{b}_{k,n}$ is subject to a block decision, typically a hard-decision.

Considering that MRC is performed, the impulse response of the RAKE receiver becomes

$$h_R(\tau,t) = \sum_{j=1}^{J} \hat{\alpha}_j(t) e^{-j\hat{\theta}_j(t)} \delta(\Delta - t + \hat{\tau}_j) \tag{7.57}$$

which is the complex conjugate of the channel impulse. Δ is some nominal delay that is required to ensure physical implementation of the RAKE, whose value should be higher than the delay corresponding to the last multipath captured by the RAKE, corresponding to the last finger. Its frequency response becomes

$$H_r(f,t) = \sum_{j=1}^{J} \hat{\alpha}_j(t) e^{-j\hat{\theta}_j(t)} e^{j2\pi f(\hat{\tau}_j - \Delta)} \tag{7.58}$$

where $\hat{x}$ stands for the estimated value of x and $\hat{\alpha}_j(t) e^{-j\hat{\theta}_j(t)}$ represents the estimate of the complex conjugate of the multipath coefficient $\alpha_j(t) e^{j\theta_j(t)}$. For this reason, it is

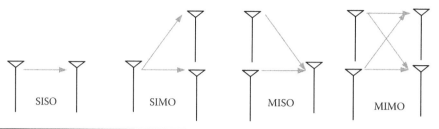

Figure 7.17   Multiple antenna configurations.

needed to make the channel estimate of amplitudes, time delays, and phase shifts of the several multipaths of the channel. The biggest limitation of the RAKE receiver comes from the fact that this estimate is never completely accurate due to channel variation (caused by the movement of the terminal, movement of the environment around terminals or changes in the atmosphere) or because it may not detect all multipaths.

7.8 Multiple Input Multiple Output

The use of multiple antennas at both the transmitter and receiver aims to improve performance or symbol rate of systems, but it usually requires higher implementation complexity. Moreover, the antenna spacing must be larger than the coherence distance to ensure independent fading across different antennas [Foschini 1996] [Foschini 1998] [Rooyen 2000].

The various configurations, shown in Figure 7.17, are referred to as multiple input single output (MISO), single input multiple output (SIMO), or MIMO. On the one hand, the SIMO and MISO architectures are a form of receive and transmit diversity schemes, respectively. On the other hand, MIMO architectures can be used for combined transmit and receive diversity, as well as for the parallel transmission of data or spatial multiplexing (SM). When used for SM, MIMO technology promises high bit rates in a narrow bandwidth, therefore, it is of high significance to spectrum users. MIMO systems transmit different signals from each transmit element so that the receiving antenna array receives a superposition of all the transmitted signals. Receiving and transmitting antennas must be sufficiently separated in space and/or polarization to create independent propagation paths.

MIMO schemes are used in order to push the performance or capacity/throughput limits as high as possible without an increase in spectrum bandwidth although there is an obvious increase in complexity [Rooyen 2000] [Hottinen 2003] [Marques da Silva 2001] [Marques da Silva 2002a] [Marques da Silva 2002b] [Marques da Silva 2003b]. Figure 7.18 presents a generic diagram of a MIMO scheme.

For M transmitting and N receiving antennas, we have the capacity equation [Telatar 1995] [Foschini 1998],

$$C_{EP} = \log_2\left(\det\left(I + \frac{\beta}{M}HH'\right)\right) \text{ bit/s/Hz} \qquad (7.59)$$

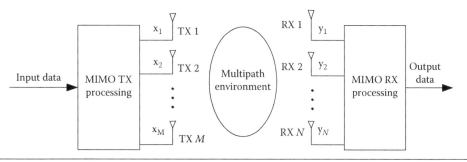

Figure 7.18   Generic diagram of a MIMO scheme.

where I_N is the identity matrix of dimension $N \times N$, **H** is the channel matrix, H' is the transpose conjugate of H, and β is the SNR at any receiving antenna. [Foschini 1996] and [Telatar 1999] demonstrated that the capacity grows linearly with $m = \min (M, N)$ for uncorrelated channels. Therefore, it is possible to employ MIMO as a multiresolution distribution system where the concurrent data streams are transmitted by M and received by N $(M \leq N)$ different antennas. The downside to this is the receiver complexity, sensitivity to interference, and correlation between antennas, which is more significant as the antennas are closer together. For a UMTS system, it is inadequate to consider more than 2 or 4 antennas at the UE/ mobile receiver.

MIMO schemes are implemented based on multiple-antenna techniques. These multiple-antenna techniques can be of different types, as follows:

- Space-time block coding
- Multilayer transmission
- Space division multiple access
- Beamforming

7.8.1 Space-Time Block Coding

Although space-time block coding (STBC) is essentially a MISO system, the use of receiver diversity makes it a MIMO, which corresponds to the most common configuration for this type of diversity. TD techniques are particularly interesting for fading channels where it is difficult to have multiple receiving antennas (as in conventional receiver diversity schemes). A possible scenario is the downlink transmission where the base station uses several transmitting antennas and the mobile terminal only has a single antenna [Alamouti 1998].

STBC-based schemes focus on achieving a performance improvement through the exploitation of additional diversity, while keeping the symbol rate unchanged [Alamouti 1998] [Tarokh 1999]. Symbols are transmitted using an orthogonal block structure, which enables simple decoding algorithm at the receiver [Alamouti 1998] [Marques da Silva 2004] [Marques da Silva 2009].

7.8.2 Open/Closed-Loop Techniques for Flat Fading

3GPP specifications define several TD schemes such as the standardized closed-loop modes 1 and 2 [3GPP 25.214-v5.5.0] or the open-loop space time TD (STTD) [3GPP 25.211-v5.2.0] for two transmitting antennas. These schemes and other transmitting schemes such as open-loop modes using space time block coding for four and eight transmitting antennas, the closed-loop selective TD (STD), and hybrid modes combining STBC and STD were analyzed.

7.8.2.1 Open-Loop Techniques

Open-loop TD schemes are performed without previous knowledge of the channel state by the transmitter. Figure 7.19 shows the general scheme of an open-loop transmitter scheme using M transmitting antennas.

STTD, also known as the Alamouti scheme, [Alamouti 1998] is a particular example of a scheme using two transmitting antennas and is already standardized for UMTS [3GPP 25.211-v5.2.0]. For this particular case, the transmission matrix representing the encoding process of the space-time block encoder is defined as follows:

$$\mathbf{G}_2(s_0, s_1) = \begin{bmatrix} s_0 & -s_1^* \\ s_1 & s_0^* \end{bmatrix} \tag{7.60}$$

Each column represents the sequence of symbols s_i transmitted by an antenna. Admitting that the channel coefficients for antennas 1 and 2 (h_0 and h_1) are constant during two symbol periods, the receiver performs the following processing:

$$\hat{s}_0 = h_0^* \cdot r(t_0) + h_1 \cdot r(t_1)^* = s_0 \cdot (h_0^2 + h_1^2) + n' \tag{7.61}$$

$$\hat{s}_1 = h_0^* \cdot r(t_1) - h_1 \cdot r(t_0)^* = s_1 \cdot (h_0^2 + h_1^2) + n'' \tag{7.62}$$

As it can be seen, the path diversity obtained is two. It is possible to use other STBC for four and eight antennas, as shown in [Tarokh 1999].

For four antennas, a code with the following generator matrix can be used

$$\mathbf{G}_4(s_0, s_1, s_2, s_3) = \begin{bmatrix} \mathbf{G}_2(s_0, s_1) & \mathbf{G}_2^*(s_2, s_3) \\ \mathbf{G}_2(s_2, s_3) & -\mathbf{G}_2^*(s_0, s_1) \end{bmatrix} \tag{7.63}$$

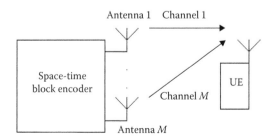

Figure 7.19   Generic block diagram of an open-loop transmitter scheme.

The decoding is performed as follows:

$$\begin{cases} \hat{s}_0 = r_0 h_0^* + r_1^* h_1 - r_2^* h_2 - r_3 h_3^* = \sum_i |h_i|^2 s_0 - c_0 s_3 + \eta_0 \\ \hat{s}_1 = r_1 h_2^* - r_0^* h_1 - r_3^* h_2 + r_2 h_3^* = \sum_i |h_i|^2 s_1 + c_0 s_2 + \eta_1 \\ \hat{s}_2 = r_2 h_0^* + r_3^* h_1 + r_0^* h_2 + r_1 h_3^* = \sum_i |h_i|^2 s_2 + c_0 s_1 + \eta_2 \\ \hat{s}_3 = r_3 h_0^* - r_2^* h_1 + r_1^* h_2 - r_0 h_3^* = \sum_i |h_i|^2 s_3 - c_0 s_0 + \eta_3 \\ c_0 = 2\operatorname{Re}\{h_0^* h_3 - h_1 h_2^*\} \end{cases} \qquad (7.64)$$

In the case of eight antennas, the generator matrix is as follows:

$$\mathbf{G}_8(s_0,s_1,s_2,s_3,s_4,s_5,s_6,s_7) = \begin{bmatrix} \mathbf{G}_4(s_0,s_1,s_2,s_3) & \mathbf{G}_4^*(s_4,s_5,s_6,s_7) \\ \mathbf{G}_4(s_4,s_5,s_6,s_7) & -\mathbf{G}_4^*(s_0,s_1,s_2,s_3) \end{bmatrix} \qquad (7.65)$$

After performing a decoding process similar to the case of four antennas, the following symbol estimates are obtained

$$\begin{cases} \hat{s}_0 = \sum_i |h_i|^2 s_0 - c_0 s_3 - c_1 s_5 + c_2 s_6 + \eta_0 \\ \hat{s}_1 = \sum_i |h_i|^2 s_1 + c_0 s_2 + c_1 s_4 + c_2 s_7 + \eta_1 \\ \hat{s}_2 = \sum_i |h_i|^2 s_2 + c_0 s_1 - c_1 s_7 - c_2 s_4 + \eta_2 \\ \hat{s}_3 = \sum_i |h_i|^2 s_3 - c_0 s_0 + c_1 s_6 - c_2 s_5 + \eta_3 \\ \hat{s}_4 = \sum_i |h_i|^2 s_4 - c_0 s_7 + c_1 s_1 - c_2 s_2 + \eta_4 \\ \hat{s}_5 = \sum_i |h_i|^2 s_5 + c_0 s_6 - c_1 s_0 - c_2 s_3 + \eta_5 \\ \hat{s}_6 = \sum_i |h_i|^2 s_6 + c_0 s_5 + c_1 s_3 + c_2 s_0 + \eta_6 \\ \hat{s}_7 = \sum_i |h_i|^2 s_7 - c_0 s_4 - c_1 s_2 + c_2 s_1 + \eta_7 \\ c_0 = 2\operatorname{Re}\{h_0^* h_3 - h_1 h_2^*\} \\ c_1 = 2\operatorname{Re}\{h_0^* h_5 - h_1 h_4^*\} \\ c_2 = 2\operatorname{Re}\{h_0^* h_6 - h_1 h_7^*\} \end{cases} \qquad (7.66)$$

Clearly, in both cases the decoding does not achieve the maximum possible path diversity because in the decoding of all symbols, there is always intersymbolic interference from one symbol in the case of four antennas (c_0) and from three symbols in the case of eight antennas (c_0, c_1, c_2).

It has been shown that the STBC constructed using an orthogonal design can achieve the maximum possible rate of one for every number of transmitting antennas using any arbitrary real constellation and for two transmitting antennas using any arbitrary complex constellations (Alamouti) [Tarokh 1999]. However, any orthogonal code of rate 1 does not exist using more than two antennas [Liang 2003] – if orthogonality is essential (fully loaded systems with significant interference levels), a code with $R < 1$ should be employed for such cases.

Although this scheme is a MISO, by adopting receiving diversity, this can be viewed as a MIMO system.

7.8.2.2 Closed-Loop Techniques The description of the space diversity component of the diversity scheme based on the STD is presented next. The block diagram of this scheme is depicted in Figure 7.20.

In the case of the downlink, the STD scheme has a low-rate feedback link from the receiver (MS) telling the transmitter (BS) which antenna should be used in transmission. There is a common or dedicated pilot sequence, which is transmitted. Different antennas with specific pilot patterns/codes enable antenna selection. Then, the transmitter has a link quality information about the M (number of transmitting antennas) links and it transmits a single symbol stream over the best antenna. The receiver is supposed to reacquire the carrier phase $\theta_k(t)$ after every switch between antennas. The antenna switch (AS) has the capability to switch for every slot duration.

The decision for the antenna selection of the STD using WCDMA signals combined with the RAKE receiver should select the signal from the transmitting antenna whose multipath diversity order is higher. In other words, it should select the transmitting antenna whose channel i with L multipaths maximizes the expression $\max\left[\sum_{l=1}^{L}|h_{i,l}|^2\right]$.

This is similar to the selective combiner that can be employed in receiving diversity. The selection combining algorithm was already described in Section 7.6.1.

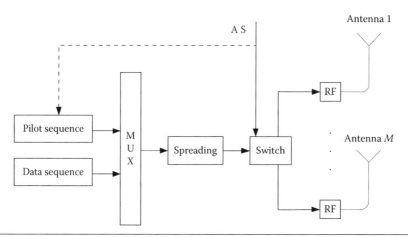

Figure 7.20   Scheme of a selective transmit diversity with Feedback Indication.

7.8.3 Multilayer Transmission

Multilayer transmission and SDMA belong to the same group, entitled spatial multi-plexing, whose principles are similar but purposes are quite different.

The goal of the MIMO based on multilayer transmission scheme is to achieve higher data rates in a given bandwidth, whose increase rate corresponds to the number of transmitting antennas [Foschini 1996] [Foschini 1998] [Nam 2002]. An example of the multilayer transmission scheme is the vertical–Bell Laboratories layered space-time (V-BLAST). In this case, the number of receiving antennas must be equal or higher than the number of transmitting antennas. The increase of symbol rate is achieved by "steering" the receiving antennas to each one (separately) of the transmitting antennas, in order to receive the corresponding data stream. This is achieved through the use of the nulling algorithm. With the sufficient number of receiving antennas, it is possible to resolve all data streams, as long as the antennas are sufficiently spaced so as to minimize the correlation [Marques da Silva 2005b]. The V-BLAST scheme is described in the following.

The $M \times N$ MIMO scheme is spectral efficient and resistant to fading, where the BS uses M transmitting antennas and the MS uses N receiving antenna [Marques da Silva 2004]. As long as the antennas are located sufficiently far apart, the transmitted and received signals from each antenna undergo independent fading.

As depicted in Figure 7.21, two different V-BLAST MIMO schemes are considered: scheme 1 and scheme 2 [Marques da Silva 2004]. Scheme 1 directly allows an increase of the data rate whose increase rate corresponds to the number of transmitting antennas. Scheme 2 allows the exploration of a higher order of diversity, without increasing data rate. The TD combining is performed using any combining algorithm, preferably the MSE-based combining algorithm. In case of the scheme 2 (depicted in Figure 7.21b), antenna switching is performed at a symbol rate, where the red dashed lines represent the signal path at even symbol periods, in the case of two transmitting antennas. Output signals are then properly delayed and combined to provide diversity.

7.8.3.1 System Description of the V-BLAST Scheme Applied to WCDMA Signals The generic diagram of the V-BLAST scheme is depicted in Figure 7.22. It is equipped

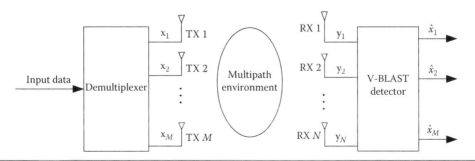

Figure 7.21   Generic diagram of the $M \times N$ V-BLAST MIMO scheme.

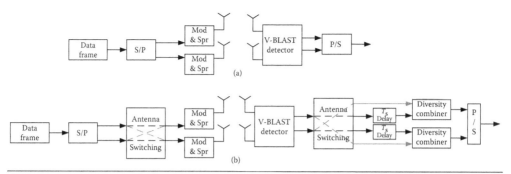

Figure 7.22   Diagram of the **2×2** V-BLAST MIMO alternatives (a) scheme 1 and (b) scheme 2.

with M transmitting and N receiving antennas ($N \geq M$). In the transmitter, the data stream is demultiplexed into M independent substreams, and then each substream is encoded (modulated and spreaded) and transmitted by M antennas. Each spread data symbol x_m is sent to the mth transmitting antenna. It is considered that the radiated signals are WCDMA signals, whose environment is typically defined by multipath frequency-selective fading channels. At the receiver, the detector estimates the transmitted symbols from the received signals at N receiving antennas.

It is assumed that all spreading sequences are unit-norm complex random binary sequences with value $(\pm 1 \pm j)/\sqrt{2}$, where $j = \sqrt{-1}$. For the scheme 1, the low-pass equivalent transmitted signals at the M antennas are respectively given by

$$
\begin{cases}
x_1(t) = \sum_{\mathrm{sf}=-\infty}^{+\infty} \left[\sqrt{E_c} \cdot \sum_{k=1}^{K} b^{(1)}(k, \lfloor \mathrm{sf}/S_F \rfloor) \cdot S_d(k, \mathrm{sf}) \right] \cdot g(t - \mathrm{sf}T_c) \\
\;\;\vdots \\
\;\;\vdots \\
\;\;\vdots \\
x_M(t) = \sum_{\mathrm{sf}=-\infty}^{+\infty} \left[\sqrt{E_c} \cdot \sum_{k=1}^{K} b^{(M)}(k, \lfloor \mathrm{sf}/S_F \rfloor) \cdot S_d(k, \mathrm{sf}) \right] \cdot g(t - \mathrm{sf}T_c)
\end{cases}
\tag{7.67}
$$

where E_c is the chip energy of data channel. The operation $\lfloor \cdot \rfloor$ stands for the integer part of operand. $S_d(k, \mathrm{sf})$ represents the spreading sequence of the kth user and sf represents the chip index order ($\mathrm{sf} : 1...S_F$). S_F is the spreading factor, T_c is the chip interval, and $g(t)$ is the chip waveform. $b^{(1)}(k, \lfloor \mathrm{sf}/S_F \rfloor)$ and $b^{(M)}(k, \lfloor \mathrm{sf}/S_F \rfloor)$ are the data symbols in the 1st and Mth transmitting antenna. The baseband equivalent of the N-dimensional received vector $\mathbf{y} = \left[y_1 y_2 ... y_N \right]^T$ at sampling instants is expressed by

$$
\mathbf{y} = \sum_{l=1}^{L} \mathbf{H}_l \mathbf{x} + \mathbf{z}
\tag{7.68}
$$

where $\mathbf{x} = \left[x_1 x_2 ... x_M \right]^T$ denotes the transmitted symbol vector with each element having the unit average power, and $\mathbf{H}_l : \mathbf{H}_1 \mathbf{H}_2 ... \mathbf{H}_L$ denotes the $N \times M$ channel matrix, whose element $h_{m,n}$ at the nth row and mth column is the channel gain from the mth transmitting antenna to the nth receiving antenna, that is, $(h_{m,n})_l = (\alpha_{m,n} e^{j\theta_{m,n}})_l \delta\left(t - \left(\tau_{m,n}\right)_l\right)$. L is the number of multipaths of the channel and the index l stands for the multipath order. It is considered a discrete tap-delay-line channel model where the channel from mth transmitting antenna to the nth receiving antenna comprises discrete resolvable paths, expressed through the channel coefficients. The $N \times M$ sets of temporal multipaths corresponding to paths between the several transmitting antennas and the several receiving antennas experience independent but i.i.d. Rayleigh fading. The elements of the N-dimensional noise vector $\mathbf{z} = [z_1 z_2 \ldots z_N]^T$ are also assumed to be i.i.d. complex Gaussian random variables with zero mean and variance σ_n^2.

7.8.4 Space Division Multiple Access

The goal of the space division multiple access (SDMA) scheme is to improve the capacity (more users per cell), while keeping the spectrum allocation unchanged. It is usually considered in the uplink, where the transmitter (UE) has a single antenna whereas the receiver (BS) has several antennas. Figure 7.23 depicts an SDMA configuration applied to the uplink. SDMA assumes that the number of antennas at the receiver is higher than the number of users that share the same spectrum. With such

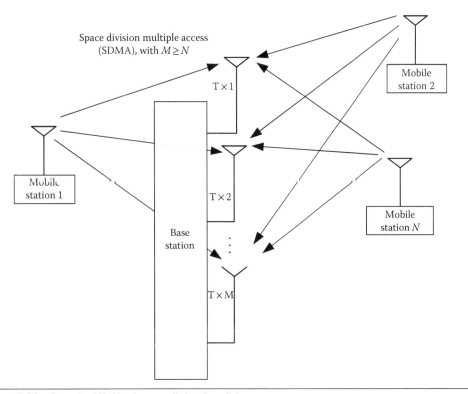

Figure 7.23 Example of SDMA scheme applied to the uplink.

approach, the receiver can decode the signals from each transmitter, while avoiding the signals from the other transmitters. Similar to the decoding performed in multilayer transmission, this can be achieved through the use of the nulling algorithm.

In SM, the symbol with the highest SNR is first detected using a linear nulling algorithm such as zero forcing (ZF) or MMSE [Foschini 1996]. The detected symbol is regenerated and the corresponding signal portion is subtracted from the received signal vector using typically a successive interference canceller (SIC). This cancellation process results in a modified received signal vector with fewer interfering signal components left. This process is repeated until all symbols are detected. According to the detection-ordering scheme in [Foschini 1996], the detection process is organized so that the symbol with the highest SNR is detected at each detection stage.

Clearly, the processing of SDMA is almost the same as that of the SM.

7.8.5 Beamforming

In STBC and SM MIMO schemes, the antenna elements that form an array are usually widely separated in order to form a TD array with low correlation among them. On the other hand, the beamforming is implemented by antenna array with certain array elements at the transmitter or receiver being closely located to form a beam array with antenna elements spacing typically half wavelength. This scheme is an effective solution to maximize the SNR, as it steers the transmitting (or receiving) beam toward the receiver (or transmitter) [Marques da Silva 2009]. As a result, an improved performance or coverage is achieved with beamforming, that is, the number of required cells to cover a given area is reduced.

Figure 7.24 depicts a transmitting station transmitting a signal using the beamforming, generated with the uniform linear antenna array (ULA). As can be seen, the beamforming allows transmitting a higher power directed toward the desired station, while minimizing the transmitted power toward the other stations.

It consists of M identical antenna elements with 120° half-power beamwidth (HPBW) [Schacht 2003]. Each antenna element is connected to a complex weight $w_m, 1 \leq m \leq M$. An analogy can be done with a receiving array antenna. The weighted elementary signals are summed together making an output signal as follows [Schacht 2003]:

$$y(t,\theta) = \sum_{m=1}^{M} w_m x \left[t - (m-1)\frac{d}{c}\sin\theta \right] \tag{7.69}$$

where $x(t)$ is the transmitted signal at the first element, d is the distance between elements (here considered $\lambda/2$), c is the propagation speed, and θ is the direction of arrival (DOA) for the main sector, $-60° < \theta < 60°$. The other two array sectors are essentially the same as in the first one, and therefore, attention is paid only to the main sector. In the frequency domain, the earlier equation can be written as:

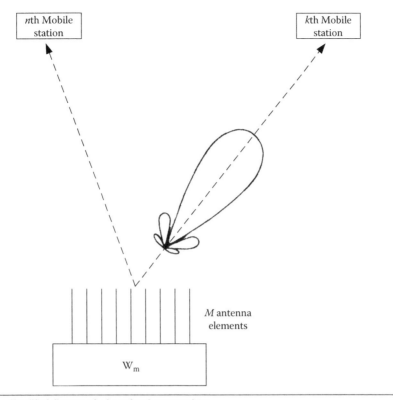

Figure 7.24   Simplified diagram of a beamforming transmitter.

$$Y(f,\theta) = \sum_{m=1}^{M} w_m X(f) e^{-j2\pi f(m-1)\frac{d}{c}\sin\theta} \tag{7.70}$$

which makes

$$H(f,\theta) = \frac{Y(f,\theta)}{X(f)} = \sum_{m=1}^{M} w_m e^{-j2\pi f(m-1)\frac{d}{c}\sin\theta}. \tag{7.71}$$

For narrowband beamforming, f is constant and θ is variable. For the beam to be directed toward the desired direction θ_1, we have

$$w_m = e^{j2\pi f(m-1)\frac{d}{c}\sin\theta_1} \tag{7.72}$$

in which, for the case of $d = \lambda/2$, (7.72) results in [Schacht 2003]

$$w_m = e^{j\pi(m-1)\sin\theta_1}. \tag{7.73}$$

In other words, for $\theta = \theta_1$, $H(f,\theta_1)$ is reduced to

$$H(f,\theta_1) = M, \tag{7.74}$$

which is the maximum attainable amplitude by beamforming.

7.9 Multiresolution Transmission Schemes

The introduction of multiresolution in a broadcast cellular system deals with source coding and the transmission of the output data streams. In a broadcast cellular system, there is a heterogeneous network with different terminal capabilities and connection speeds. For the particular case of video, a common strategy presented in the literature to adapt its content within a heterogeneous communications environment is scalable video [Li 2001] [Liu 2003] [Vetro 2003] [Dogan 2004] [Holma 2007].

A flexible common channel, suitable for point-to-multipoint transmissions is already available in UMTS networks, namely the forward access channel (FACH), which is mapped onto the secondary common control physical channel (S-CCPCH).

If we do not have macrodiversity in [3GPP TR 25.803-v6.0.0], about 40% of the sector total power has to be allocated to a single 64 Kbps MBMS if full cell coverage is required. This makes MBMS too expensive since the overall system capacity is limited by the power resource. To make MBMS affordable for the cellular system, its power consumption has to be reduced. Inner-loop power control (also called fast closed-loop power control) is not implemented in S-CCPCH channels because multiresolution schemes and/or macro diversity allow using less transmitting power. It is worth remembering that inner-loop power control in the downlink consists of the ability of the BS to adjust its output power in accordance with one or more commands received from the UE, in order to keep the received SNR at a given threshold.

Assuming that macro diversity combining is not used, extra power budget has to be allocated to compensate for the receiving power fluctuations. A common approach consists of considering MBMS video streaming as scalable, with one basic layer to encode the basic quality and consecutive enhancement layers for higher quality. Only the most important stream (basic layer) is sent to all users in the cell to provide the basic service. The least important streams (enhancement layers) are sent with less amount of power or coding protection and only the users who have better channel conditions are able to receive that additional information to enhance the video quality. Using this methodology, the transmission power for the most important MBMS stream can be reduced because the data rate is reduced, and the transmission power for the less important streams can also be reduced because the coverage requirement is relaxed.

Scalable video, depicted in Figure 7.25, provides a base layer for minimum requirements [Li 2001] and one or more enhancement layers to offer improved quality at increasing bit/frame rates and resolutions. Therefore, this method significantly decreases the storage costs of the content provider.

Besides being a potential solution for content adaptation, scalable video schemes may also allow an efficient usage of power resources in MBMS, as suggested in [Cover 1972]. This is depicted in Figure 7.25, where two separate physical channels are provided for one MBMS service (e.g., @ 256 Kbps) – one for the base layer, at half bit rate of the total bit rate (128 Kbps), and with a power allocation that

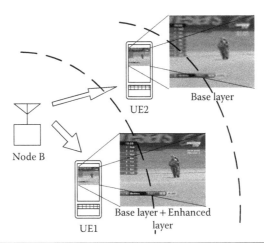

Figure 7.25 Scalable video transmission.

can cover the whole cell range; another one for the enhanced layer, also at half bit rate of the total bit rate (128 Kbps), but with less power allocation than that of the base layer.

The system illustrated in Figure 7.25 consists of two QoS regions, where the first region receives all the information whereas the second region receives the most important data. The QoS regions are associated to the geometry factor that reflects the distance of the UE from the base station antenna.

Scalable video transmission can also be implemented using different techniques. One example consists of the use of hierarchical constellations. In case the channel conditions are not above a certain threshold, since the modulation considered is QPSK (instead of 16-QAM) the received bit rate is reduced to half. In addition, hierarchical constellations may be combined with different channel coding rates, which correspond to the concept of adaptive modulation and coding [Souto 2007b].

Another possibility to implement the scalable video transmission is the use of SM MIMO technique, where each transmitting antenna sends a different data stream. The first data stream (most powerful) may include the base layer, whereas the enhanced layer may be sent by a second antenna (a less powerful data stream). Depending on the power and channel conditions, a certain UE may receive successfully either the two streams or only the base layer.

7.9.1 Hierarchical QAM Constellations

The limited spectrum resources available are one of the major restrictions for achieving high bit rate transmissions in wireless communication networks. Multilevel quadrature amplitude modulation (M-QAM) is considered an attractive technique to achieve this objective due to its high spectral efficiency and has been studied and proposed for wireless systems by several authors [Webb 1994] [Webb 1995] [Goldsmith 1997]. In

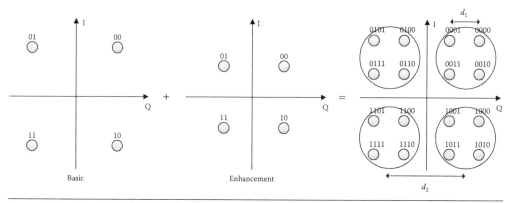

Figure 7.26   Signal constellation for 16-QAM Nonuniform modulation.

fact, 16-QAM modulation has already been standardized for the HSDPA mode of the UMTS by 3GPP.

M-QAM constellations can be constructed in a hierarchical way so as to provide multiresolution and improve the efficiency of the network in broadcast/multicast transmissions, as discussed in [Cover 1972]. In this case, the constellations can be referred to as hierarchical, embedded, or multiresolution M-QAM (we will denote them as M-HQAM (Hierarchical QAM)).

To present an example of that, 16-QAM hierarchical constellations can be constructed using a main QPSK constellation where each symbol is, in fact, another QPSK constellation, as shown in Figure 7.26. This construction procedure results in two classes of bits with different error protection, which can be modified by using nonuniformly spaced signal points.

The bits used for selecting the symbols inside the small inner constellations are called weak bits and the bits corresponding to the selection of the large outer QPSK constellation are called stronger bits. The idea is that the constellation can be seen as a 16-QAM constellation if the channel conditions are good or as a QPSK constellation otherwise. In the latter situation, the received bit rate is reduced to half. Some alterations to the physical layer of the UMTS system to incorporate these modulations have already been proposed in [Souto 2005a] [Souto 2005b] [Souto 2007b]. Figure 7.27 shows an example of the usage of a 64-HQAM constellation in a cellular system. Depending on the position in the cell, the users will demodulate the received signal as 64-QAM, 16-QAM, or QPSK.

These techniques are interesting for applications where the data being transmitted is scalable, that is, it can be split in classes of different importance. For example, in the case of video transmission, the data from the video source encoders may not be equally important. The same happens in the transmission of coded voice. Several authors have studied the use of hierarchical constellations for this purpose. In [Ramchandran 1993] and [Wei 1993] hierarchical QAM constellations were employed for the transmission of digital high definition television

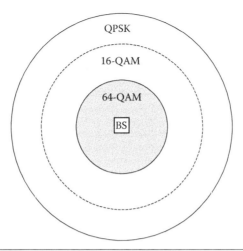

Figure 7.27 Example of the type of demodulation used inside a cell for transmission of a 64-QAM hierarchical constellation.

signals. Furthermore, [Engels 1998] compares the performance of 64-QAM and 64-DAPSK hierarchical constellations, whereas [Pursley 1999] analyzes the use of M-PSK hierarchical constellations in multimedia transmissions. Hierarchical 16-QAM and 64-QAM constellations have been incorporated into DVB-T standards [ETSI 2904].

7.9.2 Macrodiversity

There are two types of networks to be considered: the multifrequency networks (MFN) and the single-frequency networks (SFN). Macrodiversity refers to the transmission of the same information by different Node Bs to the UE in the downlink. The Node Bs, to which a terminal is linked to, are referred to as the active set. Macrodiversity aims at supplying additional diversity in situations where the terminal is far from the Node Bs, in order to compensate the path loss affecting the transmission to a UE located at the edge of the cell and to reduce the amount of transmit power needed to reach a distant receiver, thus increasing network capacity.

It is worth noting that in broadcast, the global CIR is longer due to the distance between the transmitter (BS) and the several different receivers (UEs). Nevertheless, if the cyclic extension is long enough, the global CIR will be the sum of the independent CIRs, which enables the SFN concept, allowing the macrodiversity effect. This can be seen in Figure 7.28.

Therefore, macrodiversity is used during soft handover in order to ensure smooth transitions between two cells or two sectors of the same cell, thus reducing the risk to drop the call.

Note that the performance gain brought by macrodiversity depends on the diversity order of the channel, that is, a two-path channel benefits more from macrodiversity than a six-path channel because the latter exhibits already a high multipath

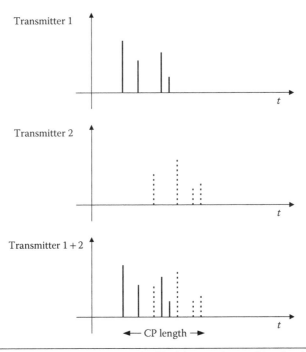

Figure 7.28   Global CIR (c) is composed of the sum of the CIR of the several transmitters (a + b).

diversity order. In opposition to the dedicated physical channel, macrodiversity for MBMS does not consume network resources, as MBMS is broadcasted simultaneously in several cells. In the case of MFN, the UE is required to estimate the carrier of each Node B it is linked to. This increases its power consumption. Moreover, the signals received from different Node Bs (especially far ones) may be significantly delayed with regard to those received from near Node Bs. This requires extra memory at the terminal in order to store the received signals for further combining or an additional synchronization procedure between the Node B's transmitters. In the downlink, the combining takes place at the mobile, which has to demodulate and then combine the signals received from the different Node Bs in the active set. The extra complexity added by macrodiversity then depends on the receiver type. In the case of an equalizer, one has to be set up and operated for each Node B the UE is linked to. Moreover, the UE must estimate one transmission channel per Node Bs. In the special case of OFDM, two main cases for macrodiversity can be distinguished:

1. Node Bs are synchronized, at least to allow UE's receiving signals from two or more Node Bs with a time difference smaller than the CP.
2. Node Bs are not synchronized.

In the first case, the Node Bs can transmit identical signals to the terminal on the same time-frequency resource. This is possible because the signals will superpose within the CP: no ISI occurs as long as the sum of the time differences plus the maximum delay of the CIRs is shorter than the CP. In this case, the terminal can employ a single receiver to demodulate the superimposed signals. This means that it will perform a unique DFT. In this case, macrodiversity behaves just like TD (from a unique transmitter with multiple spaced antennas). When different Node Bs transmit the same data over the same subcarriers, the resulting propagation channel is equivalent to the allegorical sum of all propagation channels, which increase the diversity gain. When Node Bs send the same data over different subcarriers, the maximum diversity can be achieved since each data symbol benefits from the summation of the propagation channel powers. It could also be possible to form a MIMO scheme but, in that case, the data and distinct pilot signals have to be sent on orthogonal time-frequency resources, and several propagation channels have to be estimated. If the Node Bs cannot be assumed to be synchronized, the terminal will need separated receiver chains to demodulate the signals from the distinct Node Bs. Moreover to avoid interference, orthogonal time-frequency resources have to be allocated to the different Node Bs. This is still very complex to fulfill, thus in the general case, interference will occur.

Fast cell selection is one option for macrodiversity for unicast data. Intra-Node-B selection should be able to operate on a subframe basis. An alternative intra-Node-B macrodiversity scheme for unicast is a simultaneous multicell transmission with soft combining. The basic idea of multicell transmission is that, instead of avoiding interference at the cell border by means of intercell-interference coordination, both cells are used for transmission of the same information to a UE thus reducing intercell interference as well as improving the overall available transmit power. Another possibility of intra-Node-B multicell transmission is to explore the diversity gain between the cells with space-time processing (e.g., by employing STBC through two cells). Assuming Node-B-controlled scheduling and that fast/tight coordination between different Node B is not feasible, multicell transmission should be limited to cells belonging to the same Node B. For multicell broadcast, soft combining of radio links should be supported, assuming a sufficient degree of inter-Node-B synchronization, at least among a subset of Node Bs.

7.9.3 Multihop Relays

One key to improve the coverage and capacity for high quality multimedia broadcast and multicast transmissions in mobile networks is to increase the data rates for the MSs at the cell edge. MSs at the cell edge suffer from high propagation loss and high intercell interference from the neighbor cells. Other MSs reside in areas that suffer from strong shadowing effects. Thus, the overall goal of this activity is to bring more power to the cell edge and into shadowed areas while inducing minimal additional interference for neighbor cells. There are several measures to implement multihop relays [Sydir 2009]. The next paragraphs describe some of these methods.

Adaptive Relaying: As stated before, MSs at the cell edge suffer from high propagation loss and high inter-cell interference from the neighbor cells. Other MSs reside in areas that suffer from strong shadowing effects. Thus the overall goal of adaptive relaying is to bring more power to the cell edge and into shadowed areas while inducing minimal additional interference for neighbor cells. The obvious solution for this would be to decrease cell sizes by installing additional base stations which would, of course, mean to increase the network infrastructure costs.

In opposition to conventional repeaters working with the amplify-and-forward strategy, adaptive relays are understood to work in a decode-and-forward style. By doing this, relays amplify and retransmit only the wanted component of the signal they receive and suppress the unwanted portions (they do not amplify unwanted signals). The disadvantages of relays compared to simple repeaters are the additional delay that they introduce into the transmission path between BS and UE and, depending on the algorithms, a possible signaling overhead.

The gain is based on the fact that the transmission path is split up into smaller parts that can reduce propagation loss.

Fixed relay stations (RS) positioned at a specified distance from the BS (Figure 7.29) could help to increase the probability that a MS receives enough power from several base stations. This deployment concept would sectorize the cell in an inner region, where the UEs (e.g., UE 1 in Figure 7.29) can receive their signal from the base station plus some RSs and an outer region where only the signal from the relay stations is strong enough (e.g., UE 2 in Figure 7.29).

Configurable virtual cell sizes: Kudoh and Adachi [Kudoh 2003] proposed a wireless multihop virtual cellular network. It consists of a so-called central port that corresponds to the base station, which acts as a gateway to the core network. The so-called wireless ports correspond to the RSs that communicate with the MSs and relay the signal from and to the central port. The wireless ports that are communicating directly with the MSs are called end wireless ports. The wireless ports are stationary and can act together with the central port as one virtual base station. The central port and the

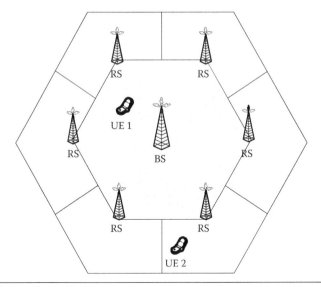

Figure 7.29   Two-hop relaying architecture.

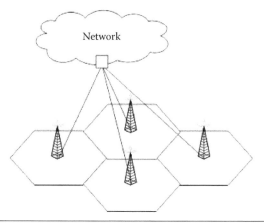

Figure 7.30 Present cellular network.

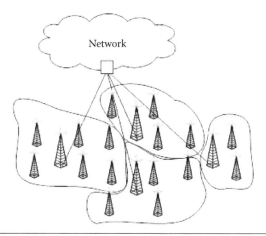

Figure 7.31 Virtual cellular network.

end wireless ports introduce additional diversity into the cell, so that the transmit power may be reduced, which means also a reduced interference for other virtual cells. The differences between present cellular networks and virtual cellular networks are illustrated in Figures 7.30 and 7.31.

From the perspective of multimedia broadcasting and multicasting, configurable virtual cell sizes could be used to adapt the cell size to the user distribution and their service needs can be very different in spatial or temporal dimensions.

End of Chapter Questions

1. What are the differences among FDMA, TDMA, and CDMA?
2. What are the differences between FDMA and OFDM?
3. What is the Rake receiver used for?
4. Which type of transmission technique is employed associated to the Rake receiver?

5. Which kinds of MIMO do you know? What are the differences among them?
6. What is the difference between beamforming and SDMA?
7. What does multiresolution technique stands for?
8. Which types of multiresolution techniques do you know? How do they work?
9. What is the difference between spread spectrum and DS-CDMA?
10. Which types of CDMA do you know?
11. What is the difference between closed-loop and open-loop MIMO techniques?
12. What are the differences between OFDMA and SC-FDE? Are these techniques employed in the uplink or in the downlink? Why?
13. Which types of diversity combining techniques do you know? What are their differences?
14. What is the difference between Alamouti-like and V-BLAST MIMO techniques?
15. What is the IB-DFE used for? How does it work?

8

CELLULAR COMMUNICATIONS

The recent need to be in permanent touch with others facilitated an enormous growth of wireless communication, both in terms of offer and service demand. In the past, cellular phones were used mainly for voice communication. A constant evolution of services allowed the massification of services such as short message services (SMS), multimedia messaging service (MMS), video call, and so on. Recently, mainly as a result of the implementation of the universal mobile telecommunication system (UMTS) by cellular operators, Internet access became possible, allowing a new myriad of services such as e-mail, video streaming, and web browsing, among others. In fact, the evolution from the second-generation cellular systems (2G) into the third generation (3G) was the main driver to allow a sudden increase of traffic due to the new services.

Cellular coverage and capacity as well as electrical consumption are key aspects in cellular systems. The cellular coverage is related to the necessary balance made by operators between costs and quality/service availability. Currently, there are still some geographic zones where, due to difficult propagation conditions or low population density, the coverage is weak or even inexistent.

8.1 Cellular Concept

The cellular network concept was the result of the need to develop a higher capacity for the mobile service. It consists of the use of several low-power transmitters, typically with less than 100 W. Figure 8.1 shows a typical cellular network structure, where a base station (BS) is placed at the center of each cell. This BS is used by mobile stations (MS) located in the corresponding cell in order to allow them to gain access to the network. Note that different BSs are typically interconnected with guided medium transmission systems (e.g., optical fibers, terrestrial microwave system, etc.).

As depicted in Figure 8.2, in order to allow a connection establishment between the MS1, located in left cell, and MS2, located in another cell, the network performs the following steps: (1) MS1 establishes a link over the air with the corresponding BS (BS1), (2) then, the existing static connection between the source BS (BS1) and destination BS (BS2) allows the call being forwarded to the destination BS, and (3) finally, BS2 establishes a link over the air with the MS2. Using these three independent links, a connection can be established between two different users located in different cells.

The areas to be covered by a cellular network are distributed among cells, where each one is allocated a set of frequency bands to be used by the BS and MS. The use

217

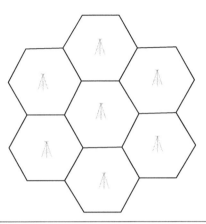

Figure 8.1 Cellular structure.

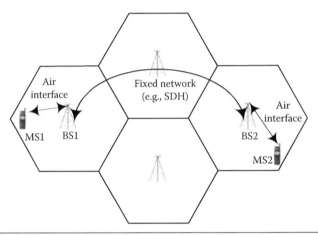

Figure 8.2 Example of interconnection between two mobile stations using the corresponding base stations.

of the same frequency bands by adjacent cells results in co-channel interference (see Chapter 3). This type of interference is not experienced when adjacent cells do not use the same frequency bands, as the propagation losses assure the necessary isolation. Consequently, the geometric pattern utilized by cellular configuration is depicted in Figure 8.3.

The cell geometry consists of a hexagon. Instead of considering a cell consisting of a circle around the center (BS), the hexagon is adopted. With such configuration, the antennas from different BSs are equidistant. In Figure 8.3, each letter represents a different set of frequency bands assigned to each cell. As can be seen, in order to avoid co-channel interference, the same bands are not assigned to adjacent cells. This leads to the reuse factor seven, where each group of seven cells use a group of seven sets of frequency bands, and the repetition of frequency bands only occurs in groups of seven cells. In fact, as described in Chapter 3 for the definition of co-channel interference, code division multiple access (CDMA) networks (such as UMTS) make use of all bands in all cells, leading to reuse factor one. This is depicted in Figure 8.4. The resulting residual interference is mitigated by employing multiuser detectors.

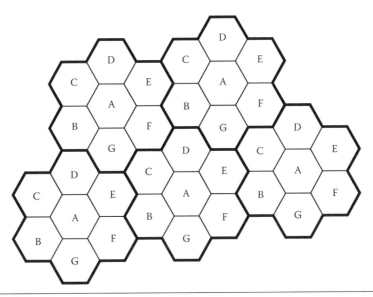

Figure 8.3 Typical cellular network structure with reuse factor seven.

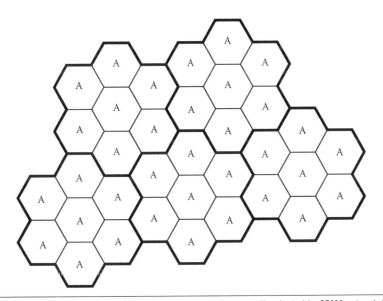

Figure 8.4 Typical cellular network structure with reuse factor one (normally adopted by CDMA networks).

Cellular coverage is related to capacity. The cellular capacity is viewed as the number of simultaneous calls within a cell. Regardless of the cell dimension, a cell accommodates a certain number of simultaneous calls. Increasing the cell dimension (e.g., macrocell, instead of a microcell) allows accommodating a lower number of calls per square kilometer while corresponding to a higher cellular coverage. This can be seen in Figure 8.5.

The decision on whether or not to implement a lower hierarchy cell relies on the rate of calls expected. Allocating M carriers per cell allows accommodating $M \times N$

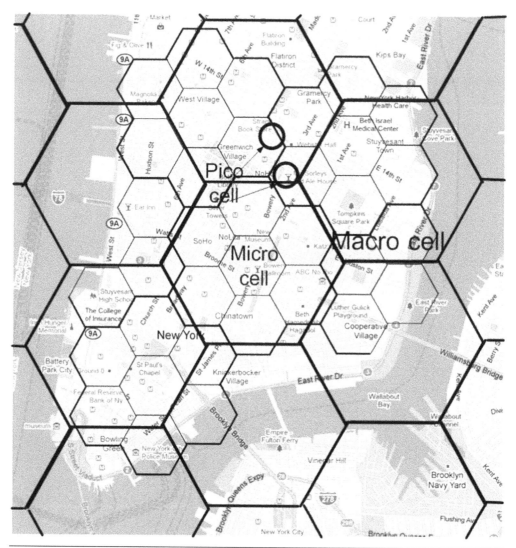

Figure 8.5   Hierarchical cellular structure.

calls, with N the number of calls per carrier.* Note that M, which relates to the number of carriers allocated per cell, and the resulting $M \times N$ maximum simultaneous calls are independent on whether it is a macrocell, microcell, or picocell. The above equivalence assumes time-division duplexing (TDD).† In case of frequency-division duplexing (FDD),‡ the number of calls per cell is halved.§

---

* GSM accommodates eight TDMA calls per carrier. The number of calls accommodated in each wideband UMTS carrier depends on the CDMA spreading factor.
† Time-division duplexing allows full-duplex operation by allocating uplink and downlink channels in a single-carrier frequency but using different time slots.
‡ Frequency-division duplexing allows full-duplex operation by allocating uplink and downlink channels in two different carrier frequencies.
§ Half of the carriers are used for the uplink and the other half for the downlink.

In 3G networks, the coverage is also related to the throughput available to each user. A lower dimension cell typically corresponds to a better signal quality due to the lower path loss. A better signal quality (higher signal strength and lower noise and interferences) allows typically a higher throughput.

Although this is not a problem inherent of cellular telecommunications, electrical consumption has been, lately, an issue to which the technological research is looking at. In the telecommunications field, this problem can be mitigated by using smart and adaptive antennas, efficient power control (see Section 8.1.5), or with the implementation of hierarchical cellular structures (e.g., macrocells, microcells, picocells, femtocells, etc.). In this sense, a lower dimension cell, where a mobile station is at a shorter distance from a BS, corresponds to energy saving.

8.1.1 Macrocell

A macrocell is the initially designed type of cell. In order to maximize the coverage, the antenna's BS is placed in geographically high locations (e.g., on the top of buildings or on the top of hills). The area of a macrocell varies from few kilometers to around 30 kilometers (rural environments). The dimension of the cell to be covered depends on the population density and on the propagation environment. The coverage area of a cell in rural environment tends to be higher than in urban scenarios, as the expected rate of calls is lower in rural areas. Note that the fading typically experienced in rural areas is modeled by a Ricean distribution. As detailed in Chapter 5, it consists of a Rayleigh fading model (Rayleigh distribution) to which a line-of-sight component is summed. Due to the existence of line-of-sight, the shadowing effect is normally not experienced, resulting in a lower rate of path loss* in rural areas as compared to urban environments. Consequently, a rural environment is much easier to cover and the cell size is typically higher than in urban areas. On the other hand, an urban environment is typically characterized by the absence of line-of-sight component, where the shadowing effect is normally experienced and whose path loss rate is higher. Note that due to the expected low rate of calls in rural areas, other lower size cells (e.g., microcells and picocells) are normally not implemented.

8.1.2 Microcell

A microcell is normally implemented in addition to the macrocell (see Figure 8.5). This cell corresponds to an area from few hundreds of meters to around 2 km. In this case, a MS may switch between a macro- and microcell, depending on the signal's quality and service desired. The shadowing effect is normally not present in microcells

---

* The free space path loss has a distance decay rate 2, whereas in real scenarios, this value varies between 3 and 5. Rural scenarios present typically a distance decay rate of the order of 3, whereas in urban scenarios, this distance decay rate varies typically between 4 and 5.

as the line-of-sight propagation component normally exists between transmitting and receiving antennas. When shadowing is present, its effect is typically of reduced consequences. Note that a macrocell is more subject to a weak coverage (e.g., due to shadowing effects and to a wider area) than lower dimension cells. Furthermore, some specific highly populated areas, and the corresponding rate of calls, justify the implementation and investment in microcells. The signal-to-noise ratio (SNR) made available by microcells is typically better than that of macrocells. Consequently, in the case of 3G (CDMA networks), the available user throughput in microcells is normally higher than in macrocells.

8.1.3 Picocell

Picocells are designed for indoor or outdoor highly populated environments such as hotels, offices, or shopping centers. In addition, the area to be covered varies from few meters to around 200 meters. Since the line-of-sight component is normally present in picocells, the type of fading experienced is typically the multipath Ricean fading.

In picocells, the SNR is normally high and the available throughput is maximized. Consequently, the number of calls per square area is maximized. The implementation of a picocell requires a high investment in infrastructures, which is only justified in highly populated areas. Since walls originate high attenuation levels, the implementation of an indoor picocell solves the problem of covering indoor environments from outdoor BS. In order to avoid co-channel interference, the spectrum made available in lower dimension cells must be different from that in higher dimension cells. Finally, due to the low distances involved, the battery use in picocells is much less than in higher order cells, which translate into energy saving.

8.1.4 Femtocell

Femtocells make use of the existing infrastructures, while solving coverage problems, as well as allowing a reduction in the electrical consumption [Zhang 2010]. It is viewed as economically more effective than picocells. A femtocell is designed for indoor and small-size environment. It makes use of a small-size BS, which typically interconnects with the cellular network through xDSL or cable modem. Although a low-cost solution, it enables a high SNR in the interior of domestic houses, offices, or other indoor buildings, resulting in high throughput available to the end user and a wide range of services.

Using femtocells, the services of a cellular operator can be viewed as an alternative to the fixed telephone operator as well as to the Internet service provider. Femtocells were not included in the 3GPP specifications for UMTS but are considered for the LTE. Similarly to picocells, since the penetration rate of electromagnetic waves over walls is limited, a femtocell solves the problem of covering indoor environments from outdoor BS. Note that, due to the low distances involved, the battery use in femtocells is minimized, representing another important advantage of this cell configuration.

8.1.5 Power Control

Power control is the adjustment in the transmitting power in order to optimize the performance of communications. In cellular communications, power control is used to maximize the SNR of signals in receivers. This is performed using fast and slow power control.

Fast power control intends to mitigate the effects of fast fading caused by the multipath channel, in the uplink as well as in the downlink. On the other hand, slow power control intends to compensate for the received power from stations far from the receiver (e.g., at the edge of a cell), in order to compensate for the near–far problem. In addition, slow power control also intends to mitigate the effects of slow fading (shadowing).

The purpose of power control is to make the received SNR as constant as possible, and consequently, to keep the bit error probability stationary. The overall maximization of SNR is performed in such a way that the transmitting power is the minimum necessary to originate a signal at the receiver with the necessary SNR, whereas the power is not higher than this value in order to minimize the level of interference generated to other users. Note that a user's signal represents interference to the others, especially in a cellular network whose air interface is based on CDMA technology.

Uplink power control improves the performance due to the following reasons: It equalizes the power received from multiple MS (avoiding the near–far problem) and compensates for fading. Moreover, besides life batteries are increased, interferences from adjacent cells (co-channel interference) are decreased. Note that co-channel interference is more important in CDMA networks, as the reuse factor normally adopted is one. Power control is always employed in CDMA networks, as this technology is highly sensitive to variations in the received power.[*] This is valid even when the CDMA network adopts multiuser detectors at receivers, whose purpose is the mitigation of the level of multiple access interference.

In the downlink, as the BS sends all the signals to all mobiles in a synchronized manner, all signals are received by all MS with the same power. Therefore, in this case, the use of power control is only considered for minimizing interferences caused in adjacent cells[†] (especially in CDMA networks) and to compensate for interferences received from adjacent cells.[‡]

---

[*] A user's signal received with a higher power represents a higher level of multiple access interference to the others.

[†] Transmit with the minimum power to accommodate each user, without generating a high level of interference in adjacent cells.

[‡] Increase the power in order to increase the value of SNR. This is normally necessary for users located farer from the reference BS, that is, at a lower distance from an adjacent BS.

Power control is performed in a dynamic way, in open or closed-loop. Open-loop power control measures the interference conditions, adjusting the transmitted power to avoid these effects. However, since in networks using different transmitting and receiving frequency bands (frequency-division duplexing) the fast fading does not present correlation between uplink and downlink, open-loop power control is not an effective mechanism. In this case, closed-loop power control is normally adopted. With closed-loop power control, the receiver measures the SNR and sends a command to the transmitter to adjust its power, in order to keep the destination's SNR at a desired level, avoiding fluctuations.

As detailed in Chapter 7, multiple input multiple output (MIMO) system is another technique that can be employed to maximize the SNR, without having to increase the transmitting power.

8.2 Evolution of Cellular Systems

First generation (1G) of cellular networks (1980–1992) was analogue (NMT, AMPS, TACS, etc.). These systems were of low reliability, low capacity, low performance, and without the ability to perform roaming between different networks and countries. The multiple access technique adopted was frequency-division multiple access (FDMA), where different signals were transmitted in different (orthogonal) frequency bands.

The second generation (2G), like the global system for mobile communications (GSM)–(1992–2002/3), introduced the digital technology in the cellular environment, with a much better performance, better reliability, higher capacity, and even with the ability to perform roaming among operators, due to its standardization and advanced technology. The multiple access technique used by GSM was time-division multiple access (TDMA).

The need to optimize the scarce spectrum and deliver higher throughputs, as well as new services, drove the development of the third generation cellular system (3G). Such services include web browsing, e-mail access, video streaming, file transfer, and so on.

8.3 UMTS

Third generation cellular systems (3G), based on wideband code division multiple access (WCDMA)* radio access technology, have been globally deployed. It was specified by third generation partnership project (3GPP), which is a group within the International Telecommunications Union (ITU). Although 3G is the global designation, the name internationally adopted was international mobile telecommunications 2000 (IMT-2000). Nevertheless, some differences exist between the systems adopted

---

* Note that 3GPP also specifies the air interface using TD-CDMA (Time Division-CDMA) and (TD-SCDMA) Time Division-Synchronous Code Division Multiple Access.

by various nations. The European implementation of 3G was designated by European Telecommunications Standardization Institute (ETSI) as UMTS. Outside Europe, it is normally known as WCDMA cellular system.

With all areas of the wireless industry developing rapidly, it has become necessary for the UMTS industry and its constituent members to provide their vision for the medium- and long-term evolution of UMTS networks and services. 3GPP has been active to provide a vision of this evolution through its activities. Important milestones were achieved in 3GPP Releases 5 and 6 with IP multimedia subsystem (IMS), namely the high-speed downlink packet access (HSDPA; enhancements enabling data transmission speeds of up to 14.4 Mbps per user), the high-speed uplink packet access (HSUPA), the incorporation of wireless local area network (WLAN) concept, and the integration of the IP protocol as a transport protocol, in addition to circuit switching. High-speed packet access (HSPA) is a generic term adopted by the UMTS forum to refer to improvements in the UMTS radio interface in the Releases 5 and 6 of the 3GPP standards. HSPA refers to both the improvements made in the downlink of UMTS, often referred to as HSDPA, and the improvements made in the uplink, often referred to as HSUPA, but also referred to as enhanced dedicated channel (E-DCH). HSPA defines a series of straightforward upgrades to UMTS Release 99 (R99) networks which offers improvements by a factor of ten in the speed of service delivery, improvements by a factor of five in network capacity and a significant improvement in service latency. Table 8.1 summarizes the differences between the several different 3GPP releases.

New terminals are required to support these capabilities and the last terminals for HSPA were made available at the end of 2008; they are category-6 terminals, capable of supporting downloads up to 14.4 Mbps and uploads up to 5.76 Mbps [Marques da Silva 2010].

Table 8.1 Comparison between Several Different 3GPP Releases

FDD	UMTS	HSPA	HSPA+	LTE	LTE/IMT
TDD	TD-SCDMA	TD-HSDPA	TD-HSUPA	TD-LTE	ADVANCED
3GPP release	99	5/6	7	8	11/12
Downlink data rate	384 kbps	14.4 Mbps[a]	28 Mbps[a]	>160 Mbps[b]	100 Mbps mobility
Uplink data rate	128 kbps	5.76 Mbps[a]	11 Mbps[a]	>60 Mbps[b]	1 Gbps nomadic
Switching	Circuit + packet switching	Circuit + packet switching	Circuit + packet switching	IP based (packet switching)	IP based (packet switching)
Transmission technique	WCDMA/ TD-SCDMA	WCDMA/ TD-SCDMA	WCDMA/ TD-SCDMA	Downlink: OFDMA Uplink: SC-FDMA	Downlink: OFDMA Uplink: SC-FDMA
MIMO	No	No	Yes	Yes	Yes
Multihop relay	No	No	No	No	Yes
AMC	No	Yes	Yes	Yes	Yes
Deployment	2003	2006/8	2008/9	2010	

[a] Peak data rates
[b] Assuming 20-MHz bandwidth and 2 x 2 MIMO.

Both HSDPA and HSUPA were implemented in the standard 5-MHz carrier of UMTS networks (Release 6), coexisting with the first generation of UMTS networks based on the 3GPP R99 standard. As HSPA standards refer uniquely to the radio access network, no change of the packet-switching core network beyond handling higher traffic was required.

HSPA offered a cost-effective wide-area broadband mobility and played a significant role in stimulating the demand for data services, whether they be consumer multimedia and gaming, or corporate email and mobile access. By comparing HSPA with WiMAX (described below), it is clear the higher penetration rates of the former system related to the latter. For the time being, HSPA connections are around 150 million globally, whereas WiMAX connections do not go beyond 3.5 million.

8.4 Long-Term Evolution

Emergent services are demanding more and more from the wireless infrastructures. 3GPP is deeply involved in specifying and defining the architecture of the 3G evolution. A few years ago, 3GPP has launched the study item evolved UTRAN (E-UTRAN), which aims at defining the new air interface of the 3GPP long-term evolution (LTE) [Holma 2007]. 3GPP LTE will be the natural evolution of the UMTS standard, in order to face the latest demands for voice and data services. In 2008, the increase of data rate was up to a factor of five [Bogineni 2009].

The LTE air interface has been added to the specification in Release 8 and its initial deployment began in 2010. Goals include improving spectral efficiency by a factor 2–4 compared to HSPA Release 6 and 7 [Marques da Silva 2010], making use of new spectrum and reframed spectrum opportunities, as well as better integration with other open standards. [3GPP 25.913-v7.0.0] [NGMN 2006] [NGMN 2007]. The LTE air interface is a completely new system based on orthogonal frequency-division multiple access (OFDMA) in the downlink and single carrier–frequency domain equalization (SC-FDE; also referred to as single carrier-frequency division multiple access (SC-FDMA) by many authors) in the uplink. SC-FDE is the selected transmission technique for the uplink due to its lower peak to average power ratio, as compared to OFDM. In addition, LTE comprises MIMO as an option in order to achieve peak data rates in downlink exceeding 300 Mbps and uplink peak data rates of approximately 75 Mbps [Dahlman 2008]. Four different types of MIMO have been considered in 3GPP Release 8, each one can be selected depending on the objective, namely: space-time block coding (STBC) for improved performance; spatial multiplexing for increased data rate; space-division multiple access (SDMA) for increased cell capacity; and beamforming for increased coverage [Astély 2009]. The modulation comprises QPSK, 16-QAM or 64-QAM. To handle occasional retransmission, LTE considers hybrid-automatic repeat request (H-ARQ) in the medium access control (MAC) sublayer.

Another important modification is that LTE is based on all-IP architecture (i.e., all services are carried out on top of IP), instead of the previous UMTS circuit plus packet-switching combined network. In fact, it is for use over any IP network, including WiMAX and Wi-Fi, and even wired networks [Dahlman 2008]. Therefore, an important focus of this evolution is on enhancements for packet-based services.

3GPP community has been working on LTE and various contributions were made to implement Multimedia Broadcast and Multicast Service (MBMS) in LTE [Astély 2009]. In the 3GPP LTE project, two types of transmission scenarios exist [Marques da Silva 2010]:

1. Multicell transmission: Multimedia broadcast over a single-frequency network (MBSFN) on a dedicated frequency layer or on a shared frequency layer
2. Single-cell transmission: Single cell–point to multipoint (SC PTM) on a shared frequency layer

Multicell transmission in single-frequency network (SFN) area is a way to improve the spectral efficiency, since all MBMS cells transmit the same MBMS session data, the signals can be combined for a UE located at a cell boundary. Furthermore, the multicell transmission may be provided over a cell group, which comprises cells that transmit the same service. In contrast, the single-cell transmission covers only one cell.

The concept of dynamic MBSFN area is introduced where the MBMS transmission is switched off in some cells of the MBSFN area when a certain MBMS is not required. In some cases, the released resource can be reused for other MBMS or unicast services.

By considering MBSFN, LTE will allow delivering services such as Mobile TV. It is worth noting that LTE standardization already reached a state where specification changes are limited to corrections and bug fixes.

OFDM is an attractive choice to meet requirements for high data rates, with correspondingly large transmission bandwidths and flexible spectrum allocation [Liu 2005]. OFDM also allows for a smooth migration from earlier radio access technologies and is known for high performance in frequency-selective channels. It enables further frequency-domain adaptation, provides benefits in broadcast scenarios, and suits well for MIMO processing.

The possibility to operate in vastly different spectrum allocations is essential. In addition, different bandwidths are realized by varying the number of subcarriers used for transmission, while the subcarrier spacing remains unchanged. In this way, operation in spectrum allocations of 1.4, 2.5, 5, 10, 15, and 20 MHz can be supported, where the later is required to provide the highest LTE data rates [Astély 2006] [Astély 2009]. The draft document [3GPP 25.814-v1.2.2], edited by radio access network 1 (RAN1) working group within 3GPP, defines and describes the potential physical layer evolution under consideration for evolved UTRA and UTRAN. Due to the fine frequency granularity offered by OFDM, a smooth migration of, for example, 2G spectrum is possible. A 2G GSM operator can in principle migrate on a 200-KHz GSM carrier-by-carrier basis by using only a fraction of the available OFDM subcarriers. FDD, TDD, and combined FDD/TDD are supported to allow for operation in paired as well as unpaired spectrum.

By allocating the proper time slots and carrier frequencies, LTE provides intra-cell orthogonality between users in both uplink and downlink. Nevertheless, inter-cell interference is a major problem, as compared to WCDMA/HSPA, especially for users at the edge of the cell. Several measures can be implemented in order to minimize this problem, as power control and intercell interference coordination or advanced interference cancellation schemes [Marques da Silva 2000a] [Marques da Silva 2000b] [Marques da Silva 2003d] [Marques da Silva 2003e] [Marques da Silva 2004b] [Marques da Silva 2004c] [Astély 2009].

8.5 WiMAX-IEEE802.16

WiMAX, standardized by the Institute of Electrical and Electronics Engineers (IEEE) as IEEE 802.16, was initially created in 2001 and updated by several newer versions. It consists of a technology that implements a wireless metropolitan area network (WMAN) using the IP [Eklund 2002] [Andrews 2007] [Peters 2009]. WiMAX means worldwide interoperability for microwave access and allows fixed and mobile access. The basic idea is to provide wireless Internet access to the last mile, with a range of up to 50 km. Therefore, it can be viewed as a complement or competitive to the existing asynchronous digital subscriber line (ADSL) or cable modem, providing the service with the minimum effort in terms of required infrastructures. On the other hand, fixed WiMAX can also be seen as backhaul for Wi-Fi, cellular BS, or mobile WiMAX. As the standard only defines the physical layer and MAC sublayer, it can be used by either IPv4 or IPv6 (or even ATM).

In order to allow the operation of WiMAX in different regulatory spectrum constraints faced by operators in different geographies, this standard specifies channel sizes ranging from 1.75 MHz to 20 MHz, using either TDD or FDD, with many options in between [Yarali 2008].

The initial version of WiMAX was upgraded several times to

- IEEE 802.16-2004, also referred to as IEEE 802.16d. This version only specified the fixed interface of WiMAX, without providing any support for mobility [IEEE 802.16-2004]. This version of the standard was adopted by ETSI as a base for the high-performance metropolitan area network (HiperMAN).
- IEEE 802.16-2005, also referred to as IEEE 802.16e. It consists of an amendment to the earlier version. It introduced support for mobility, handover, and roaming, among other new capabilities [IEEE 802.16e]. In addition, MIMO schemes were improved in order to achieve better performances.
- Relay specifications are included in IEEE 802.16j amendment. The incorporation of multihop relay capability in the foundation of mobile IEEE 802.16-2005 is a way to increase both the available throughput by a factor of 3 to 10 and/or coverage (and higher channel reuse factor), or even to fill the "coverage hole" of indoor coverage [Oyman 2007] [IEEE 802.16] [Peters 2009].

It is worth noting that, as referred earlier, multihop relay capability is being considered as an option for inclusion in IMT-Advanced [Astély 2009]. For backward compatibility, 802.16e subscriber station (SS) is able to benefit from this capability without awareness of the presence of the relay station. In addition, a distributed MIMO system can be established using both the BS and the relay station transmitting antennas on one side and receiving antennas of the mobile station on the other side. Alternatively, various adjacent BS and RS antennas can cooperatively perform space-time coding to send data packets.

In addition to these versions, requirements for the next version of Mobile WiMAX entitled IEEE 802.16m [IEEE 802.16m] were completed by the end of 2007, and implemented in the beginning of 2011. The goal of IEEE 802.16m version is to reach all the IMT-Advanced requirements as proposed by ITU-R in [ITU-R M.2133], making this standard a potential candidate for the 4G. Advances in IEEE 802.16m include wider bandwidths (up to 100 MHz, shared between uplink and downlink), an adaptive and advanced TDMA/OFDMA access schemes, advanced relaying techniques (already incorporated in IEEE 802.16j), advanced multiple-antenna systems, adaptive modulation schemes as hierarchical constellations and Adaptive Modulation and Coding (AMC) frequency adaptive scheduling, and so on [ITU-R M.2133]. In order to preserve investments, IEEE 802.16m will keep backward compatibility with earlier WiMAX versions.

WiMAX forum is also making some adaptations to IEEE 802.16e in order to allow its use in mobile satellite services coupled with an ancillary terrestrial component, creating the so-called S-WiMAX [Ansari 2009].

The original version of the standard specified a physical layer operating in the range 10–66 GHz, based on OFDM and TDMA technology. IEEE 802.16-2004 added specifications for the range 2–11 GHz (licensed and unlicensed), whereas IEEE 802.16-2005 introduced the scalable OFDMA (SOFDMA) with MIMO (either Space-Time Block Coding (STC) based, spatial multiplexing based, or beamforming) or advanced antenna systems (AAS) [IEEE 802.16e], instead of the simple OFDM with 256 subcarriers considered by the earlier version.

The introduction of Scaling OFDMA (SOFDMA) that is, scaling the discrete Fourier transform (DFT) to the channel bandwidth is very effective, as it keeps the carrier spacing constant across different channel bandwidths. This results in higher spectral efficiency and improved resistance to multipath interference in a wide range of delay spreads. In addition, with the IEEE 802.16-2005 the number of subcarriers considered may vary from 128 to 2048, instead of the rigid number of 256 subcarriers considered by the IEEE 802.16-2004. Finally, IEEE 802.16-2005 introduced the hybrid automatic repeat request (H-ARQ) in between the SS and the BS, AMC technique (BPSK, QPSK, 16-QAM, or 64-QAM, with modulation and coding rate chosen dynamically), as well as turbo coding and low-density parity check (LDPC), greatly enhancing performance and allowing a new capability for nonline-of-sight

(NLOS) and for mobility using Time Division Duplexing (TDD) [IEEE 802.16e] [Yarali 2008]. In opposition, the earlier version only supported fixed profile in line-of-sight (LOS) mode, using either TDD or Frequency Division Duplexing (FDD). Finally, the additional Quality of Service (QoS) made available with the enhanced real-time polling service brought by the 2005 version allowed WiMAX standard becoming much more appropriate for MBMS applications [IEEE 802.16e].

Wi-Fi MAC algorithm based on Carrier Sence Multiple Access with collision avoidance (CSMA-CA) is not effective for MAN coverage, due to the high probability of distant SS to be interrupted by closer SS, reducing their throughputs. This would degrade the QoS of services as voice over Internet protocol (VoIP) and Internet protocol based television (IPTV). Therefore, the WiMAX working group has decided to use TDMA (2004 and 2005 versions) and/or OFDMA (2005 version), which is connection-oriented MAC. With such access technique, the SS needs to compete only once, after which it is allocated a slot (time or frequency subcarriers) by the BS. In addition, based on the throughput requirements, the slot can enlarge or contract in order to provide the required QoS. WiMAX runs on a connection-oriented and contention-based MAC, which allows providing QoS for a higher number of simultaneous users than Wi-Fi. WiMAX has been designed to scale from one user to hundreds of users within one radio-frequency (RF) channel.

In terms of throughputs and coverage, these two parameters are subject to a trade-off [IEEE 802.16e]: typically, mobile WiMAX provides up to 10 Mbps per channel (symmetric), over a range of 10 km in rural areas (LOS environment) or over a range of 2 km in urban areas (NLOS environment) [Ohrtman 2008]. With the fixed WiMAX, this range can normally be extended. Mobile version considers an omni-directional antenna, whereas fixed WiMAX uses a high-gain antenna. Throughput and ranges may always change. Nevertheless, by enlarging one parameter the other has to reduce, otherwise the bit error rate (BER) will experience degradation. In the limit, WiMAX may deliver up to 70 Mbps per channel (in LOS, short distance and fixed access) and may cover up to 50 km (in LOS for fixed access) with a high-gain antenna [Ohrtman 2008], but not both parameters simultaneously, and in this case, a single SS may be supported by the BS. In opposition to UMTS, where handover is detailedly specified, mobile WiMAX has three possibilities but only the first one is mandatory: Hard handover (HHO), fast BS switching (FBSS), and macro diversity handover (MDHO). FBSS and MDHO are optional, as it is up to the manufacturers to decide about their implementation. Therefore, there is the risk that handover is not possible between two BS from different manufacturers. Another drawback on the use of WiMAX is still the speed allowed in mobility, which is limited to 60 Km/h. For higher speeds, the user experiences a high degradation in performance.

In terms of physical layer characteristics, when comparing WiMAX against UMTS, it is clear that the former standard uses more advanced transmission techniques (e.g., OFDMA vs. WCDMA), which is incorporated in the LTE. The WiMAX version currently available (IEEE 802.16-2005) already incorporates most of the techniques not considered by UMTS but specified for LTE (from 3GPP). Examples of such techniques are OFDMA, MIMO, advanced turbo coding, all over IP architecture, and so on. In addition, the inclusion of multihop relay capabilities (IEEE 802.16j), which aims to improve the speed of service delivery and coverage by a factor of 3 to 10, will make the WiMAX standard closer to what is expected for the IMT-Advanced. Recall that this latter standard foresees a target of 100 Mbps mobile and 1 Gbps nomadic. Therefore, the future of WiMAX may be a potential candidate for the IMT-Advanced 4G. Although most of the techniques foreseen for IMT-Advanced are already integrated in the earlier WiMAX versions, IEEE 802.16m will integrate and incorporate several advancements in these transmission techniques in order to meet all the IMT-Advanced requirements as defined by ITU-R [Osseiran 2009].

8.6 Fourth Generation of Cellular Communications

Future mobile cellular communication systems beyond 3G aim at allowing subscribers ubiquitous access to a virtually unlimited amount of multimedia content with a guaranteed QoS. This constitutes extensive requirements for improvements to present mobile cellular communication systems to provide a greater throughput to the subscribers. As described in Table 8.1, the expected data rates for the fourth generation cellular system (4G) are in the range of 100 Mbps for vehicular mobility to 1 Gbps for nomadic access (in both indoor and outdoor environments, and in both uplink and downlink).

The 4G is also called international mobile telecommunications advanced (IMT-Advanced), as defined by the ITU–Radio communications (ITU-R) [ITU-R M.2133]. Going back to IMT2000 technology, it is an umbrella term covering several standards, namely IEEE 802.16e/WiMAX, CDMA2000, WCDMA, and so on. In some bibliography, 4G is also referred to as IMT-Advanced. In this sense, both LTE Advanced (to be standardized by 3GPP) and IEEE 802.16m (standard from IEEE) are candidates for IMT-Advanced (i.e., 4G).

It is expected that, in future mobile radio networks, multihop relaying will be introduced [Sydir 2009]. Therefore, new topological approaches like multihop or distributed antenna solutions and relaying allow an increased coverage of high rate data transmission, as well as improved system performance and capabilities [3GPP 36.913-v8.0.0]. Standardization is expected to be finalized in 3GPP Release 10. Nevertheless, for the time being, the work on LTE-Advanced within 3GPP is still in an early phase, with several discussions taking place on the technologies to use in order to achieve the requirements.

Within 4G, voice, data, and streamed multimedia will be delivered to the user based on an all over IP packet switched platform, using IP version 6 (IPv6) [3GPP 36.913-v8.0.0]. The goal is to reach the necessary QoS and data rates in order to accommodate the emergent services as mobile TV, high definition television (HDTV), DVB, MMS, video chat, and so on. [3GPP 36.913-v8.0.0]. All of these services must be delivered in the concept of "anywhere" and "anytime."

Potential technologies for the air interface of 4G include the following options [Astely 2009] [Boudreau 2009]:

- Carrier aggregation composed of, for example, multiple components of 20 MHz in order to support transmission bandwidths of up to 100 MHz.
- AAS increasing the number of downlink transmission layers to eight and the number of uplink layers to four.
- Multihop relay (adaptive relay, fixed relay stations, configurable cell sizes, hierarchical cell structures, etc.) in order to improve coverage and data rates [Marques da Silva 2010].
- Advanced intercell interference cancellation schemes [Boudreau 2009] [Marques da Silva 2008a].
- Multiresolution schemes (hierarchical constellations, MIMO systems, OFDM transmission technique, etc.) [Marques da Silva 2010].

A candidate for IMT-Advanced is the IEEE 802.16m standard, which is expected to be composed of several advances to be implemented over the earlier WiMAX version (entitled IEEE 802.16j, which already includes multihop relay and advanced MIMO schemes), in order to meet all the IMT-Advanced requirements [ITU-R M.2133] as defined by ITU-R. Another solution is to use 3GPP LTE (R8) as a basis for additional advances, in order to implement 4G.

Due to the improvements in terms of space address with the 128 bits made available by IPv6, multicast and broadcast applications will be easily improved, as well as the improved security, reliability, intersystem mobility, and interoperability capabilities. In addition, as the 4G concept consists of a pool of wireless standards, this can be efficiently implemented using the software defined radio (SDR) platform, being currently an interesting R&D area for many industries worldwide.

End of Chapter Questions

1. What is the difference between a macrocell, a microcell, and a picocell?
2. What are the advantages of using a hierarchical cellular structure?
3. What is the typical propagation environment experienced in urban environments?
4. What is the typical propagation environment experienced in rural environments?
5. What is the typical propagation environment experienced in indoor environments?
6. What is the difference between a picocell and a femtocell?
7. What is a femtocell? Which services can be provided by such type of cell?
8. What is the relationship between a cell size and the cell capacity? Why?
9. What is the reuse factor adopted in different types of cellular networks?
10. Assuming M carriers in the cell and N calls per carrier, what is the maximum number of simultaneous calls in a cell using FDD?
11. What does multihop relay stands for?

12. What are the advantages of the UMTS, relating to the GSM?
13. What are the differences between LTE and UMTS?
14. What are the differences between LTE and 4G?
15. In which cellular generations is the OFDM transmission technique employed?
16. What are the differences between LTE and WiMAX?
17. Which cellular generations only consider IP-based network (i.e., all over IP)?
18. Which cellular generations consider MIMO systems?
19. Which cellular generations consider multihop relay systems?
20. Which potential technologies are expected to be implemented by the 4G in order to provide 100 Mbps in mobility and 1 Gbps nomadic?

TRANSPORT NETWORKS

Transport networks consist of MAN or WAN networks, being commonly employed to interconnect different LAN, MAN, or telephone PABX. These types of networks may belong to either a telecommunications operator or be propriety of a certain organization.

Figure 9.1 depicts an example of a transport network (WAN or MAN type) using a ring topology, which is employed to interconnect different LAN. Depending on the type of technology, it may present different topologies. An important advantage of the ring topology relies on its redundancy, which translates in resistance from failures.

The transport networks can be grouped into the following categories:

- Permanent circuit: The main advantage of this type of connectivity relies on its dedicated and guaranteed bandwidth, without sharing with other users. In addition, the available bandwidth is typically high. The disadvantage is the high cost that is normally associated to this type of connectivity. A permanent circuit is normally leased from a telecommunications operator. Alternatively, a circuit may be implemented and established by an operator or institution as a request to a specific need (using, e.g., terrestrial microwave systems or optical fibers). Logical link control (LLC) protocols that can be ran over this type of connectivity include the synchronous data link control (SDLC), high-level data link control (HDLC), and the point-to-point protocol (PPP).
- Circuit switching* (ISDN or PSTN): This type of connectivity is the traditional way circuits are shared among different users. Thus, it leads to a cost reduction, as compared to permanent circuits. Nevertheless, the available bandwidth is typically limited. In case it is used for data communications, the PPP and the link access procedures over D channel (LAPD) are among the most used LLC protocols.
- Packet switching (X.25, Frame relay, MPLS, etc.): As defined earlier for packet switching (see Chapter 1), its main advantage relies on the statistical multiplexing performed as a function of the instantaneous available resources and of the users' need. In addition, while the available bandwidth is typically higher than that of circuit switching,† the cost is typically kept within reduced limits, as opposed to the high cost associated to permanent circuits. The most

---

* In case this type of circuits are used for data communications, dial up or xDSL modems can be employed.
† But typically based on the best effort, that is, not guaranteed.

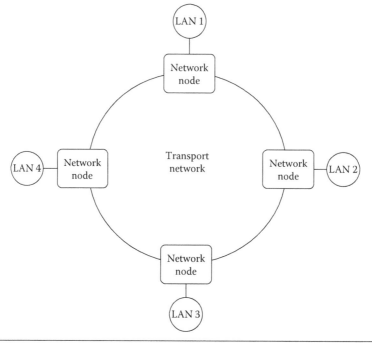

Figure 9.1   Example of a transport network (ring topology) used to interconnect different LAN.

used LLC protocols employed in packet switching include the link access procedures balanced (LAPB) and the Internet engineering task force (IETF).

- Cell switching (ATM): This corresponds to a variation of the packet switching. The reason for calling it cell switching, instead of packet switching, relies on the use of fixed size packets (cells*), instead of variable size packets (MPLS, IP, etc.).

When the service consists of analog or digital telephony, the traditional used transport network is of circuit-switching type. Contrarily, packet switching is normally employed for data communications.

Initially, transport networks consisted of point-to-point connections among different PABX, using coaxial cables. Currently, single-mode optical fibers are mostly employed, using a ring topology (with two or four optical fibers). Synchronous digital hierarchy (SDH) is normally implemented over optical fibers. Moreover, the asynchronous transfer mode (ATM) or the multiprotocol label switching (MPLS) is often implemented over SDH.

9.1 Circuit-Switching Transport Networks

As described in Chapter 1, this type of switching considers the establishment of a physical path between the origin and the destination of a communication. Although it

---

* The ATM uses fixed size packets of 53 octets, where 48 octets correspond to the payload data and 5 octets to the header.

Table 9.1   European FDM Carrier Standards

FDM HIERARCHY		
DESIGNATION	NUMBER OF CHANNELS	BANDWIDTH
Group	12	60–108 kHz
Supergroup	60	313–552 kHz
Master group	300	812–2044 kHz
Supergroup master	900	8516–12338 MHz

requires the previous establishment of a connection, it allows a synchronous exchange of data, with a delay only due to the propagation of signals. Consequently, this mode is ideal for delay sensitive media such as telephony or video teleconference.

9.1.1 FDM Hierarchy

Frequency division multiplexing was earlier used in transport networks to support the exchange of voice channels among different PABX.* As can be seen from Table 9.1, these hierarchies comprise different tributary orders, with the multiplexing of different number of voice channels. Coaxial cables and terrestrial microwave systems are among the most used transmission mediums. Each voice channel comprises the bandwidth of 300–3400 kHz, while the frequency carriers are 4 kHz spaced apart to avoid adjacent channel interference.

Modern telephone networks do not comprise FDM transport networks. Nevertheless, FDM is currently being used in cable television distribution using hybrid networks† (some parts are composed of optical fiber, whereas the final distribution is typically implemented using coaxial cable).

9.1.2 Plesiochronous Digital Hierarchy

Plesiochronous digital hierarchy (PDH) is a family of circuit-switching transport networks implemented using time-division multiplexing (TDM). The perfect synchronism among different tributaries is not achieved in PDH because, although they present the same nominal rate, they do not have exactly the same phase and data rate. This results from the fact that different tributaries are synchronized by different and independent clocks and their accuracy is not sufficient. Consequently, the maximum throughput possible to be exchanged in PDH networks is limited. PDH tributaries are referred to as plesiochronous and the corresponding hierarchy is entitled plesiochronous digital hierarchy.

As described in Chapter 6, TDM requires that the Nyquist sampling theorem is followed, with a sampling period of $T_a \leq 1/(2B)$. Moreover, the TDM requires

---

* The reader should refer to Chapter 6 for the description of FDM multiplexing and demultiplexing.

† In some cases, cable television distribution is already being implemented using only optical fiber cables.

that the demultiplexer is synchronized with the multiplexer. Such synchronization is achieved using a synchronism signal, which is transmitted in one of the time slots, in parallel with the transported signals. The whole signal is designated as frame and the synchronism signal is referred to as framing.

Similarly to the FDM hierarchy, the transported signals consist of voice channels typically in the range of 300–3400 kHz. Using TDM, the analog channels need to be encoded with an analog coding system such as PAM, PDM, or PPM (see Chapter 6). Nevertheless, due to inability to perform regeneration of signals, analog voice is not well fitted for long-range transmission. Consequently, the TDM is normally employed to transport digital voice, using pulse code modulation (TDM-PCM). In this case, since the sampling frequency is 8 ksamples/s, the sampling period results in 125 μs, which also corresponds to the period of the frame repetition.

The European PDH system is entitled Conference of European Post and Telecommunications (CEPT), followed by one figure which represents the hierarchy having been normalized by the ITU as [ITU-T G.732]. On the other hand, the PDH system used in the North America is entitled digital signal (DS), having also been normalized by the ITU as [ITU-T G.733]. Moreover, the Japanese PDH system consists of a variation of the DS used in North America. Figure 9.2 depicts the generation of different PDH hierarchies. As can be seen, all of the hierarchies are composed of a number of ITU-T G.711 voice channels (TDM-PCM), which are properly multiplexed. A higher hierarchy is generated using a multiplexer. The generation of a nth order hierarchy requires a stack of n multiplexer. Similarly, the extraction of a nth order tributary requires a stack of n (de)multiplexers.

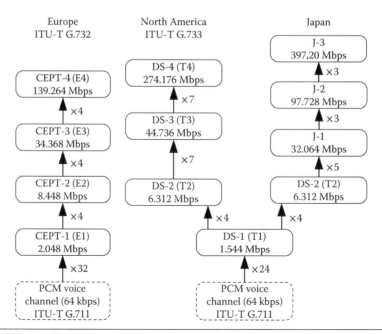

Figure 9.2   PDH hierarchies.

Table 9.2 Generic Characteristics of the Primary PDH Hierarchy

	PDH HIERARCHY	
DESIGNATION	CEPT-1 [ITU-T G.732]	DS-1 [ITU.T G.733]
PCM law	A (A = 87.6)	μ (μ = 255)
Number of time slots	32	24
Number of voice channels	30	24
Number of bits per frame	$32 \times 8 = 256$	$24 \times 8 + 1 = 193$
Frame throughput	$256 \times 8 = 2.048$ Mbps	$193 \times 8 = 1.544$ Mbps
Framing	In block, using 7 bit words in the time slot 0 of odd frames	Distributed, using the sequence 101010..., composed of the 193rd bit in odd frames
Signaling	In time slot 16, at a rate of 4 bits per channel, split in 16 frames (multiframe)	In the eight bit of each channel in one frame out of six

The primary European multiplexing hierarchy is called CEPT-1 [ITU-T G.732], whereas the North American and Japanese versions are referred to as DS-1 [ITU-T G.733]. Note that the interface provided by a CEPT-1 is referred to as E1, whereas the interface provided by a DS-1 is named as T1.[*] From the transportation of data point of view, the nomenclature CEPT-x or DS-x is employed, whereas the nomenclature employed in interfaces is E-x, T-x, or J-x.

Both CEPT-1 and DS-1 use word interposition, with 8 bits per word, and a rate of 8000 frames per second (see Table 9.2). The voice codec comprised by the CEPT-1 is the A law and its frame has 32 time slots. Similarly, the DS-1 considers the μ law and its frame has a total of 24 time slots.

From the 32 time slots[†] of the CEPT-1, 30 are used to transport 64 kbps telephone channels. In addition, the time slot 0 of odd frames is employed for framing and the time slot 16 is used for signaling.[‡] The CEPT-1 comprises a cumulative throughput of $32 \times 8 \times 8000 = 2.048$ Mbps.

The DS-1 frame comprises a total of $24 \times 8 + 1 = 193$ bits, corresponding to twenty-four 64 kbps telephone channels and an additional bit (per frame) for framing purposes.[§] The cumulative throughput of the DS-1 frame is $193 \times 8000 = 1.544$ Mbps. The signaling is transported in the data time slots. Specifically, the signaling bits are transmitted from the 6th to 12th frame of each multiframe,[¶] using the 8th bit of each time slot.[**] This results in a PCM word of 7 bits, which results in a small degradation of the PCM voice channels.

---

[*] The same concept applies to other hierarchies.

[†] Numbered from 0 up to 31.

[‡] At a rate of 4 bits per channel, split over 16 frames.

[§] The F bit is used in odd frames for framing purposes. It presents the pattern 101010... This corresponds to a distributed framing, instead of block framing employed in the CEPT-1.

[¶] A multiframe is composed of 12 frames.

[**] The signaling information refers to the corresponding transported channel in the time slot.

Higher order tributaries are obtained from the multiplexing of a number of immediately lower order tributaries. With the exception of the primary multiplexing tributary order,* all upper multiplexing tributary orders are multiplexed using bit interposition.

It is worth noting that the throughput of a higher hierarchy is higher than the number of lower order tributaries multiplied by their elementary throughputs.[†] This results from the fact that an additional overhead is necessary for framing and for frame justification.

Frame justification consists of the addition or removal of some bits, in order to allow the correct operation of multiplexers and demultiplexers when the different tributaries rate is subject to fluctuations, and therefore it differs from the nominal rate. When the rate is higher than the nominal, the justification comprises the addition of a bit without information (from time to time), in order to adjust the rates. Some bits are preallocated in frames, for justification purposes. In addition, when bit justification is employed, this needs to be properly signalized using a justification indication bit, which is also a preallocated bit in the frame composition.

9.1.3 Synchronous Digital Hierarchies

The inaccuracy of independent clocks employed in PDH did not allow throughputs higher than 140 Mbps, which was not enough to face the new information exchange requirements. This was the main motivation for the development of a synchronous hierarchy, optimized for optical fiber transmission mediums.[‡]

Two different synchronous circuit-switching transport network systems were developed [Sexton 1992]: The SDH was employed in Europe and standardized by ITU-T as [ITU-T G.707] and the synchronous optical network (SONET) was used in North America and standardized by ANSI as [ANSI T1.105]. These systems consider atomic[§] clocks that allow the exchange of throughputs much higher than those possible with PDH systems. Moreover, the higher level of standardization comprised by SDH/SONET[¶] among equipment of different manufacturers lead to improved interoperability.

Another important innovation of synchronous transport network systems relies on the ability to insert or extract a tributary from any other tributary of any other order.

---

* Which uses word interposition (of PCM channels).

[†] As an example, while the throughput of the CEPT-2 is 8.448 Mbps, the cumulative throughput of four CEPT-1 tributaries is 4×2.048 Mbps = 8.192 Mbps (which is lower than 8.448 Mbps).

[‡] This represents an important improvement, as compared to PDH. Nevertheless, it is possible to find equipments that implement SDH/SONET over other transmission technologies such as terrestrial microwave systems.

[§] The central (master) atomic clocks use cesium and rubidium to achieve a high level of accuracy. The other slave clocks are periodically synchronized making use of the information contained in the SDH frame.

[¶] Including the standardization of new operation, administration, and maintenance (OA&M) capabilities.

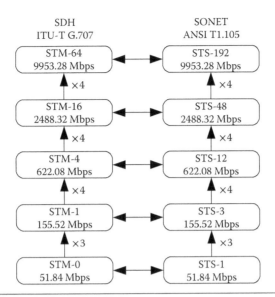

Figure 9.3 SDH and SONET hierarchies.

Figure 9.3 shows different tributaries of both SDH and SONET hierarchies. As can be seen, different SDH tributaries are referred to as synchronous transport module (STM), whereas SONET tributaries are named as synchronous transport signal (STS). There is an equivalence between the tributaries' rates of different hierarchies. It is also worth noting that, above STM-1/STS-3, the multiplexing of higher hierarchical levels is performed with four lower order hierarchical tributaries.

The different SDH/SONET tributaries can be either used to transport multiplexed ITU-T G.711 voice channels, PDH tributaries or packets belonging to packet-switching networks (such as ATM or MPLS). Figure 9.4 shows examples of PDH encapsulation into SDH/SONET. In Figure 9.4, boxes with straight lines refer to SDH/SONET hierarchy and boxes with dash lines refer to PDH hierarchy. Note that Figure 9.4 only depicts some possible encapsulations. Nevertheless, a wide variety of PDH tributaries mixes can be jointly used to generate any SDH/SONET tributary.

9.1.3.1 SDH/SONET Network A SDH/SONET network comprises a number of elementary devices that are used to allow the multiplexing/demultiplexing and transport of data. Such devices are the following:

- Line terminal multiplexer (LTM): It accepts lower order tributaries to generate a higher order tributary. This type of device is employed at the beginning and at the end of a path. This means that all multiplexed tributaries end in this device.
- Add and drop multiplexer (ADM): It accepts lower order tributaries to generate a higher order tributary. Nevertheless, contrarily to the LTM, the ADM can be used in the middle of a path in order to insert or remove some of the lower order tributaries.

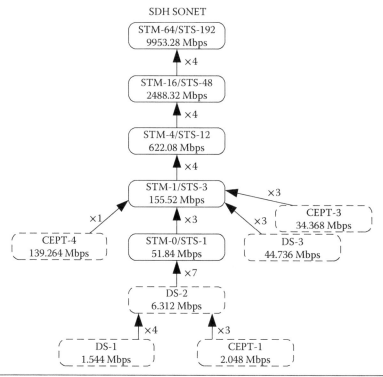

Figure 9.4 PDH encapsulation into SDH/SONET.

- Synchronous digital cross connect (SDXC): It is used to perform semipermanent switching along the SDH/SONET network, in order to provide "permanent circuits" to a customer who requires it.
- Regenerator (REG): It is used to perform the regeneration of signals, mitigating the effects of channel impairments. SDH/SONET networks make use of regenerators, typically, every 60 km of optical fibers.

Figure 9.5 shows an example of two SDH networks with interconnection. As can be seen, an STM-16 hierarchy uses ADM-16 to multiplex up to four STM-4 tributaries. In the example depicted in Figure 9.5, an STM-4 tributary can be imported from a STM-4 network. Moreover, the STM-16 ring and the STM-4 ring are interconnected in two points in order to achieve interconnection redundancy. This results in an overall mesh topology.

It is worth noting that the elementary physical topology employed in SDH networks is the ring. Since optical fibers are unidirectional, a cable composed of, at least, two optical fibers (one pair) is employed to allow simultaneous bidirectional communications. As depicted in Figure 9.6, in case of two optical fibers, both are employed for operation and backup using different time slots. In the case that the optical fiber cable is cut, such impairment is detected by the two closest ADM and bridges are introduced in those ADM. Consequently, a switch to the backup time slot is performed by the network.

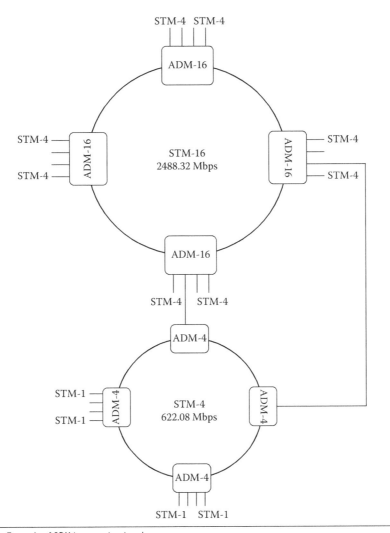

Figure 9.5 Example of SDH transport networks.

The plotted example refers to a STM-16 ring. Nevertheless, the same principle applies to other hierarchies. The same concept is also applicable to SONET networks.

As shown in Figure 9.7, in case the SDH ring is implemented with four optical fibers, all time slots of two optical fibers are allocated for operation, whereas the other two optical fibers (one pair) are reserved for backup. In this case, redundancy is not assured with TDM but with an extra optical fiber pair. Similar to the previous case, the closest ADM inserts bridges, but those bridges are responsible for forwarding the signals to the redundant fibers in the opposite direction.

Both SDH and SONET networks comprise different layers. Each layer refers to the communication among certain type of devices, which includes some functionality. The description of different layers and their functionalities is included in Table 9.3. Figure 9.8 plots an example of a network with the identification of different layers using both SDH and SONET terminologies.

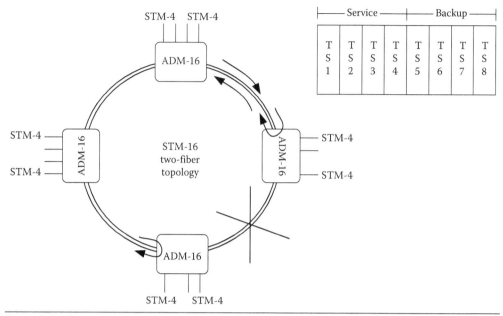

Figure 9.6   Failure procedure in a two optical fibers cable.

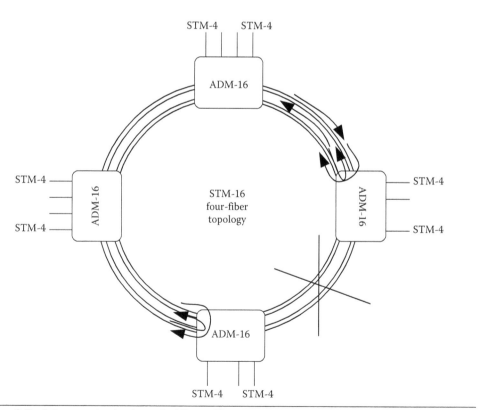

Figure 9.7   Failure procedure in a four optical fibers cable.

Table 9.3 SDH/SONET Multilayer Architecture

SDH/SONET MULTILAYER ARCHITECTURE			
LAYER	SDH TERMINOLOGY	SONET TERMINOLOGY	FUNCTIONALITIES
4	Path	Path	Definition of the end-to-end transported tributary, throughput, and so on.
3	Multiplexing	Line	Synchronism, multiplexing, switching, OA&M, type of protection from failures, and so on.
2	Regeneration	Section	Regeneration distance, electrical to optical conversion, and so on.
1	Physical	Photonic	Definition of the type of optical fiber, light wavelength, transmit power, and so on.

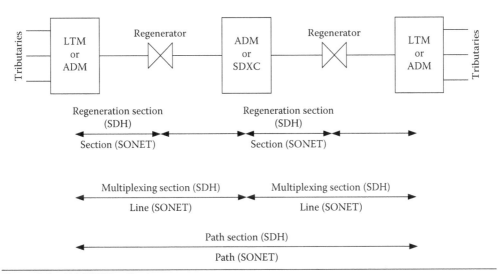

Figure 9.8 SDH/SONET architecture.

Each layer has its own overhead type, which is processed by the corresponding device. As can be seen from Figure 9.8, the path is considered end-to-end. Consequently, the path overhead is inserted by the initial multiplexer (LTM or ADM), being removed by the final multiplexer (LTM or ADM). The multiplexing section consists of the parts of the path among adjacent multiplexers, including those intermediates (ADM or SDXC). Consequently, the path of the example plotted in Figure 9.8 is composed of two multiplexing sections. In this case, the intermediate multiplexer (ADM or SDXC) removes the initial multiplexing section overhead at its input and inserts another multiplexing section overhead at its output, corresponding to the second multiplexing section. Identical rational applies to the regeneration section overhead. Note that a regenerator only processes the regeneration section overhead (removed at the input of the device and inserted at its output), whereas an intermediate multiplexer processes both the regeneration section overhead and the multiplexing section overhead.

9.1.3.2 SDH/SONET Frame Format   The SDH/SONET frame comprises a different header type for each different layer (except for the physical layer), as follows:

- The path overhead: It is used to manage the end-to-end path,[*] inserted at the beginning and removed at the end of the path. This is used to manage the end-to-end path between extreme devices (LTE or ADM). The end-to-end transport of a PDH tributary into SDH/SONET is performed by the path layer and managed by the path overhead. In addition, this layer also implements error protection mechanisms and provides engineering orderwire communication channels[†] (at path level).
- The multiplexing section (line) overhead: It includes functions such as multiplexing of lower order tributaries, frame synchronization, switching information, error protection, and engineering orderwire communication channels, being processed and used to manage multiplexers.
- The regeneration section (section) overhead: It allows the alignment of the frame, assures error protection, and provides engineering orderwire communication channels. This overhead is processed and used to manage the communication among different regenerators and multiplexers.

Different SDH and SONET hierarchies are formed by octet interposition of the lower hierarchy tributaries. In addition, a STM-1 frame can be formed by octet interposition of three STS-1 frames (STM-0). Due to this reason, the following frame description focuses on STS-1, whereas the composition of STM-1 frame can be deducted in a straightforward manner.

The basic structure of an STS-1 frame is depicted in Figure 9.9. It is composed of a total of 810 octets, split into 9 rows and 90 columns. The frame transmission is performed starting from the 1st up to the 90th octet of each row, from row to row. This structure is transmitted every 125 µs,[‡] while another frame with the same structure is transmitted in an equal period. The transmission of $810 \times 8 = 6480$ bits in 125 µs results in a throughput of 51.84 Mbps. Since each octet corresponds to 8 bits which is repeated every 125 µs, this corresponds to 64 kbps.[§]

The frame comprises two main blocks: the header and the synchronous payload envelope (SPE).[¶] The section header[**] comprises three columns and three rows, the

---

[*] End-to-end path from the SDH communication point of view.

[†] Note that error protection and engineering orderwire communication channels are provided by all different layers and managed by the corresponding overheads.

[‡] This is the sampling period, which corresponds to the inverse of the 8 ksamples/s sampling rate employed in voice.

[§] 64 kbps corresponds to the throughput necessary to support one PCM voice channel. Nevertheless, one STS-1 octet can be used to transport different type of data.

[¶] In the SDH terminology, the SPE is referred to as virtual container (VC).

[**] In the SDH terminology, the section header is referred to as regeneration header.

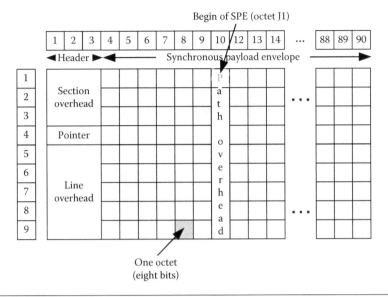

Figure 9.9 STS-1 frame format.

1	2	3	4	5	6	7	8	9	10	11	12	13	14	15	16	17	18	19	20	21	22	23	24
0	1	1	0	1	0	Pointer (number between 0 and 782)											Space used for negative justification							

Figure 9.10 STS-1 pointer.

pointers correspond to three columns and one row, and the line overhead* is composed of three columns and four rows. The SPE comprises the octets used to transport the payload data and the path overhead (transported in the tenth column).

The pointer is employed to identify the beginning of the SPE within the payload area and to accommodate justification bits. It is worth noting that both SDH and SONET comprise a concept where a SPE/VC does not occupy a rigid position within the payload. On the contrary, it may fluctuate within the payload area. The initial octet of the path overhead is entitled J1 and marks the beginning of the SPE/VC. Although Figure 9.9 depicts the octet J1 at the top of the path overhead column, due to load fluctuations this octet may be located anywhere in the payload area.

This transportation concept can be viewed as the payload of a truck. We could have the furniture of ten houses to be transported at the same time between two different cities. Each family's furniture requires one standard truck. At principle, one truck would be employed for each family's furniture. Nevertheless, a different method where each truck could be used to transport part of a family's furniture and another part of another family's furniture. This is the concept employed in SDH and SONET. To implement such concept, there is the need to have a pointer toward the beginning of the SPE/VC.

As depicted in Figure 9.10, the STS-1 pointer uses three octets (24 bits). The logic states of the initial 6 bits are fixed, whereas the 10-bit pointer is used to quantify a

---

* In the SDH terminology, the line header is referred to as multiplexing header.

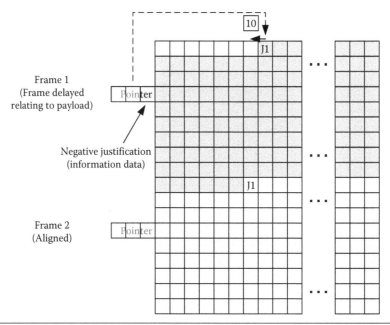

Figure 9.11   Frame delayed relating to the SPE (STS-1).

value between 0 and 782 octets.* This value corresponds to the octet order where the SPE/VC begins, within the payload area.

Due to successive insertions and removals of headers in different devices (regenerators, multiplexors, etc.), the payload (SPE/VC) travels faster than the corresponding frame.[†] Another problem that causes similar effects is when the incoming clock is slower than the outgoing clock. In order to overcome such fluctuation, an extra octet is inserted into the last octet of the pointer.[‡] This is known as negative justification. In this case, the value of the pointer needs to be decreased by one. The negative justification is depicted in Figure 9.11.

Contrarily, when the incoming clock is faster than the outgoing clock, the frame flows faster than the SPE. In order to overcome such fluctuation, a stuff octet[§] is transmitted after the pointer. This is known as positive justification. This can be seen from Figure 9.12.

As can be seen from Figure 9.13, the basic structure of a STM-1 frame is composed of a total of 2430 octets, split into 9 rows and 270 columns, formed by octet interposition. Consequently, the STM-1 header comprises 9 columns (instead of 3 columns, as considered for the STS-1).

---

* In case of STM-1 (SDH), nine octets are employed for similar purposes, instead of three.
† The frame is delayed by successive overhead processing.
‡ In case of STM-1 (SDH), instead of using a single octet for negative justification (STS-1/SONET), a total of three octets are employed.
§ An octet without information.

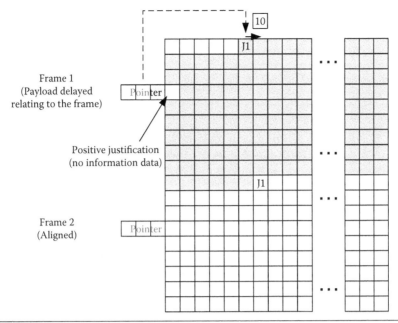

Figure 9.12 SPE delayed relating to the frame (STS-1).

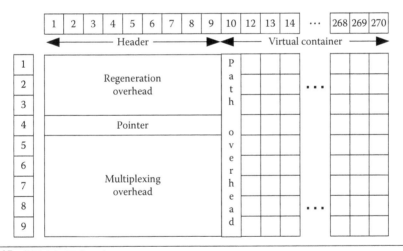

FIGURE 9.13 STM-1 frame format.

9.1.4 Digital Subscriber Line

The digital subscriber line (DSL) is not exactly a transport network. Nevertheless, it consists of a transmission technique that allows the transport of data over conventional installed copper twisted pairs (analog transmission medium). In addition, the PPP protocol (see Chapter 10) is normally implemented over DSL. Most common uses of DSL modems are made over circuit-switching networks.

Different versions of DSL exist (generically referred to as xDSL*), and each version makes use of different transmission techniques to allow the exchange of digital data

---

* For example, for x = A, the xDSL becomes asymmetric DSL (ADSL).

Figure 9.14 Block diagram of a communication chain using DSL modems.

with different characteristics (throughput, reliability, bandwidth, modulation scheme, etc.). The xDSL transmission technique is implemented making use of modems at both ends of the transmission medium (e.g., at home and at the ISP site).

As described earlier, a modem consists of a device that allows the exchange of digital data over analog transmission mediums. It implements the modulation at the transmitter side, whereas the demodulation process is carried out at the receiver side. The modulation involves the process of encoding one or more source bits into a modulated carrier wave. Moreover, the modem also implements an error control technique. An important advantage of using a modem, instead of line coding, relies on the fact that the transmitted signal is carrier modulated (bandpass), instead of transmitted in the baseband. This allows selecting the transmission bandwidth where the channel impairments (e.g., attenuation and distortion) are less destructive. Figure 9.14 depicts a block diagram of a communication chain using xDSL modems.

Note that the communication implemented by the modems is bidirectional. In addition, today's xDSL communications are full duplex. This is possible because modems use, typically, two different frequency bands (employing the FDD technique), one for transmission and another for reception of signals. As depicted in Figure 9.15, a lower bandwidth is normally utilized for uplink, whereas an upper bandwidth is employed for downlink. Since data communications tend to require a higher throughput in the downlink than in the uplink, most xDSL modems allocate a higher bandwidth for the downlink. In this case, the communication established by the modem is referred to as asymmetric. Moreover, a requirement for xDSL modem is the ability to keep the analog telephony in simultaneous with the exchange of data. This is possible by keeping a baseband bandwidth allocated for conventional analog telephony, with a typical reserved band of 0–20 kHz. Typical uplink frequency band is 25 kHz to 140 MHz, whereas the typical downlink band is 150 kHz to 1 MHz. Latest xDSL standards utilize higher frequencies. An example is the very high data rate DSL (VDSL) whose frequency band utilized for data goes from 25 kHz up to 12 MHz.

In order to mitigate the effects of the channel impairments, most xDSL modems have embedded an equalization module, which is an effective measure to minimize the effects of intersymbol interference. This kind of interference is also mitigated by using higher order modulations (e.g., M-QAM) or by employing the orthogonal frequency-division multiplexing (OFDM) transmission technique. Some types of xDSL modems use a variation of the OFDM transmission technique in the downlink (more demanding direction), entitled discrete multitone (DMT). Figure 9.16 depicts the generic spectrum of a modem employing DMT transmission technique. The DMT

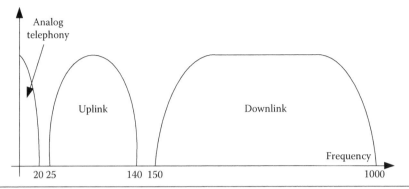

Figure 9.15 Generic use of spectrum by a xDSL modem.

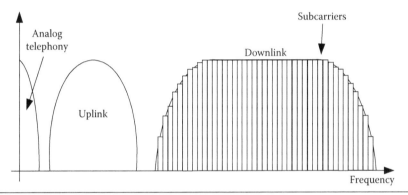

Figure 9.16 Generic spectrum of a xDSL modem that uses DMT.

is very similar to the OFDM technique described in Chapter 7, with the difference that some subcarriers are removed and the modulation order of different subcarriers can be different (as a function of the channel impairments experienced by each subcarrier). Similar to the OFDM technique, each DMT subcarrier suffers from flat fading (i.e., nonfrequency-selective fading). In order to mitigate the remaining fading effects, each subcarrier is subject to an equalization process at the receiver side. This results in a better signal quality, which translates in a more efficient use of the spectrum (expressed in bit/s/Hz).

An alternative to the preallocation of different uplink and downlink bandwidths (i.e., frequency division duplexing [FDD]) consists of using a bandwidth simultaneously for the uplink and downlink. In such case, the downlink bandwidth includes the uplink bandwidth, and an echo cancellation (EC) mechanism* is employed to avoid interferences between the uplink and the downlink spectrum. Moreover, some xDSL modems use a single frequency band for both uplink and downlink. In this case, the full-duplex operation is possible by employing Time Division Duplexing (TDD).

---

* Implemented using an adaptive equalizer to reject the nondesired signal.

Table 9.4 xDSL Standards and Characteristics

DESIGNATION	xDSL MODEM				
	ADSL	ADSL2	HDLC	VDSL	VDSL2
Standard	ANSI-T1.413-1998	ITU-T G.992.3	ITU-T G.991.1	ITU-T G.993.1	ITU-T G.993.2
Transmission technique	CAP–QAM	DMT–M-QAM	Digital baseband transmission using line coding 2B1Q	DMT–M-QAM	DMT
Maximum uplink data rate	1 Mbps	1.3 Mbps	1.544 (T1) or 2.048 Mbps (E1)	16 Mbps	100 Mbps
Maximum downlink data rate	8 Mbps	12 Mbps	2.048 Mbps	52 Mbps	100 Mbps
Range	4 km	5 km	4 km	1.5 km	300 m
Observations	One twisted pair (TP)	One TP	HDLC uses two TP. HDLC2 have same performance with single TP	One TP. Supports HDTV	One TP. Supports HDTV

Table 9.4 shows most common xDSL standards alongside with their important characteristics. As can be seen, the initial asymmetric DSL (ADSL) does not employ the DMT transmission technique. Instead, the FDD is utilized, using carrierless amplitude phase modulation (CAP) associated to QAM modulation scheme. Since the CAP suppresses the carrier, the required power for the signal transmission is reduced. The ADSL2 allows higher uplink and downlink data rates, as well as an extended range. This results from the use of the DMT transmission technique. On the other side, the high bit rate DSL (HDSL) transmits signals in baseband employing the 2B1Q line coding technique. Since digital transmission is performed in baseband, cohabitation with analog telephony is not possible. Finally, the very high data rate DSL (VDSL) supports higher data rates at the cost of lower distances. The extreme data rates provided by both VDSL and VDSL2 allow their use for the provision of cable TV. This makes the twisted pair act as a competitor to coaxial cables, typically employed in cable TV.

9.1.5 Data over Cable Service Interface Specification

Data over cable service interface specification (DOCSIS) consists of a set of standards that allows high-speed data communications over coaxial cables, typically installed for the provision of television service. The DOCSIS can be viewed as an alternative to xDSL, but using coaxial cables as transmission medium (instead of twisted pairs). Alternatively, hybrid fiber-coaxial may also be utilized as transmission medium.

Figure 9.17 depicts the block diagram of a communication chain using DOCSIS modems.

DOCSIS is currently widely employed by ISPs to provide Internet access to domestic users. This service is normally provided together with the cable television. When telephony service is also provided, the three services are commonly known as triple

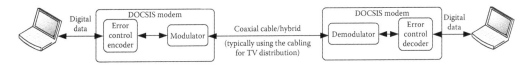

Figure 9.17 Block diagram of a communication chain using DOCSIS modems.

Table 9.5 DOCSIS Standards and Characteristics

	DOCSIS MODEM		
DESIGNATION	DOCSIS 1.0	DOCSIS 2.0	DOCSIS 3.0
Standard	ITU-T J.112 (Annex B)	ITU-T J.122	ITU-T J.222
Transmission technique	FDMA (TDMA in uplink)	FDMA (CDMA in uplink)	FDMA (CDMA in uplink)
Uplink channel bandwidth/ modulation	200 kHz to 3.2 MHz using QPSK or 16-QAM	200 kHz to 6.4 MHz using 8- to 128-QAM	200 kHz to 6.4 MHz using 8 to 128-QAM
Downlink channel bandwidth/modulation	6-MHz channels using 64- or 256-QAM	6-MHz channels using 64- or 256-QAM	6-MHz channels using 64- or 256-QAM
Maximum uplink data rate	10.24 Mbps	30.72 Mbps	122.88 Mbps (in four channels (typical value))
Maximum downlink data rate	42.88 Mbps	42.88 Mbps	171.52 Mbps (in four channels (typical value))

play. In this case, the telephony service is provided using voice over IP over DOCSIS modems (employed between the subscriber home and the ISP).

As can be seen from Table 9.5, the DOCSIS modem shares the spectrum with regular analog or digital video distribution, using FDMA (downlink). The uplink channels are implemented using TDMA (DOCSIS 1.0) or CDMA (DOCSIS 2.0 and 3.0). In the uplink, the higher modulation order and bandwidth of DOCSIS 2.0 and 3.0 allow a higher data rate per channel, as compared to DOCSIS 1.0. In the downlink, all three DOCSIS versions allow the same data rate per channel. Nevertheless, an important difference of the DOCSIS 3.0 as compared to DOCSIS 2.0 relies on its ability to aggregate multiple channels (in both uplink and downlink), which results in a data rate that is N times higher. Typically, four channels are aggregated in DOCSIS 3.0, which results in a data rate of 122.88 Mbps in the uplink and 171.52 Mbps in the downlink.

9.2 Packet-Switching Transport Networks

Packet-switching networks are of lower costs than circuit switching and are ideal for the exchange of data. Packet-switching networks intend to provide a higher bandwidth (based on the best effort) but at costs close to circuit-switching networks. The network resources are made available as a function of each users need and as a function of the instantaneous network traffic. Therefore, it is normally stated that a packet-switching network performs statistical multiplexing, as opposed to synchronous access control mechanisms (frequency or time multiplexing) typically employed in circuit-switching networks.

Packet switching involves the segmentation of a message into small pieces of data, and each piece is switched independently by the network nodes. Each piece of data is referred to as a packet. In order to allow its routing by the network nodes, each packet contains additional information (overhead) in a header. Moreover, each node of a packet-switching network is able to store packets, in case it is not possible to send it due to temporary congestion. In this case, it is not guaranteed the time for message transmission, but this value is kept within reasonable limits, especially if quality of service (QoS) is offered.

There are different packet-switching protocols such as Internet protocol, asynchronous transfer mode, multiprotocol label switching, frame relay, X.25, and so on. Although the IP protocol is typically employed by the network to interact directly with users (providing services), the ATM, MPLS, Frame Relay, and X.25 protocols are mostly employed in transport networks.

9.2.1 Asynchronous Transfer Mode

The integrated services digital network (ISDN) [Prycker 1991] was developed and standardized in the eighties in order to become the most important domestic and commercial data network, providing both voice and data services. ISDN was also expected to be used as a circuit-switching transport network (using, e.g., a primary access of 2.048 Mbps). Nevertheless, its implementation required, in most cases, the replacement of the existing voice-graded UTP cables, which was an important limitation. On the other hand, the rapid development of the Ethernet technologies based on installed UTP cables, as well as the requirements of the new services, became important obstacles for the development of the ISDN standard. The ATM is a protocol intended to fit these new requirements, while providing much higher throughputs necessary to support new multimedia services, support of services which require variable throughputs, delay sensitive services, and support of services sensitive to errors [Prycker 1991] [Handel 1991]. Note that the ATM is associated to a concept entitled Broadband-ISDN (B-ISDN). It is important referring that while the ATM is a protocol, the B-ISDN is a concept, which is materialized with a reference model, as depicted in Figure 9.18.

Although the ATM can be employed to connect end users, its most common applicability is as a transport network. In addition, although the ATM protocol can be directly implemented as a baseline protocol of a transport network, its most common implementation is above a SDH or SONET circuit-switching transport network. In this case the SDH/SONET network is only used to transport the ATM cells, whereas the data of different services is typically encapsulated into the ATM cells.

The ATM is a special type of packet-switching protocol. In fact, some authors refer to it as a cell switching or as cell relay. The reason comes from the fact that the ATM packets are of fixed size (cells), as opposed to variable size packets adopted by most of the packet-switching protocols. The ATM protocol is based on virtual circuits, instead of datagrams. There is the need to previously establish the circuit,* and the cells are always routed by the same intermediate nodes, until the circuit is terminated.

---

* Consequently, the ATM protocol is connection oriented.

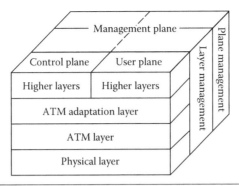

Figure 9.18 B-ISDN reference model.

Header	Payload data
5 octets	48 octets

Figure 9.19 ATM cell format.

An important advantage of the ATM relies on its faster switching, which results from the use of fixed routes and from the reduced level of processing and decision by intermediate nodes. When a network node receives a cell, it checks the route to which it belongs (such information is contained in a label, which is part of the ATM cell header) and, consulting the routing table, it verifies which is the output interface to use in order to forward the cell.

Due to fast switching and to small size of cells, the ATM is very well fitted to support delay sensitive services. This makes the ATM protocol well fitted to support voice and video services.

The cell label has a local meaning. This means that the label is removed by each node and another label is reinserted. Moreover, since the multiplexing performed by ATM does not follow any time synchronous mechanism (such as time multiplexing), the ATM is referred as an asynchronous mode.

As can be seen from Figure 9.19, an ATM cell has a fixed length of 53 octets. Its header has a length of 5 octets and the payload data is 48-octet long.

An important characteristic of the ATM relies on the ability to support different classes of traffic, while compatible with their different QoS requirements. The ATM protocol presents the following advantages:

- Support constant or variable data rate services
- Support services from very low to high data rate requirements
- Support symmetric and asymmetric communication services
- Support different services in a single network (such as voice, video, multimedia, etc.)
- Support delay sensitive services (e.g., telephony, audio, or video streaming)
- Support error-sensitive services (e.g., file transfer)

9.2.1.1 The B-ISDN Reference Model According to the [ITU-T I.121] recommendation, the ATM is the protocol used to implement the B-ISDN concept. The B-ISDN reference model comprises different layers and planes, as depicted in Figure 9.18.

The B-ISDN reference model includes different classes of services, grouped as follows:

- Class A: Circuit-switching emulation, audio and video of constant bit rate
- Class B: Compressed audio and video of variable bit rate
- Class C: Connection-oriented data (such as file transfer of web browsing)
- Class D: Connectionless data (such as network management protocols or DNS)

In order to provide different services, the B-ISDN reference model is split into different layers* with the following functionalities [Costa 1997]:

- Physical Layer: It includes two different groups of functionalities:
  - Physical medium: It includes functions that are dependent of the physical medium.
  - Transmission convergence: It takes care of the generation and recovery of frames, header error control (HEC), cell rate decoupling, and so on.
- ATM layer: It is responsible for the multiplexing and demultiplexing of cells, flow control, insertion and removal of headers, and so on. It encapsulates blocks of data with different sizes, generated by the ATM adaptation layer (AAL), into cells (fixed size).
- AAL: It adapts the requirements of the different classes of services provided by the higher layers to the ATM layer. Consequently, the AAL is service dependent. The rate of generated blocks of data and their corresponding sizes is service dependent, being performed by the AAL. Then, these blocks of different size are encapsulated into cells by the ATM layer. In order to support different services, with different requirements, some service dependent and service independent functions have to be provided. Consequently, the AAL is split into two sublayers (see Figure 9.20): Convergence sublayer (CS), which provides services to the upper layer, and segmentation and reassembling sublayer (SAR), which segments the data received by the CS, in order to allow its encapsulation into ATM cells. To allow these functionalities, the AAL provides different types of service:
  - Type 1 AAL: It supports constant bit rate services (e.g., circuit-switching emulation) from the higher layer (class A), keeping the synchronism information between the source and the destination. In addition, it manages errors (lost cells, corrupted cells, cells in wrong sequence, duplication of cells, etc.).
  - Type 2 AAL: It multiplexes low data rate channels, such as mobile communications.

---

* Different layers are interfaced using service access points (SAP).

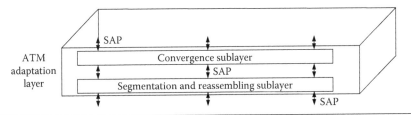

Figure 9.20 AAL sublayers.

- Type 3/4 AAL: It supports classes C and D variable bit rate services with error detection.
- Type 5 AAL: While the type 3/4 AAL considers the multiplexing of small blocks of data, the type 5 AAL groups the data into large blocks, allowing a more efficient multiplexing (reducing the overhead). Consequently, the type 5 AAL is more efficient for data communications.
- Higher Layers: They interface with different classes of service and provide them the required services.

As can be seen from Figure 9.18, the management plane comprises the independent management of each layer (management layer), as well as the management of the B-ISDN reference model as a whole.

9.2.1.2 ATM Network As described earlier, the ATM network can be employed as a transport network or can be employed to provide different services to the end user. In the former case, there are only interfaces between two adjacent network nodes (network–network interface), whereas the latter case comprises interfaces between users and network nodes (user–network interface). The B-ISDN reference model comprises the adaptation of the ATM network to the different services generated by the users.

The transport of data is performed by the ATM network using two different basic layers:

1. Physical layer: It comprises the exchange of bits within different types of devices.
2. ATM layer: It comprises the layer 3 switching performed by the network nodes.

Similarly to the SDH layers, depending on the type of devices, the physical layer is split into different sublayers, each one with each own overhead:

- Transmission path sublayer: It corresponds to the SDH path layer and comprises the ATM end-to-end exchange of data, from the local where the data is encapsulated into cells up to the local where those bits are removed from cells.
- Digital section sublayer: It corresponds to the exchange of data between, for example, two adjacent SDH multiplexers.
- Regeneration section sublayer: It corresponds to the exchange of data between two adjacent regenerators.

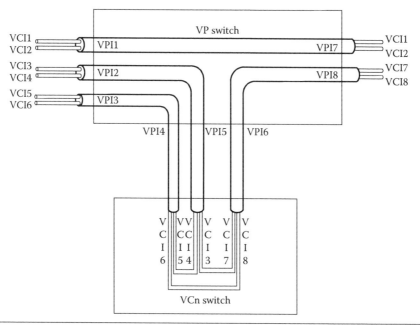

Figure 9.21   Virtual channel and virtual path switch.

Similarly, the ATM layer also comprises two hierarchical sublayers defined as [ITU-T I.113] (see Figure 9.21) [Costa 1997]:

1. Virtual channel (VCn) sublayer: It comprises the layer 3 switching of a group of different cells, which are identified by a common VCn Identifier (VCI).
2. Virtual path (VP) sublayer: It comprises the layer 3 switching of a group of different channels, which are identified by a common VP Identifier (VPI).

A VP switch performs the switching of different VPs, translating the incoming VPIs into another outgoing VPIs. Note that a VP switch performs the switching of multiple VCs whose inputs and outputs are common. Similarly, a VCn switch performs the switching of different VCns, translating the incoming VCIs into another outgoing VCIs.

9.2.1.3 ATM Cell Format   As can be seen from Figure 9.19, the ATM cells are 53-octets long, being composed of a 48-octet-long payload data field and of a 5-octet-long header field.

Depending on the type of interface where cells are employed, the headers present two different formats. Those different cell formats are depicted in Figure 9.22, where UNI stands for user–network interface and NNI for network–network interface. The contents of different header's fields are as follows:

- Generic control field (GCF): with the default value 0000
- VPI: composed of 8 bits (UNI) or 12 bits (NNI), as above described

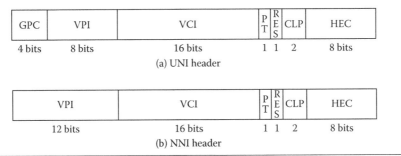

Figure 9.22   Cells header format of (a) UNI and (b) NNI.

- VCI: composed of 16 bits, as above described
- Payload type (PT): to signalize the type of data transported in the cell
- Reserved (RES): for future use
- Cell loss priority (CLP): in case of congestion, cells with this bit active are discarded first
- HEC: composed of 8 bits and used to allow checking the presence of errors in the cell header

9.2.2 Multiprotocol Label Switching

The MPLS protocol is employed in packet-switching transport networks (MAN/WAN). It was designed to support any layer 3 protocol (IPv4, IPv6, IPX, ATM, etc.) [RFC 3031] [RFC 3270]. Moreover, it can be implemented over any type of layer 2 network such as SDH, SONET, Ethernet, and so on. Consequently, it is normally stated that the MPLS protocol belongs to the layer 2,5 of the Open System Interconnection (OSI) reference model (see Figure 9.23). It is worth noting that most common MPLS implementations consider a SDH or SONET circuit-switching network below it.

9.2.2.1 The MPLS Network   The MPLS protocol is a connection-oriented protocol, whose forwarding of packets is performed using the virtual circuit method (similar to the ATM protocol). This method allows a high routing speed and the delivery of packets in the correct sequence, which results in good QoS.

The MPLS protocol was developed taking the weaknesses of the ATM protocol. An important limitation of the ATM protocol relies on the fact that the overhead is high (5 octets), as compared to the cells' size (48 octets), which results in a high overhead and in a low efficiency.

The currently existing physical layer protocols allow the exchange of data at a much higher rate than before. This allows exchanging longer frames without introducing delays to the supported services. Consequently, contrarily to the ATM, the MPLS protocol allows the exchange of connection-oriented packets with different sizes.

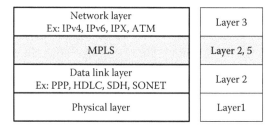

Figure 9.23 Location of the MPLS protocol in the OSI reference model.

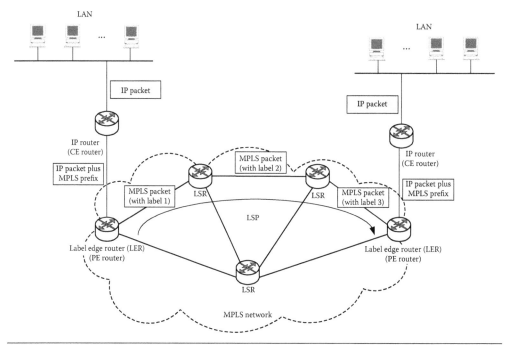

Figure 9.24 Example of a MPLS network.

The routing of MPLS packets is performed making use of a label* contained in the packet header. The label identifies the virtual circuit to which the packet belongs. Due to this reason, the MPLS protocol is referred to, by some authors, as label-switching protocol.

An example of a MPLS network is depicted in Figure 9.24. As can be seen, the MPLS consists of a WAN or MAN, which is used to transport large amount of data among multiple LAN. The different MPLS routers are called label-switching router (LSR). Nevertheless, the LSR routers that are located at the edge of the MPLS cloud (network), interfacing with the customer's equipment router (CE Router), are referred to as label edge router (LER) or simply as edge LSR. The LER is a provider equipment router (PE Router), as it is owned by the transport service provider.

The CE Router consists of, for example, an IP router. However, it has knowledge that the following router is a MPLS router. The packets received from its LAN and

---

* The label is similar to the VPI/VCI identifiers employed in the ATM protocol.

forward to the MPLS router (LER) are added with a prefix. The prefix consists of a code that identifies the virtual circuit that is to be employed in order to forward the packet into its destination. The table where the prefixes are stored is entitled VPN* routing and forwarding (VRF). The VRF is provided by the LER to the CE router.

Since the MPLS uses the virtual circuit method, the routing path is calculated in advance, before the data is transmitted through the network. Such path calculation is performed taking into account the required QoS and the traffic conditions. The MPLS virtual circuit is designated, in the MPLS world, as label-switching path (LSP), as depicted in Figure 9.24. Moreover, as a preventive measure, the MPLS performs traffic engineering [RFC 2702]. This consists of load balancing of different traffic flows over different paths.

The IP packet is received by the LER router with its prefix. This router translates the IP packet into the MPLS format and performs the conversion from prefix into MPLS label. It is important noting that, similar to the VPI/VCI identifiers employed in ATM networks, the label has a local meaning. This means that each LSR performs the removal of the label corresponding to the previous point-to-point MPLS connection, adding a new label corresponding to the following point-to-point connection. This function is referred to as label swap, whereas the addition or removal of a label performed by LER routers is referred to as push or pull, respectively. The calculation of the output interface to use by a MPLS router in order to forward a packet is performed taking into account the incoming label, as well as the information contained in the routing table.

The list of labels, employed by different MPLS routers, is kept in a database entitled label forwarding information base (LFIB). When a new LSP is created, the corresponding labels are added to the LFIB. Label distribution protocol (LDP) is employed to perform the discovery of paths (LSP) and to perform the resulting distribution of labels by different routers. In order to assure the reliability of data, the LDP runs over the TCP protocol. Note that the LDP is accompanied by bandwidth reservation for specific LSPs, and therefore, also contributes for the MPLS to be a QoS enabler protocol.[†]

A certain LSP can be used to transport different types of traffic (voice, video streaming, file transfer, etc.). For the sake of QoS provisioning, different traffic types must be properly identified with a different forward equivalent class (FEC), also included in the MPLS label. As can be seen from Figure 9.25, a label identifies a pair of LSP and a FEC. When a new FEC is created, it is assigned to a certain LSP. Note that different FECs of a communication between the same two end points may belong to the same LSP or to different LSP. In the example depicted in Figure 9.25, the FEC1 and FEC2 belong to LSP1, whereas the FEC3 and FEC4 belong to LSP2.

---

* Virtual private network.
† Together with the low latency introduced by this protocol.

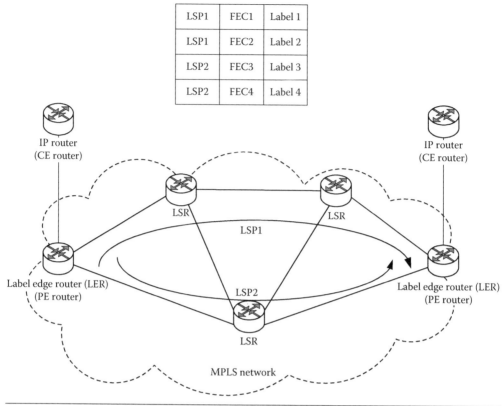

LSP1	FEC1	Label 1
LSP1	FEC2	Label 2
LSP2	FEC3	Label 3
LSP2	FEC4	Label 4

Figure 9.25   LSP versus FEC.

MPLS shim header

Label	CoS	S	TTL
20 bits	3	1	8 bits

Figure 9.26   MPLS shim header format.

The MPLS routers process differently the packets with FEC that identify traffic with higher priority or with higher sensitivity to loss of data (packet discard).

9.2.2.2 MPLS Packet Format   As described earlier, the MPLS is a protocol used to transport layer 3 packets. Contrarily to the ATM protocol that comprises a certain packet (cell) format, the MPLS limits to add a certain overhead to the transported layer 3 packets (ATM, IPv4, IPv6, IPX, etc.). Alternatively, the overhead is added between the layer 2 header and the layer 3 header. The MPLS overhead is referred to as shim header and comprises the following different fields (see Figure 9.26):

- Label: It is employed to identify the LSP (VP), as well as the FEC (type of traffic being transported in the packet).
- Class of service (CoS): It is used for QoS provisioning and for explicit congestion notification purposes.

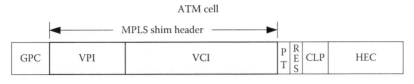

Figure 9.27 Location of the MPLS shim header.

- Stack (S): In case multiple labels are inserted into a frame, it indicates a hierarchy. The value one means that the label is the last of the stack.
- Time to live (TTL): It corresponds to the time-to-live field of IPv4 packets, being assigned to the last label of a stack.

When the transported protocol has fields that identify virtual circuits, the MPLS shim header is directly inserted into the corresponding fields. This is the case of ATM and frame relay protocols:

- ATM: The shim header is inserted into the VPI/VCI fields (see Figure 9.27a).
- Frame relay: The shim header is inserted into the data link channel identifier (DLCI) field.

Alternatively, the shim header is placed between the layer 2 and the layer 3 headers. This can be seen from Figure 9.27b). This latter procedure is employed in protocols such as Ethernet, PPP, Token Ring, and so on.

End of Chapter Questions

1. What is the difference between ATM and B-ISDN?
2. What is the ATM adaptation layer used for?
3. What are the different layers of the B-ISDN reference model?
4. Define the functions of the different B-ISDN layers.
5. Why is ATM referred to as cell switching, instead of a packet-switching network?
6. In the scope of the ATM protocol, what is the difference between a virtual path and a virtual circuit?
7. In the scope of the ATM protocol, which sublayers are included in the physical layer? What are their functionalities?
8. Which types of ATM cell headers exist. Define the different header fields.
9. What is the difference between an ATM virtual circuit switch and an ATM virtual path switch?
10. Which types of ATM classes of services exist?
11. Which AAL sublayers do you know? Define the functionalities provided by each one?

12. Which type of services can the AAL provide? Define each AAL type.
13. What does FEC consist of?
14. How does the FEC concept implemented by the MPLS protocol can contribute to provide QoS?
15. What are the advantages of the MPLS protocol relating to the ATM protocol?
16. What is the MPLS packet format?
17. In which location is the MPLS shim header introduced in the packets/frames?
18. What does LSP stands for?
19. Which mechanisms are implemented by the MPLS protocol that make it well fitted for providing QoS?
20. Which type of information is contained in a label?
21. Which mechanism is employed by different LSR in order to calculate the output interface to forward a certain packet?
22. What is the difference between a LSR and a LER?
23. What is the difference between a CE router and a PE router?
24. What is a MPLS prefix?
25. Describe the working functionalities of a MPLS network.
26. Describe the MPLS shim header format.
27. To which OSI reference model layer does the MPLS protocol belong?
28. Define the SDH architecture.
29. Define the SONET architecture.
30. What are the reasons that may originate fluctuations between a SPE/VC and a frame?
31. Which mechanisms can be implemented to counteract the fluctuations between a SPE/VC and a frame?
32. What is the difference between the ITU-T G.732 and the ITU-T G.733 standard?
33. Which type of interposition is employed in SDH and SONET hierarchies?
34. Which type of interposition is employed in PDH hierarchy?
35. What are the advantages of the synchronous hierarchies (SDH/SONET) relating to plesiochronous hierarchies?
36. Which type of protection from failures can be employed in a SDH ring using two optical fibers? How does it work?
37. Which type of protection from failures can be employed in a SDH ring using four optical fibers? How does it work?
38. How are different SDH headers managed by different SDH devices?
39. How do DOCSIS 1.0 modems allow the exchange of digital data over coaxial cables?
40. Which type of transmission techniques can be employed in xDSL modems?
41. What is the difference among DOCSIS 1.0, 2.0, and 3.0?
42. What is the throughput per channel made available by different DOCSIS modems?

43. How does the DOCSIS 3.0 provide a higher throughput than DOCSIS 2.0?
44. Which uplink/downlink bandwidths and modulation schemes are employed in different DOCSIS standards?
45. Which studied xDSL version performs the transmission in baseband? What is its line coding technique and the corresponding code efficiency?
46. What is the difference between a xDSL modem and a DOCSIS modem?
47. Which versions of xDSL do you know and what are their differences?
48. What is the difference between discrete multitone (DMT) and OFDM?
49. Which types of transmission techniques are employed by xDSL modems?
50. How does discrete multitone (DMT) tend to allow the exchange of higher data rates?
51. What does echo cancellation stands for?
52. What is the typical spectrum utilized by DSL modems?

DATA LINK LAYER

The physical layer is responsible for allowing the exchange of bits between two adjacent nodes* of a network. However, these bits are subject to channel impairments, such as noise, interference, or distortion. These channel impairments originate errors in the exchanged bits. As described in Chapter 3, the bit error probability increases with

- The increase of the power of noise, interference, and distortion
- The increase of distance without regeneration
- The decrease of the transmitted power
- The decrease of the reliability of the transmission medium. A radio transmission medium is typically less reliable than an optical fiber. Consequently, the bit error probability of the former transmission medium tends to be worse than the latter.

As can be seen from Figure 10.1, the data link layer is directly above the physical layer, being responsible for allowing a reliable exchange of data between two adjacent nodes. The data link layer is composed of two sublayers:

- Logical link control (LLC), which deals with flow control and error control. These functionalities are implemented using protocols such as the IEEE 802.2, HDLC, point-to-point protocol (PPP), serial line Internet protocol (SLIP), and so on.
- Medium access control (MAC), which determines when a host is allowed to transmit within the shared medium of a local area network (LAN) or a metropolitan area network (MAN). Note that this sublayer does not exist in WAN networks. This sublayer is implemented using protocols such as the IEEE 802.3, IEEE 802.5, IEEE 802.11, and so on.

The network layer makes use of the services provided by the data link layer. In fact, the network layer is an extreme-to-extreme layer, responsible for allowing data being routed along the several network nodes, such that it reaches the final destination. Therefore, the network laycr makes use of a concatenation of data link layers, a different one for each different network hop.† A certain hop can be a satellite link, using a certain data link layer protocol, whereas another link can be an optical fiber link, using the same or another data link layer protocol.

---

* For example, between adjacent routers or between a computer and a router.
† A link between two adjacent nodes.

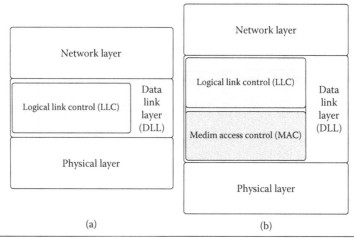

Figure 10.1   Data link layer of (a) WAN and (b) local area network/metropolitan area network.

10.1 LAN Devices

Before the description of error control and flow control is performed, it is worth describing the devices used within a LAN or a MAN.

10.1.1 The Hub

A hub is a network device that retransmits the bits present at one of its input into all output interfaces.* In addition, it performs regeneration of signals, which mitigates the channel impairments (noise, interference, distortion, etc.). The hub also acts as a repeater, which is an important functionality when the network length is longer than the maximum segment size. Nevertheless, the hub is not able to detect errors.[†] The star is the most common physical topology of a network, which employs a hub. Using a hub (repeater) as a central node, the logical topology of the network becomes a bus (see Figure 10.2). By definition, a bus consists of a network that makes use of a common transmission medium.

It is worth noting that the hub deals with bits, and therefore, it works at layer 1 of the open system interconnection (OSI) model.

10.1.2 The Bridge

Contrarily to the hub, which is normally used as the central node of a network, the bridge is typically used to interconnect two network segments. Moreover, these two segments may use the same MAC sublayer protocol or different MAC sublayers protocols,[‡] as long as the LLC sublayer protocol is the same. As defined further, both

---

* Contrarily to the switch.
† Contrarily to some types of bridges and switches.
‡ As long as the bridge interfaces are compatible with these two specific types of MAC sublayer protocols.

the IEEE 802.3 and IEEE 802.5 MAC sublayer protocols make use of the common IEEE 802.2 LLC sublayer protocol, and therefore, a bridge may be used to interconnect these two types of networks.

Another common application of the bridge relies on its ability to achieve a segmentation of a collision domain. The bridge breaks a unique collision domain into two smaller collision domains (see Figure 10.3). Naturally, this reduces the number of hosts that share the medium and to where the carrier sense multiple access with collision detection (CSMA-CD) mechanism is applied. Consequently, the network

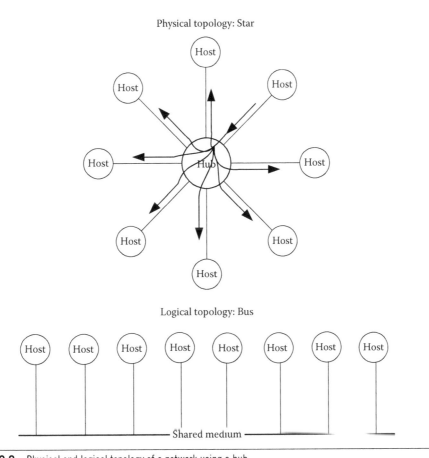

Figure 10.2 Physical and logical topology of a network using a hub.

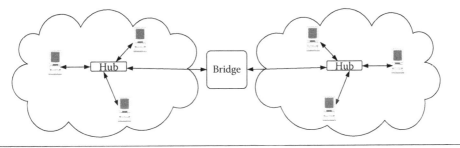

Figure 10.3 Network segmentation performed with a bridge.

performance tends to be improved. Moreover, since a certain network host stops receiving all the network frames, this represents an advantage from the security point of view.

In any of the above described bridge applications, when a frame arrives a bridge port, it must decide about whether the frame has to be forward to the other network segment. Such decision is taken based on the destination's MAC address of the frame, that is, using the layer two of the OSI model.

The mapping between the output port and the MAC address is performed based on a switching table,* which is stored in the bridge. Initially, such a table is empty. Consequently, when a switch receives a frame, it forwards to all output interfaces (ports). As the bridge receives frames, it registers the port and the corresponding source MAC address of the frame, and registers such mapping in the switching table. After having exchanged a certain number of frames, the table stored in the bridge has the full knowledge about the network segments where each of the MAC address of the network nodes is located.

There may be multiple paths between two network points. Multiple paths exist in the case of a network with a mesh topology (see Chapter 1). In this case, the bridge has to decide which one to use. Such discovery and decision is performed making use of the spanning tree protocol (IEEE 802.1d). The obtained information is used to remove loops and to keep information about redundant paths. In case a path is interrupted, the bridge switches to the backup path. Moreover, there are situations where a bridge uses two or more paths simultaneously to perform load balancing.

Contrarily to the hub, which performs the repetition of bits by hardware, the bridge operation is performed by software. Consequently, the delay introduced by such network device is typically much higher.

The bridge works using the store-and-forward mode, defined as follows: it accepts a whole received frame and stores it in memory. Then the cyclic redundancy code (CRC) calculation is performed and the output interface is calculated using the switching table. It still verifies the frame delay between the origin and destination, before it forwards it to the output interface.

10.1.3 The Switch

Similar to the bridge, the switch performs the segmentation of a network into smaller collision domain segments. In fact, in case each segment connects a single host, the collision domains are avoided, and consequently, the CSMA-CD mechanism described below is not applicable. However, while the bridge is typically equipped with only two interfaces (which can be of different MAC sublayer types), the number of interfaces of the switch is typically high. Moreover, these interfaces are typically of

---

* A switching table is similar to a routing table. While the routing table performs the mapping between a destination's IP address of a packet and an interface, the switching table performs the mapping between a destination's MAC address of the frame and a port.

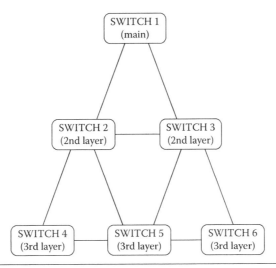

Figure 10.4 Example of network with loops.

the same type (e.g., IEEE 802.3/Ethernet protocol). Nowadays, the switch tends to be the central node of the network, instead of the previously used hub. Moreover, the bridge has been replaced by the switch.

It is possible to have different switches linked in cascading (see Figure 10.4).* Typically, a IEEE 802.3 network is implemented using a star as the physical topology. Nevertheless, while the corresponding logical topology of a network using a hub is the bus, the star is the logical topology that results from the use of the switch as the central node. This results in a high improvement of the network performance. A network using a hub as a central node only allows one host transmitting at a time. Contrarily, a switch allows up to half of the network hosts transmitting to the other half of the network (half duplex).† Assuming that a LAN works at 10 Mbps in half-duplex mode and that it has a total of ten hosts, we may have up to five hosts transmitting at 10 Mbps to the other five hosts (see Figure 10.5). This results in a cumulative network throughput of 50 Mbps. In case the network works in full-duplex mode, then the maximum cumulative network throughput becomes 100 Mbps.

It is worth noting that most of the switches currently available in the market present autosensing. This is a capability that allows a switch detecting and adapting to different throughputs in different interfaces, as well as to half or full duplex. Note that the switch may connect with different devices at different speeds. It may connect with hosts at 100 Mbps, whereas the connectivity with a server may be at 10 Gbps.

Similar to the bridge, the switch performs the forwarding of frames based on the destination's MAC address. Therefore, this device works at layer two of

---

* For example, a main switch may be used to serve a whole organization, while a second layer of switches may be used for different departments.

† In case the network is full duplex, it is possible to have all network hosts transmitting to all network hosts, simultaneously, at the maximum data rate.

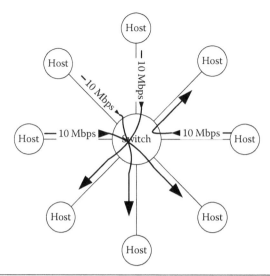

Figure 10.5 Example of half-duplex transmission in a local area network with switch.

the OSI model.* The forwarding of frames is performed based on a switching table, which maps destination MAC addresses into output interfaces (ports). Similar to the bridge, the switching table is filled in as the switch receives frames from the corresponding interface ports. Moreover, there may be more than one path between two network nodes. The spanning tree protocol (IEEE 802.1d) normally resolves that by using a metric based on the lower number of hops, while the other paths are kept as redundancy. Load balancing is also possible in high dimension LANs.

Contrarily to the bridge, which performs the switching of frames by software, the switch performs its functionality through hardware. Consequently, the latency introduced by a switch is typically much lower than that of a bridge.

Depending on the type of switch, the forwarding of frames is performed using one of the following modes:

- Store-and-forward: it accepts a whole received frame and stores it in memory. Then the CRC is computed and the output interface is calculated using the switching table. It also verifies the frame delay between the origin and destination, before it forwards it into the calculated output. This is the mode that achieves the best level of integrity. Nevertheless, the delay introduced with this mode tends to be higher than in the case of the other modes.
- Cut-through: it only reads the initial octets of the frame header, up to the destination address field (to be able to compute the output port), before the frame is forward. The advantage of this mode relies on the maximization of the throughput, while the disadvantage relies on the risk of forwarding corrupted frames.

---

* In any case, it is currently possible to find network devices that perform both layer 2 (switch) and layer 3 (router) switching.

- Fragment free: it only accepts the initial 64 octets of a frame, before the computation of the output port is determined, and the switching path is established. This is used to allow the detection of a collision, as established by the CSMA-CD mechanism. Switches, using this mode, are normally connected to a hub,* and therefore, its current application is limited.

10.1.4 Spanning Tree Protocol

As previously described, the switch and the bridge perform layer two switching based on the destination MAC address contained in a frame and taking into account the information present in a switching table. A switching table maps MAC addresses into output ports. Initially, this table is empty. Consequently, the device forwards the frames received in one of the ports through all ports. As long as a device receives frames from the different ports, the corresponding source MAC addresses are registered and associated to certain ports, resulting in the construction of the switching table.

Small size networks have typically a physical topology consisting of a single layer (e.g., IEEE 802.3 networks). Medium to high dimension networks have typically a topology with several layers of switches (see Figure 10.4). To improve the resistance from failures, it is a good choice to keep redundant paths in a network. This results in a network with a mesh configuration, where there exist multiple paths between different network points. In this case, a bridge or a switch has to decide about which output port to use, among several possibilities. These decisions are taken by switches and bridges using the spanning tree protocol, which results in an efficient manner of building switching tables. The IEEE 802.1d is a spanning tree protocol standard.

Figure 10.4 shows an example of a mesh network with several layers of switches, namely a main switch (organizational switch), second layer of switches (e.g., branch switches), and third layer of switches (e.g., division switches). As can be seen, the communication between switch 1 and switch 5[†] can be established through switch 2 or through switch 3. In addition, it can also be established through switch 4 or switch 6, although with a higher number of hops.

The spanning tree protocol avoids loops by spanning the mesh network in a tree. When a switch or a bridge is established, an exchange of data among these devices is performed, and a sequence of opening and closing of ports is executed. During this phase, the leds normally present at the device interfaces show the orange color. In the steady state, the loops are avoided by disabling those links between any two points that are not part of the selected path between those two points. Once the spanning tree protocol finalizes its transient phase, the leds of the device present the green color.

The spanning tree protocol avoids loops in the network, while keeping information about redundant paths. The existence of loops results in flooding and network

---

* This is the reason why the CSMA-CD mechanism is applicable.

† In fact, the purpose is to establish a communication between switch 1 and a host connected to switch 5, not to switch 5 itself. The switch 5 is only an interim network device.

overload, which represents a performance degradation of the network. In case a path is interrupted, a bridge or a switch automatically selects the backup path. Moreover, in high dimension networks, a bridge or a switch may use two or more paths simultaneously to perform load balancing.

10.2 Logical Link Control SubLayer

Similar to the upper layers that make use of the services made available by the lower layers, the DLL makes use of the service provided by the physical layer. Since the physical layer performs the nonreliable exchange of bits, the data link layer needs to add reliability. This is performed by the LLC sublayer. Consequently, the LLC sublayer is responsible for allowing a reliable exchange of data between two adjacent nodes of a network. This is achieved by implementing the following functionalities:

- Error control
- Flow control
- Grouping isolated bits into frames

Note that the LLC sublayer is established using one of the following modes:

- Connectionless and nonconfirmed
- Connectionless and confirmed
- Connection oriented

As described in Chapter 1, the connection-oriented mode requires the previous setup of the connection before data is exchanged. In addition, it considers the connection termination after the exchange of data. Since the connection-oriented mode is always confirmed, it uses either error detection with retransmission of frames or error correction. Moreover, for the sake of error control and to allow the delivery of frames to the network layer in the correct sequence, the frames are numbered.

The nonconfirmed connectionless mode is used in scenarios where the error probability is reduced (e.g., the transmission of bits in an optical fiber) or in scenarios where an upper layer has the responsibility of performing error control. This mode does not assure reliability of data, that is, there is not feedback from the receiver to the transmitter about whether it was correctly received.

In case confirmation is used, error control can be implemented using either error detection with retransmission or error correction. In case of error detection with retransmission, there is a feedback link informing the transmitter about whether or not the data was received free of errors. Using error detection or error correction,[*] there is an additional level of processing, overhead, and delay in signals.

---

[*] Connection-oriented or connectionless and confirmed modes.

Normally, the transmitter can send data quicker than the receiving entity is able to receive. To avoid lost of bits, the receiver needs to send feedback (control data) to the transmitter about whether it is ready to receive more data. This is achieved through flow control.

The data link layer creates groups of bits to which the corresponding overhead (redundant bits to allow error control and flow control) is added. This group of bits, with a specific format depending on the data link layer protocol, is referred to as a frame. A frame consists of a group of bits necessary to allow the implementation of error control and flow control.

Figure 10.6 shows the decomposition of a frame, as composed of the payload data (data received from the network layer for transmission) plus this layer overhead.

The frame overhead consists of the start of frame, the destination and source address, the control bits, the redundant bits for error control, and the end of frame. The start and end of frame is used to allow frame synchronization (i.e., layer 2 synchronization), that is, for the receiver to understand when the frame starts and when it terminates. In addition, in case the link is asynchronous, the start of frame may also be used to allow bit synchronization (i.e., layer 1 synchronization). The control bits are used for the management of the flow control and error control. The redundant bits are used to allow the detection of errors in a frame or to implement the error correction.

The start and end of frame can be signalized using different procedures, namely

- Delimiting character string
- Flag
- Violation of the line coding mechanism

Delimiting character string: in this case, the receiver detects the start of the frame through the reception of a sequence of two predefined characters. As can be seen from Figure 10.7, these two characters are the data link escape (DLE) and the start of text (STX). Moreover, the receiver detects the end of a frame by receiving another group of two predefined characters, namely the DLE and the End of Text (ETX). A limitation of this procedure comes from the fact that the bits corresponding to the DLE character may be part of the payload data. In this case, to avoid that the receiver becomes confused, instead of sending a single DLE in the payload data, the transmitter sends twice the DLE character, and the receiver removes one of these characters.

Start of frame	Address	Control	Payload data	Redundant bits for error control	End of frame

Figure 10.6 Generic frame format.

DLE	STX	. . .	DLE	ETX

Figure 10.7 Delimiting character string.

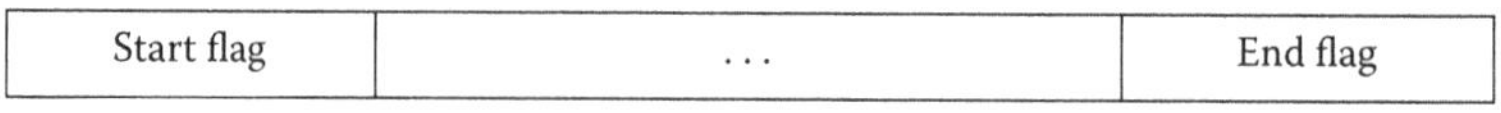

Start flag	. . .	End flag

Figure 10.8 Flag.

Flag: this is the most common frame synchronization method. In this case, the receiver detects the start and end of frame through the reception of a predefined sequence of bits (see Figure 10.8). It is composed of a sequence of bits with low probability to occur with in the payload data part of a frame. The PPP and the HDLC protocols use the sequence 01111110 as the flag. A common limitation of this procedure comes from the fact that a part of the payload data may have a sequence of bits equal to the flag. In such case the receiver could interpret it as start or end of a frame. To avoid this, a procedure known as bit stuffing is implemented, which is defined as follows for the HDLC protocol: every time the transmitter detects a sequence of five "1" logic state bits in the data, it inserts a "0'" logic state bit as the sixth bit. The receiver performs the reverse operation. With the bit stuffing procedure, it is assured that a sequence of six "1" logic state bits are never transmitted within the data field.

Violation of the line coding mechanism: in this case, the receiver detects the start and end of a frame through the reception of a predefined sequence of signal levels that is not allowed to occur within the normal transmission of data. As an example, the bi-phase Manchester line coding technique considers always a transition from 0 to 1 or from 1 to 0 at the middle of the bit period (see Chapter 6). The absence of such transition may be used as a signaling for the start and/or end of a frame.

10.2.1 Error Control Techniques

Depending on the transmission medium and data link layer protocol that is used to exchange data, error control can be performed using either error detection or error correction techniques [Benedetto 1997]. In the case of error detection, codes such as CRC or parity bits can be used to allow errors being detected at the receiver. In case an error is detected, the receiver requests a retransmission of a corrupted frame. On the other hand, in case the transmission medium is highly subject to channel impairments (e.g., wireless medium), the choice goes normally to the use of error correction, instead of error detection.[*] In this latter case, the level of overhead per frame is higher, but it avoids successive repetitions, which also translates in a decrease of overhead.

With regard to error detection and error correction codes, the code rate is an important parameter that is worth defining. The code rate R_C is defined as [Benedetto 1997]

$$R_C = \frac{n}{k} \tag{10.1}$$

---

[*] In some cases, these two techniques are used together.

where n stands for the number of information bits and k for the number of transmitted bits.

As can be seen from Figure 10.9, an error control algorithm considers a block of n information bit to which m redundant bits are added to allow the error detection or error correction capability. These redundant bits are calculated from the original n information bits. The output of the error control algorithm comprises a total of $k = n + m$ bits (codeword), which are transmitted.

Error control techniques are commonly used by modems, as well as by many protocols such as the HDLC, PPP, transmission control protocol (TCP), and so on.

10.2.1.1 Hamming Distance Before proceeding with the description of other error detection and error correction techniques, it is worth introducing the notion of Hamming distance. Regardless the specific type of error control code under consideration, let us consider the mapping between information bits and encoded bits (codewords), as depicted in Figure 10.10.

The Hamming distance is the minimum number of bits between any two codewords (blocks of encoded bits). Figure 10.11 shows the distance between any two codewords of Figure 10.10. As can be seen, the minimum distance between any two codewords is five, which is the Hamming.

The error correction and detection capabilities are deducted from the Hamming distance as follows [Benedetto 1997]:

- The maximum number of corrupted bits that can be corrected r is obtained from the Hamming distance d as $r = (d-1)/2$.
- The maximum number of corrupted bits that can be detected s is obtained from the Hamming distance d as $s = d-1$.

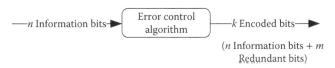

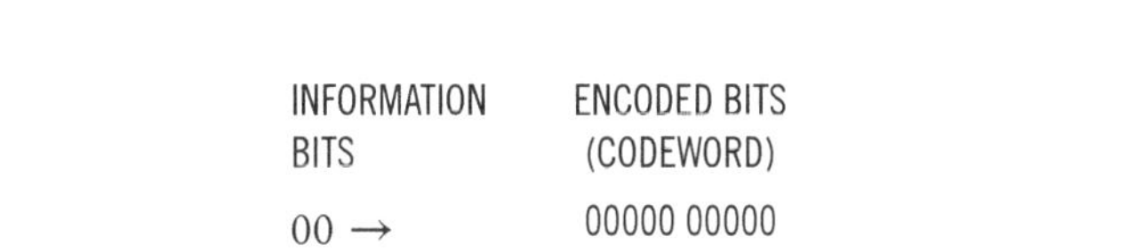

Figure 10.9   Error control encoder.

INFORMATION BITS	ENCODED BITS (CODEWORD)
$00 \rightarrow$	00000 00000
$01 \rightarrow$	01010 10101
$10 \rightarrow$	10101 01010
$11 \rightarrow$	11111 11111

Figure 10.10   Example of mapping between information bits and encoded bits.

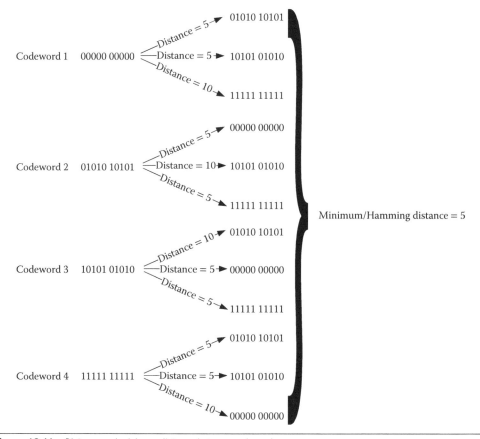

Figure 10.11   Distance and minimum distance between codewords.

Focusing again on the example of Figure 10.10 and from the above equivalences, we conclude the following:

- The maximum number of corrupted bits that can be corrected becomes $r = (5-1)/2 = 2$. Let us assume that the transmitted block is 00000 00000. If the received block is 00000 01010 (two corrupted bits), the closest codeword is 00000 00000. If the receiver is using this code as an error correction code, the receiver deducts this latter block as the estimated transmitted block. In this case, the receiver's decision is correct. On the other hand, if the received block is 00001 01010 (three corrupted bits), then the closest codeword is 10101 01010. If the receiver opt by this latter sequence as the estimated transmitted one, then the receiver makes the wrong estimate. Consequently, it is confirmed that using the proposed code as error correction code, a maximum of two bits can be corrected.

- The maximum number of corrupted bits that can be detected becomes $s = 5-1 = 4$. Assuming that the transmitted block is 00000 00000, if the received block is 00101 01010 (four corrupted bits), then the receiver concludes that such codeword does not correspond to any of the valid codewords and requests the retransmission of data or just discards it. If the received block is 10101 01010 (five corrupted bits), since this sequence of bits corresponds to a valid block, the receiver assumes this as the estimated transmitted one. In this latter case, the receiver makes a wrong estimate. Consequently, it is confirmed that using the proposed code as error detection code, a maximum of four bits can be detected.

10.2.1.2 Error Detection Codes   Once a frame is received, the host checks for errors using the redundant bits. In case an error is detected, two possibilities exist:

- The corrupted frame is requested for retransmission.
- The receiver discards the corrupted frame and an upper layer implements an error control mechanism.

In case an error is detected and the retransmission procedure is used, the frame is repeated by the transmitting host. When a frame is transmitted, the transmitting host starts a chronometer. Depending on the type of confirmation,* the procedure used by the receiving host for signalizing the transmitting one is different.

In case of positive confirmation, the receiving entity sends a feedback message in case the frame is free of errors. This procedure is commonly referred to as positive acknowledgement with retransmission (PAR) and the feedback message is known as acknowledgement (ACK). The positive confirmation procedure can be resumed as follows (see Figure 10.12):

- With errors: if the ACK is not received by the transmitting entity within a certain time period (controlled by the transmitter's chronometer), it assumes an error and retransmits the frame.
- Without errors: if the ACK is received by the transmitting entity within the expected time period, it assumes correct reception of the frame and proceeds with the transmission of the following frame.

In case of negative confirmation, the receiving entity only sends a feedback message in case of errors. This procedure is commonly referred to as negative acknowledgement (NAK). The negative confirmation procedure can be resumed as follows:

- With errors: if a feedback message is received by the transmitting entity within a certain time period informing that the frame was in the presence of errors, this entity repeats the frame.
- Without errors: if a feedback message is not received within a certain time period informing the transmitting entity about the presence of errors, this entity assumes correct reception of the frame and proceeds with the transmission of the following frame.

---

* As described in Chapter 1, the confirmation is employed in connection-oriented services or in confirmed connectionless services.

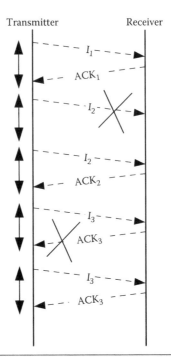

Figure 10.12   Example of positive acknowledgement with retransmission handshaking.

The advantage of negative confirmation relies on the lower amount of data exchanged, as opposed to the positive confirmation. Note that, in both cases, the frames are numbered for use by the positive or negative confirmation handshaking.

Figure 10.12 shows an example of the PAR procedure. The transmitter sends a first information frame (I_1) to the receiver. This frame is received free of errors, the receiving host acknowledges (ACK_1) the correct reception of such frame, and such confirmation is received within the expected time period. In case of the second frame, an error is detected at the receiving side. Consequently, this latter host does not send the corresponding acknowledgement. The emitter chronometer reaches the timeout and the second frame is retransmitted. In the case of the third frame (I_3), it is correctly received, but it is the corresponding acknowledgement (ACK_3) that does not reach the transmitter. Consequently, the transmitter assumes that the third frame was received corrupted and retransmits the frame. In this last situation, although the receiving host expects the fourth frame, it knows that this last received frame is a repetition of the previously sent because the frame is numbered and discards it. Otherwise, the receiving host would have assumed this frame as the fourth and would incorrectly deliver it to the network (upper) layer.

As previously described, when the error control mechanism is based on positive confirmation, the receiver sends only a feedback signal to the transmitter when the message is correctly received. In such case, flow control is associated to error control in the sense that the feedback signal informs the transmitter that

- The previously received message is free of errors (error control).
- The receiver is able to receive more data (flow control).

Contrarily, the negative confirmation considers a feedback signal sent by the receiver only in case an error is detected in the received message. Consequently, using the negative confirmation, flow control is not automatically associated to error control.

10.2.1.2.1 Parity Bits   A basic error detection mechanism relies on the use of parity bits. These parity bits are redundant bits used by the receiver for checking errors. The parity bits, which result from a predefined operation, are added to a frame at the transmitting side. The receiver performs the same operation and observes such redundant bits. The parity can be even or odd.

The transmitting entity applies the exclusive-OR (XOR) operation to the frame and adds the resulting parity bit to the sequence (one additional bit). For odd parity, if the number of "1" logic state bits in the frame is odd, the parity bit is set to "1." For even parity, if the number of "1" logic state bits in the frame is even, the parity bit is set to "1." Otherwise, the parity bit is set to "0." Finally, the frame and the parity bits are transmitted together. Figure 10.13 shows two examples of odd parity, while Figure 10.14 shows two examples of even parity.

The receiving entity performs the same XOR operation to the information bits and checks for the received parity bit. In case the frame is received free of errors, the result of this processing originates the same parity bit as the one received. In case an odd number of bits have been received corrupted, the result of this processing originates a parity bit different from the one received. This can be seen from Figure 10.15, where

	Original message	Odd parity	Resulting data for transmission
Example 1	1 1 0 0 1	**1**	1 1 0 0 1 **1**
Example 2	1 0 0 1 0	**0**	1 0 0 1 0 **0**

Figure 10.13   Examples of odd parity.

	Original message	Even parity	Resulting data for transmission
Example 1	1 1 0 0 1	**0**	1 1 0 0 1 **0**
Example 2	1 0 0 1 0	**1**	1 0 0 1 0 **1**

Figure 10.14   Examples of even parity.

Transmitted data	Received data	Received parity bit	Result of odd parity (except parity bit)
110011	100011	1	0

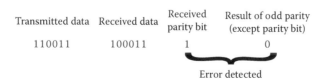

Error detected

Figure 10.15   Example of an error detected by the receiver using odd parity.

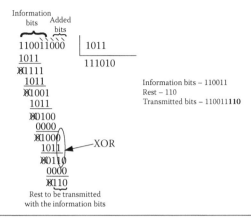

Figure 10.16 Example of cyclic redundancy code encoding.

the third bit has been received corrupted. Note that with this method, there is no way to find out which one is the corrupted bit.*

In case an even number of corrupted bits is received, the result of the parity check operation corresponds to the received parity bit. In this case, the parity operation is not able to detect the errors.

In the case of the example in Figure 10.13, which consists of error detection, the number of information bits is 5 and the number of transmitted bits is 6. From (10.1), the code rate becomes $R_C = 5/6 = 0.8333$.

10.2.1.2.2 Cyclic Redundancy Check The CRC is the most common error detection technique used nowadays. Examples of protocols that use CRC codes include the PPP, the HDLC, the Ethernet/IEEE 802.3, and the TCP. CRC codes are also used to check the integrity of stored information.

The CRC encoding considers a block of n information bits to which m redundant bits are added to allow the error detection capability. These redundant bits are calculated from the information bits using a certain predefined generator polynomial. Note that the generator polynomial is known by the transmitting and receiving entity. Finally, a total of $k = n + m$ bits are transmitted.

The processing of the transmitting entity is as follows:

- Add† a total of m zeros to the block of n information bits, where m corresponds to the degree of the generator polynomial.
- Divide the resulting block by the bits that results from the generator polynomial (see Figure 10.16). The rest of this division originates a total of m bits.
- A total of k bits, consisting of the concatenation of the n information bits with the rest of the division (m redundant bits), are transmitted.

---

* Otherwise, such bit would be corrected, and this would be an error correction code.
† Concatenate.

Based on the received block of k bits, the receiving entity performs the following processing to check the integrity of the data:

- The block composed of k received bits is divided by the bits that results from the generator polynomial. If the rest of this division is zero, then the received block is assumed free of errors. If the rest of this division is different from zero, then the received block is assumed corrupted.
- Alternatively, the receiver may extract the initial n bits of the received block composed of k bits, followed by a similar processing as that performed by the transmitting entity (add zeros and divide by the bits that result from the generator polynomial). If the obtained rest equals the received one, then the received block is free of errors. Otherwise, the received block is corrupted.

Note that the transmitted block includes the original information bits, to which some redundant bits (the rest of the division) are added. In this case, the CRC is referred to as a systematic error control code. On the other hand, error control codes whose generated codeword does not directly include the original information bits are called nonsystematic.

Let us consider an example of CRC use, where the information bits consists of the sequence 110011 and the generator polynomial is $P(x) = x^3 + x + 1$. Based on the above description, the processing of the transmitting entity becomes as follows (see Figure 10.16):

- Since the level of the generator polynomial is three, the number of zeros added to the information sequence is also three. The resulting sequence becomes 110011**000**.
- The generator polynomial $P(x) = x^3 + x + 1$ is mathematically equivalent to $P(x) = 1 \times x^3 + 0 \times x^2 + 1 \times x + 1 \times x^0$. Then, the sequence of bits that results from the generator polynomial is 1011. The division is depicted in Figure 10.16, and the rest of this operation becomes **110** (the number of bits taken from the rest of the division corresponds to the degree of the generator polynomial). Note that the binary difference operation performed to calculate the rest consists of the module 2 adder without carry.[*]
- The transmitting sequence that consists of the concatenation of the information bits with the rest comes 110011**110**.

Let us suppose that the block was received free of errors. Based on the above description, the processing of the receiving entity becomes (Figure 10.17)

- The received block of bits (110011110) is divided by the bits that result from the generator polynomial (1011). As can be seen from Figure 10.17, the rest is zero. Consequently, we may conclude that the received block is free of errors.

---

[*] A modulo 2 adder is implemented with an OR-exclusive gate (XOR), whose relationship between the input and the output bits is $0 \oplus 0 = 0$; $0 \oplus 1 = 1$; $1 \oplus 0 = 1$; $1 \oplus 1 = 0$.

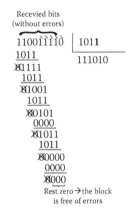

Figure 10.17   Example of cyclic redundancy code decoding without errors in the received block.

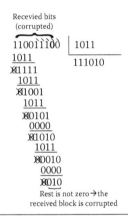

Figure 10.18   Example of cyclic redundancy code decoding with errors in the received block.

We may now consider the example depicted in Figure 10.18, where we assume that the received block is corrupted, once the received bits are 110011100, instead of 110011110.

Based on the above description, the processing of the receiving entity becomes the following:

- The received block of bits (110011100) is divided by the bits that results from the generator polynomial (1011). As can be seen from Figure 10.18, the rest is not zero (010). Consequently, we may conclude that the received block is corrupted.

10.2.1.3 Error Correction Codes   Error correction codes allow the receiving entity to correct one or more corrupted bits within a received block, without having to request a retransmission. As previously described, error correction codes are typically used in channels highly subject to impairments, that is, where the bit error rate (BER) is high. In such situations, the use of error detection and retransmission would lead to successive retransmissions of data, which would translate in excessive additional bandwidth and delay. Error correction codes are also used in channels where retransmission is not

		Odd parity
Frame 1	1 0 1 0 1	1
Frame 2	1 1 0 0 1	1
Frame 3	1 1 0 0 0	0
Frame 4	0 1 0 1 0	0
Odd parity	1 1 1 1 0	0

Figure 10.19 Example of odd parity applied in two dimensions.

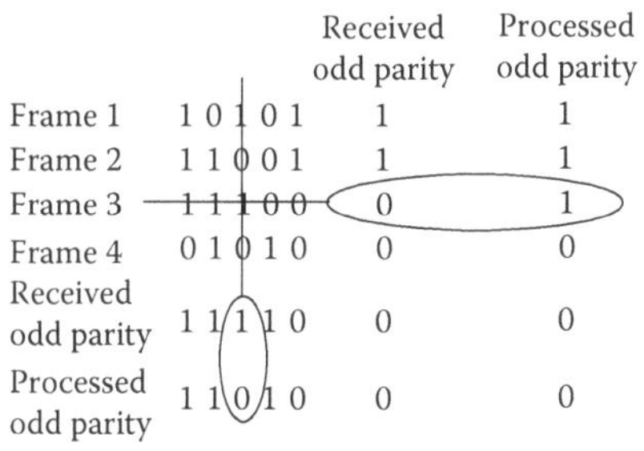

		Received odd parity	Processed odd parity
Frame 1	1 0 1 0 1	1	1
Frame 2	1 1 0 0 1	1	1
Frame 3	1 1 1 0 0	0	1
Frame 4	0 1 0 1 0	0	0
Received odd parity	1 1 1 1 0	0	0
Processed odd parity	1 1 0 1 0	0	0

Figure 10.20 Correction of a corrupted bit using odd parity in two dimensions.

possible, such as data broadcast. This capability is normally implemented in a modem, but can also be included in a protocol.

To start the description of error correction codes, we return to the previous description of the parity bits. We have seen that a traditional implementation of error detection relies on the use of parity bits. We may now consider the situation where the parity operation is performed in two dimensions. Let us consider the example depicted in Figure 10.19. In this case, the parity operation is applied independently to each frame (in rows) and, simultaneously, it is applied in columns, to a group of frames. The transmitting data consists of each frame with the corresponding parity bit, by ascending order, followed by the odd parity bits applied in columns.

Considering that the third bit in frame 3 is received corrupted, the receiver is now able to identify the location of such corrupted bit. This is done using triangulation of the row and column parity bits. This example is depicted in Figure 10.20. In this case, the receiver is now able to perform the error correction, instead of only detecting the presence of a corrupted bit.

The price to pay for this additional capability is the reduced code rate R_C* of the error correction, as compared to the code rate of error detection. We have seen that the code rate of the example in Figure 10.13, which consists of error detection, is $R_C = 5/6 = 0.8333$. On the other hand, the code rate corresponding to the error correction depicted in the example of Figure 10.20 comes $R_C = 25/36 = 0.69444$, which translates in a higher overhead associated to error correction than that of the error detection.

---

* Obtained from (10.1).

The error correction codes are generically referred to as forward error correction (FEC). There are two main types of FEC as follows:

- Convolutional codes
- Block codes

Note that in some applications, these two types of error correction codes are combined, originating the so-called concatenated codes.

10.2.1.3.1 Convolutional Codes   Convolutional codes encode an arbitrary group of input bits, using a certain logic function. The parity operation depicted in Figure 10.19 can be seen as a convolutional code. Figure 10.21 shows another example of a convolutional encoder. Since there is no output bit fed back to the input, this belongs to the group of nonrecursive codes [Benedetto 1997].

This example implements the following operations:

$$k_1 = n_1 \oplus n_0 \oplus n_{-1}$$
$$k_2 = n_0 \oplus n_{-1} \tag{10.2}$$
$$k_3 = n_1 \oplus n_0$$

Note that the symbol $\oplus$ in (10.2) stands for modulo 2 adder (XOR). The encoder is normally initialized by filling all registers of the shift register with zero logic state bits. Then, for each data bit that is fed, the shift register shifts once to the right. An important parameter of a convolutional code is its constraint length. This corresponds to the number of previous input bits that a certain output bit depends on. This constraint length equals the number of memory registers.

In the example depicted in Figure 10.21, for each input bit, three output bits are generated using the logic operations. The corresponding code rate is $R_C = 1/3$. The generator polynomial of this convolutional encoder is $G_1 = (1,1,1)$, $G_1 = (0,1,1)$, and $G_3 = (1,1,0)$, and its constraint length is 3. Moreover, since the output sequence does not include the input sequence, this convolutional code is nonsystematic.

Assuming that the first input bit is 1 and that, initially, the registers are fed with 0, the state of the shift registers is (1,0,0), respectively, for (n_1, n_2, n_3) and the output bits are (1,0,1), respectively, for (k_1, k_2, k_3). Assuming that the second input bit is 0, then the corresponding state of the shift registers is (0,1,0), respectively, for (n_1, n_2, n_3) and the output bits are (1,1,1), respectively, for (k_1, k_2, k_3).

Another way of expressing this convolutional code is using the Z transfer function. The Z transfer function of the convolutional encoder depicted in Figure 10.21 is

$$H_1(z) = 1 + z^{-1} + z^{-2}$$
$$H_2(z) = z^{-1} + z^{-2} \tag{10.3}$$
$$H_3(z) = 1 + z^{-1}$$

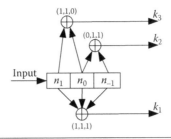

Figure 10.21 Example of a nonrecursive convolutional encoder.

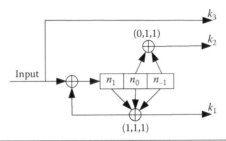

Figure 10.22 Example of a recursive convolutional encoder.

Figure 10.22 shows another example of a convolutional encoder. Since there is an output bit (k_1), which is fed back to the input, this convolutional code is considered recursive. Moreover, since the output bit k_3 corresponds to the input bit, the output sequence includes a replica of the input sequence (among other bits). Consequently, the convolutional encoder depicted in Figure 10.22 is systematic [Benedetto 1997].

The convolutional codes are normally decoded using the Viterbi algorithm.

10.2.1.3.2 Block Codes Block codes encode a fixed group of input bits to generate another fixed group of output bits. They use n input bits to generate a codeword of length k from a certain alphabet. The encoding depicted in Figure 10.10 is an example of a block code. Reed–Solomon codes are among the most used block codes [Benedetto 1997].

10.2.1.3.3 Adaptive Modulation and Coding Adaptive modulation and channel coding (AMC) considers changes to the modulation and coding rate as a function of the link conditions. If a user experiences poor link conditions, his modulation can be changed (e.g., from 16QAM to QPSK, i.e., reducing the modulation order M), reducing the required SNR level to achieve an acceptable BER performance or decreasing the coding rate. The opposite can happen when a user has very good link conditions, increasing the modulation order and/or increasing the coding rate to achieve a higher throughput.

10.2.1.3.4 Interleaving It is known that the bits are normally corrupted in bursts. This is a result of channel impairments such as a deep fading, an impulsive noise, or

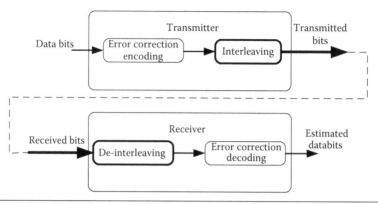

Figure 10.23   Error correction and interleaving.

even an instantaneous strong interference. As previously described for the Hamming distance, the error correction codes are able to correct up to a certain number of bits. Beyond this number, error correction codes are not able to correct those bits. To improve the error correction capability, the error correction is normally associated to interleaving. The interleaving is used to remove the sequential properties of errors and to allow the error correction codes to perform better [Benedetto 1997].

The interleaver is somewhat similar to the scrambler described in Chapter 6. Nevertheless, while the interleaver simply changes the sequential position of the bits, splitting a sequence of corrupted bits into several frames, the scrambler performs a mathematical operation using shift registers at the transmitting side. The de-interleaver located at the receiver side performs the opposite operation, repositioning the bits in the original sequence.

As can be seen from Figure 10.23, the interleaving operation is performed after the error correction-encoding algorithm is applied at the transmitting side, and the de-interleaving is performed before the error correction decoding algorithm is applied at the receiving side.

Let us focus on a certain group of encoded bits,* before interleaving. After the interleaving operation, such group of encoded bits is split among different transmitted frames. The effect of an instantaneous channel impairment will affect a certain transmitted frame, that is, a certain group of interleaved bits. Since the receiver performs the de-interleaving operation, the bits of the affected transmitted frame are split among different codewords, which will then be present at the input of the error correction decoder. Since such group of bits does not consist of a long sequence of corrupted bits anymore, the error correction can easily perform its functionality.

10.2.1.3.5 Puncturing   The puncturing operation can be applied at the output of the error correction encoder, which artificially increases the code rate of an error correction code (see Figure 10.24). This operation is achieved by periodically removing some bits from a codeword at the transmitting side, whereas the receiver inserts zero

---

* Block at the output of the error correction encoder.

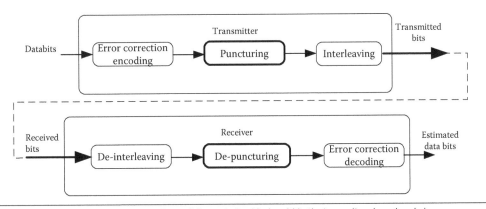

Figure 10.24   Location of the puncturing and depuncturing blocks within the transmit and receive chain.

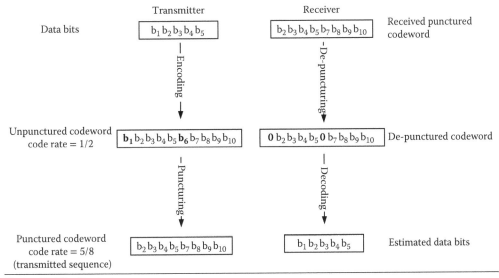

Figure 10.25   Puncturing.

value bits in the corresponding predefined positions [Benedetto 1997]. The puncturing operation should be performed in such a way that the error correction code should be able to correct the inserted zero value bits to the correct logic values.

As can be seen from the example depicted in Figure 10.25, the puncturing operation consists of removing the bits b_1 and b_6 within each ten bits codeword. The resulting punctured codeword is composed of only eight bits, instead of ten. The punctured code rate is increased from $R_C = 1/2$ (before puncturing) into $R_C = 5/8$ (after puncturing). The depuncturing operation, performed by the receiver, adds zero value bits in the positions of the removed bits, while it is expected that the error correction coding is able to correct those inserted bits into the correct logic value.

10.2.2 Automatic Repeat Request

As previously described, error detection is normally used in the transmission of data services (e.g., file transfer, web browsing, etc.) through most of the reliable transmission mediums (e.g., optical fiber, twisted pair, etc.). Moreover, when a frame is detected as corrupted, two possibilities exist: the frame is discarded and the error is handled by a higher layer, or a request for frame retransmission is sent back by the receiving entity to the transmitting entity. The latter procedure is used when the data link layer uses confirmed services.* As previously described, the retransmission procedure may use positive (PAR) or negative (NAK) confirmation, and the frames are numbered to allow accurate acknowledgements and handling of the repeated frames.

The most common data link layer retransmission protocol is the automatic repeat request (ARQ), which consists of a positive confirmation technique. Three different versions of the ARQ protocol exist, as defined in the following subsections.

10.2.2.1 Stop and Wait Automatic Repeat Request   Using the stop and wait ARQ, each frame is separately acknowledged by the receiver, using positive confirmation. When a frame is transmitted, the transmitting entity starts a chronometer and waits for the corresponding acknowledgement from the receiving entity. When a frame is received, the receiving entity uses the redundant bits to check for errors. In case the frame has been received free of errors, the entity sends the corresponding acknowledgement using a feedback channel (positive confirmation). Otherwise, in case the frame has been received corrupted, no signal is sent back to the transmitting entity. In such case, the chronometer of the transmitting entity reaches the timeout and it considers the frame as lost, proceeding with its repetition. In case the acknowledgement is received, the transmitting entity proceeds with the transmission of the following frame. Note that the acknowledgement message sent back by the receiving entity to the transmitting entity may use a dedicated frame or may be piggybacked.† In both cases, the acknowledgement (control information) is sent within the control field.

The example depicted in Figure 10.12 corresponds to the stop and wait ARQ.

As previously described, the frames are numbered to allow the frames and the acknowledgements being correctly identified. Let us suppose an example where an acknowledgement is lost. In such situation, the transmitting entity assumes that the frame was received with errors and repeats its retransmission. If the frames had no numbering, the retransmission would be processed by the receiver as a new one. Nevertheless, since the frame is numbered, the receiver is able to identify that this frame is a repetition of a frame previously correctly received and discards it.

A limitation of this protocol results from the fact that every frame is acknowledged, and the following frame is not transmitted before the previous frame has been correctly acknowledged. This represents an additional overhead and delay. Let us suppose a link using a GEO satellite. In this case, although the transmission rate can be high, the bottleneck is the propagation time that corresponds to approximately 260 ms in each direction. Consequently, even though if a frame consisting of 32 octets

---

* Connection-oriented or connectionless services.
† Sent in the control field of an information frame.

is transmitted at a rate of 256 kbps in 1 millisecond, the transmitting entity stops its transmission and waits for the acknowledgement. The acknowledgement arrives only at the instant $t = 2 \times 260 + 1 = 521$ milliseconds.[*] From this example, we verify that this version of the ARQ protocol corresponds to a very inefficient use of the expensive satellite space segment. Some advancements are achieved by the following versions of the ARQ protocol.

10.2.2.2 Go Back N Automatic Repeat Request The go back N ARQ considers the acknowledgement of a group of frames,[†] instead of each frame independently. This allows a more efficient use of the transmission medium. An important parameter is the selection of the number of successive frames that are transmitted together in a window. In the example of the GEO satellite link, by choosing 521 frames in the group (window) to be transmitted simultaneously, it is assured that the acknowledgement[‡] of the first frame is received at time $t = 2 \times 260 + 1 = 521$ milliseconds. This corresponds to the transmission instant of the 521th frame. Afterwards, the transmitting entity may send another frame (522th frame) and, simultaneously, the acknowledgement of the second frame is received. Consequently, by using the go back N ARQ, the transmitter has been avoided to stop and wait for the reception of the acknowledgement. This corresponds to a more efficient use of the satellite resources.

In the above description, all the frames were assumed free of errors. Nevertheless, in case one frame (within the whole group) is received corrupted, the receiving entity does not send the corresponding acknowledgement and the chronometer of the transmitting entity reaches the timeout. Consequently, such corrupted frame is retransmitted followed by all other frames included in the window, even those transmitted frames that were correctly received. Returning to the example of the GEO satellite link, let us suppose that the first frame is received corrupted. Consequently, since the corresponding acknowledgement is not received, at time instant $t = 2 \times 260 + 1 = 521$ milliseconds, the chronometer of the transmitter reaches the timeout and the first frame is retransmitted followed by all other 520 previously transmitted frames. Note that there is a variation of this protocol where the receiving entity sends a reject (REJ) message in case the frame is received corrupted. Therefore, using the go back N ARQ, the transmitter always sends frames in sequence and the retransmissions are signalized using the timeout or the reject message procedure. Moreover, the receiver discards all frames received after a corrupted one. Consequently, the transmitter must resend the first frame, followed by all other 520 frames that have already been transmitted.[§] In this case, while the transmitter has a window of 521 frames,[¶] since the

---

[*] A 260 milliseconds correspond to the frame propagation time, and another 260 milliseconds corresponds to the acknowledgement propagation time.

[†] Some authors refer to this group of frames as window of frames.

[‡] The acknowledgement is also referred to as "receive ready."

[§] Even though if only the first one out the 521 frames had been received corrupted.

[¶] It is allowed to send a group of 521 frames.

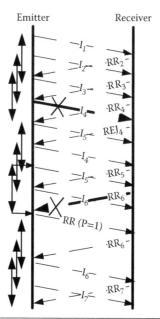

Figure 10.26   Example of go back N automatic repeat request protocol for a transmit window of $N = 3$ using the reject message.

receiver discards all frames after a certain corrupted one, it can be stated that the receiving window is one, whereas the transmitting window is N. The number of N frames in the window is such that it allows the transmitting entity not having to stop the transmission and wait for the reception of an acknowledgement. This is defined by

$$N = \frac{2 \times T_P}{T_F} + 1 \tag{10.4}$$

where T_P stands for the propagation time* and T_F stands for the frame duration† (i.e., frame transmission time).

Figure 10.26 depicts an example of the go back N ARQ protocol using the reject message. Note that instead of explicitly using the acknowledgement n (ACK) message, the receiver sends a receive ready (RR) $n + 1$ message. The meaning is similar, but instead of explicitly signalizing that the nth order frame was correctly received, it signalizes that it is ready to receive the $N + 1$th frame. By sending the RR $n + 1$, the receiver is, implicitly, informing the transmitting entity that all previous frames were correctly received. The use of the receive ready message is directly related to flow control functionality, as defined in the sliding window protocol.

Even if the reject message is used, the chronometer needs to keep being used because the RR message may be lost or corrupted. In case the timeout is reached without having received the RR message, the transmitting entity sends an RR message

---

* Approximately 260 milliseconds in the case of the GEO satellite link.
† 1 millisecond for a frame composed of 32 octets and a transmission rate of 256 kbps.

with the *P* bit active (RR *P* = 1). This corresponds to an interrogation from the transmitting into the receiving entity about which frame is it ready to receive. Then, the receiving entity responds as appropriate. This is depicted in Figure 10.26.

10.2.2.3 Selective Reject Automatic Repeat Request   The selective reject ARQ is similar to the go back N ARQ. In both cases, the transmission window is higher than one, consisting of a value that is a function of the propagation time. Nevertheless, while in the case of the go back N ARQ, the receiving window is one (the receiver discards all frames after a corrupted received frame), in the selective reject ARQ, the receiving window is such that the received frames after a corrupted one are stored in a memory. When the receiver rejects a certain corrupted frame (REJ *n*), the transmitting entity only resends a certain corrupted frame, returning to the normal transmission sequence (i.e., the transmitting entity does not retransmit all frames after a corrupted one).

The transmit window of the selective reject ARQ is $2 \times N + 1$, and the receive window is N. As in the case of the go back N ARQ, the value of N is such that it allows the transmitting entity not having to stop the transmission and to wait for the reception of an acknowledgement. The value N is defined by

$$N = \frac{T_P}{T_F} \tag{10.5}$$

Figure 10.27 shows an example of the selective reject ARQ protocol. The only difference relating to the go back N ARQ protocol (Figure 10.26) relies on the fact that the former only requires that a certain corrupted frame is resent, whereas the go back N protocol requires that all frames after a certain rejected one are retransmitted.

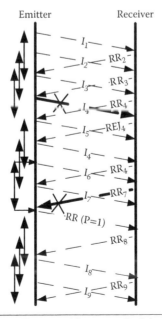

Figure 10.27   Example of selective reject ARQ protocol for a transmit window of $(2 \cdot N + 1) = 3$.

10.2.3 Flow Control Techniques

The transmitter is normally able to transmit more data than the receiver is able to receive and process. To avoid lost of data, the receiver needs to send feedback (control data) to the transmitter informing about whether it is ready to receive more data. This handshaking is known as flow control.

When the error control is based on error detection with positive confirmation (PAR), flow control is automatically associated and performed. In this case, when the receiving entity sends a feedback signal stating that the previously received frame was correctly received (error control), it is also informing the transmitting entity that it is ready to receive another frame. Otherwise, the receiver does not send the acknowledgement until the moment it is ready to receive more data. Although flow control can be jointly performed with error control procedure, it consists of a different functionality.

Naturally, when the error detection is implemented using negative confirmation, since feedback is only sent by the receiving into the transmitting entity in case of errors, the flow control functionality is not implicitly performed. The same applies when error correction is used.

Depending on the adopted ARQ procedure, there are two basic versions of flow control protocol that can be implemented, namely

- Stop and wait
- Sliding window

10.2.3.1 Stop and Wait

This type of flow control is implicitly and automatically performed when the stop and wait ARQ protocol is implemented. It is worth remembering that the ARQ uses positive confirmation. Similar to the corresponding ARQ protocol, this procedure is performed independently for every frame. The transmitter sends a single frame and, when the receiver is ready to receive more data, it sends the receive ready message, as described for the stop and wait ARQ. This can be seen from Figure 10.12, but where the acknowledgement is achieved using the receive ready message.

10.2.3.2 Sliding Window

The sliding window flow control is associated to the go back N ARQ or to the selective reject ARQ protocols. When any of these ARQ protocols are implemented, the sliding window flow control is implicitly implemented.

As previously described, the objective of the transmitting window is to allow the transmitter keep sending data (successive frames), without having to stop and wait for the reception of confirmations. Ideally, if everything occurs as expected, the transmitting entity receives confirmations while still transmitting data. This allows the maximization of the transmission medium usage.

An important aspect of this protocol relies on the fact that the frames need to be numbered. The transmitting window N is calculated by (10.4) for go back N ARQ or by (10.5) for the selective reject ARQ and is a function of the relationship between the propagation time and the frame duration. The window size N corresponds to the

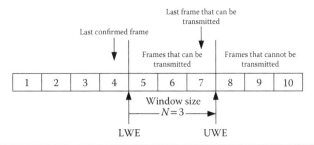

Figure 10.28 Example of sliding window flow control with $N = 3$.

maximum number of frames that can be transmitted without confirmation. If the confirmations are delayed, the transmitter may also have to delay the transmission of the following frames. Nevertheless, if such delay is within the window size, the transmitter has permission to proceed with the transmission of more data. This concept is plotted in Figure 10.28, where the lower window edge (LWE) corresponds to the position of the last confirmed frame and the upper window edge (UWE) is an upper bound for the last frame that can be transmitted. These two parameters are related by $UWE = LWE + N$.

10.3 Link and Network Control Protocols

Before the description of the LLC protocols, it is worth introducing the concepts associated to link control protocol (LCP). Due to their relationship, the concept of network control protocol (NCP) is also introduced in this chapter, instead of in Chapter 11.

NCPs run directly over the data link layer protocol, being used to negotiate options and specific parameters of the network layer (which also runs over the data link layer). Examples of NCP are the PPP Internet protocol control protocol (IPCP*), the Internet protocol version 6 control protocol (IPv6CP) over PPP, the PPP Apple talk control protocol, and so on.

For connection-oriented protocols, the NCP is responsible for performing the connection establishment and the connection termination. In addition, it is responsible for configuring and supporting the operation of the network layer.[†] As can be seen from Figure 10.29, the NCP is only initiated after the LCP of the data link layer has successfully been established. Similarly, the LCP is responsible for establishing and configuring a data link layer, as well as for operating and terminating a link. Different data link layers use different LCP, with different message formats. Note that the authentication[‡] is performed after the establishment and configuration of the LCP, being invoked by this protocol. When authentication is required, the NCP link establishment does not start before authentication process succeeds.

---

* Also referred to as IPv4CP.
† Always necessary for both connection-oriented and connectionless protocols.
‡ Such as the challenge authentication protocol (CHAP) or the password authentication protocol (PAP).

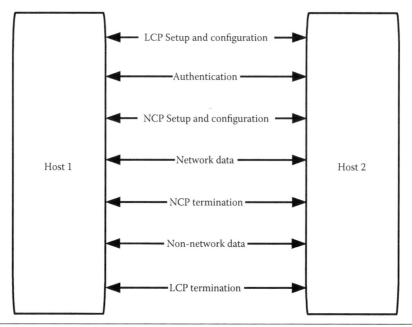

Figure 10.29 Link control protocol and Network control protocol.

The LCP does not have knowledge about the network protocol used to allow the exchange of end-to-end data. This is known, configured, and managed by the NCP. It is important to note that each different data link layer protocol or network layer protocol has as its own NCP. As an example, the IPCP consists of the NCP used by the PPP to configure and manage the IPv4. Changing either the data link layer (e.g., from PPP into HDLC) or the network layer protocol (e.g., from IPv4 into IPv6) results in a different NCP, with a different message format.

Several network protocols can be used* (even simultaneously) over a specific data link layer protocol. Consequently, a different NCP is required for each different network protocol being supported.

The following are examples of messages exchanged by both LCP and NCP protocols:

- Configure request: requests the establishment and configuration of a connection.
- Configure ACK: acceptance of a configure request, accepting all proposed options.
- Configure NAK: some proposed options in the configure request have unacceptable values.
- Configure reject: rejection of a configure request because some options are not recognized.
- Terminate request: requests to terminate a connection.
- Terminate ACK: accepts to terminate a connection.
- Code reject: an invalid or not recognized code was received.

---

* For example, IPv4, IPv6, Apple Talk, and so on.

- Echo request: a message testing the connection. It requests the counterpart node to respond with an echo reply message.
- Echo reply: the response of an echo request message, being used to test the connection.

10.4 Logical Link Control Protocols

LLC protocols are used to either interconnect different networks (WAN and MAN) or to allow the exchange of data within a LAN. While the most common LLC protocol in LAN is the IEEE 802.2, the choice of the type of LLC protocol used to interconnect different networks depends on different parameters, such as the required bandwidth, the quality of service, the cost, the reliability, the availability, and so on. These parameters are also a function of the type of connectivity,* as defined in Chapter 9.

The previous sections described which functionalities are implemented by the LLC sublayer and how these functionalities are implemented. The following subsections describe some of the LLC protocols that are used in LAN, MAN, or WAN networks.

10.4.1 High Level Data Link Control Protocol

The HDLC is a connection-oriented LLC protocol used in permanent circuits, having been standardized by [ISO 3309] and [ISO 4335]. It may establish links in either half- or full-duplex modes, and the links may be point-to-point or point-to-multipoint (one to several hosts). Moreover, it includes the synchronous transfer of data, with assured physical layer synchronism (i.e., clock synchronism). Consequently, a flag is only used at the start and end of the HDLC frame to allow the data link layer synchronism, that is, to achieve the frame synchronization.[†]

The HDLC protocol considers three different types of frames:

- Information (I): it is used to transport upper layer (network layer) information data, with error control[‡] data piggybacked (in the control field).
- Supervisory (S): it is used for error control purposes when piggybacking is not used, that is, when the receiving entity has no information data to transmit.
- Unnumbered (U): it consists of frames not numbered, being used for several functions such as connection establishment or connection termination, as well as to define the mode of the link to be established, as defined in the following.

The HDLC protocol considers three different types of stations:

- Primary: consists of a station with the ability to control the link. It can keep one or more simultaneous links and can send commands.

---

* Permanent circuit, circuit switching, packet switching, and so on.
† To signalize the start and end of a frame.
‡ This includes control data such as receive ready, receive not ready, selective reject, and so on.

Flag	Address	Control	Payload data	CRC	Flag
8 bits	8 or more	8 or 16	Variable	16 or 32	8

Figure 10.30 HDLC frame format.

- Secondary: it cannot send commands. This kind of station is limited to receive commands and to send appropriate response to these commands.
- Combined (or mixed): it can send and receive commands. It can also send the appropriate response to the received commands.

HDLC links can be established in two forms:

- Unbalanced: it is composed of a primary station and one or more secondary stations.
- Balanced: it is composed of two combined stations.

In addition, HDLC links can transfer data in three different modes:

- Normal response mode: it considers an unbalanced link, where a primary station sends commands and the secondary sends the corresponding responses.
- Asynchronous response mode: it is similar to the normal response mode, but where a secondary station may also initiate the data transfer.
- Asynchronous balanced mode: contrarily to the two previous modes, it is composed of a balanced link, where any station may initiate the data transfer. This is the most used HDLC mode of data transfer.

Figure 10.30 shows the frame format used by the HDLC protocol. This frame includes the following fields:

- Flag: consists of the bit pattern 01111110, which is used to signalize the start and end of a frame. To prevent the receiver to incorrectly consider a start or end of a frame, when a sequence of six or more one logic state bits is detected by the transmitting entity in any part of the frame other than the flag, such entity inserts an additional zero logic state bit after the fifth bit.* The receiving entity performs the inverse procedure, that is, when it receives a sequence of five one logic state bits, followed by a zero logic state bit, it removes the zero logic state bit. This procedure is known as bit stuffing.
- Address: it contains the destination address, being composed of one or several groups of eight bits. In case the MSB of each group is a one logic state bit, it indicates that this is the last group of eight bits used by the address field. Alternatively, if the MSB of the group is a zero logic state bit, it indicates that another group of eight bits is sent as address field. In addition, the bit pattern 11111111 corresponds to the broadcast address.

---

* As an example, instead of transmitting 111111111111, the sequence 11111011111011 is transmitted.

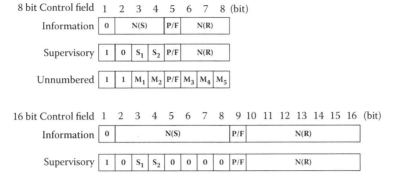

Figure 10.31 Format of the control field of an HDLC frame (8 or 16 bits format).

- Control: it is used to control the connection and can be 8 or 16 bits long. Depending on the type of frame (I, S, or U) and of the control field length, its content varies, as depicted in Figure 10.31. The use of 8 or 16 bits in the control field needs to be negotiated during setup phase (setup extended mode corresponds to using 16 bits in the control field). The 16 bit control field format is used when the sequence of the frame numbering is high (e.g., long window size in the sliding window mechanism). Note that only I and S frames can be 16 bit long (U frames are always 8 bit long). As can be seen from Figure 10.31, some bits are variable, whose meanings are defined as follows:
  - N(S): Sending sequence number. It is used to identify the sequence number of the current transmitted frame.
  - N(R): Receiving sequence number. It is used for control purposes[*] to identify the sequence number following a previously correctly received frame. This number corresponds to the sequence number of the frame that the receiving entity is ready to receive.
  - S_n: Supervisory bits, used for flow control and error control, with the following meanings for S_1 and S_2 bits:
    - 00: receive ready (RR). This corresponds to the acknowledgement (RR followed by the number of the frame N(R) that the entity is ready to receive).
    - 01: Reject (REJ). This is used to reject a frame with errors.
    - 10: Receive not ready (RNR). This is used by the receiving entity when it is not ready to receive more data.
    - 11: Selective reject (SR). This is used by the receiving entity to reject a specific frame using the selective reject ARQ protocol.
  - M_n: Unnumbered bits, used for multipurpose functions, such as setting up and finalizing a connection.
  - P/F: Pool/final bit. When the P bit is active (1), the transmitting entity is querying the receiving entity about which frame is it ready to receive. The F bit active means that it is a response to a previous query (P).

---

[*] In case of I frames, the control is performed in piggybacking mode.

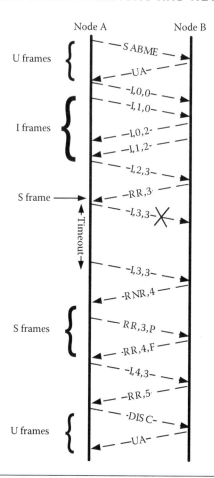

Figure 10.32   Example of HDLC handshaking using the control fields.

- Payload data: it is only used in I frames (information frames) and in some U frames. In case of I frames, it is used to transport network layer data (upper layer).
- CRC: also known as frame check sum (FCS), it consists of 16 or 32 redundant bits used for error detection. The generator polynomials used by the HDLC protocol are one of the following two*:

$$P_{16}(x) = x^{16} + x^{15} + x^2 + 1$$
$$P_{32}(x) = x^{32} + x^{26} + x^{23} + x^{22} + x^{16} + x^{12} + x^{11} + x^{10}$$
$$+ x^8 + x^7 + x^5 + x^4 + x^2 + x + 1 \tag{10.6}$$

Figure 10.32 shows an example of a handshaking performed between node A and node B, using the three types of frames (I, S, and U). The shown messages are those that are properly encoded using the frame control fields. The message SABME sent

---

* $P_{16}(X)$ is the generator polynomial used in CRC16 (16 redundant bits), whereas $P_{32}(X)$ is the generator polynomial used in CRC 32.

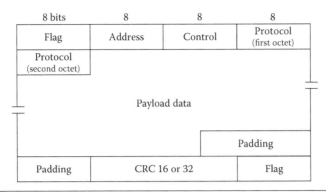

Figure 10.33 PPP frame format.

using a U frame stands for setup asynchronous balanced mode extended (extended corresponds to 16 bit control field). In addition, DISC stands for disconnect and UA for unnumbered acknowledgement.

10.4.2 Point-to-Point Protocol

The PPP is a LLC connection-oriented protocol used in either permanent circuits or circuit switching (e.g., dial-up or xDSL connection), having been standardized by [RFC 1663]. It can establish links in either half- or full-duplex modes, using synchronous or asynchronous mode, being, however, limited to point-to-point connections.

The PPP is the successor of the SLIP, which is a very simple protocol whose functionalities were almost limited to a basic framing, without advanced functionalities. Contrarily, the PPP includes error control, authentication,* compression, encryption, LCP, and NCP capabilities.

The frame format of the PPP is depicted in Figure 10.33.

The PPP frame format was developed taking the HDLC format as a baseline. Some similarities were kept, such as the content of the flag field and the generator polynomials used by CRC. The PPP frame includes the following fields:

- Flag: consists of the bits pattern 01111110, which is used to signalize the start and end of a frame.
- Address: since the PPP is only used in point-to-point, the address field is fixed to the bit pattern 11111111 (FF in hexadecimal notation). When compression is used, this field is omitted.
- Control: this field is fixed to the bit pattern 00000011 (03 in hexadecimal notation). When compression is used, this field is omitted.
- Protocol: it identifies the type of protocol being handled by the frame. Possible protocols being handled by the frame include LCP, NCP, authentication, encryption control protocol (ECP), compression control protocol (CCP), and so on.

---

* Namely the PAP and the Challenge Handshaking Protocol (CHAP). The reader should refer to Chapter 14 for the description of these protocols.

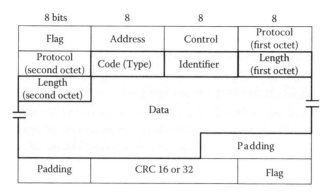

Figure.10.34   Payload field used by the protocols being managed by frames.

- Payload data: used to allow the exchange of upper layer data or the exchange of variables relating to the protocol being handled by the frame (e.g., LCP, NCP, ECP, CCP). In the latter case, some of the payload data is used to transport the following fields (see Figure 10.34):
  - Code (type): identifies the type of message being exchanged.
  - Identifier: a response includes a copy of the identifier used in the corresponding query.
  - Length: consists of the length of the code, identifier, length field, and data field, expressed in octets.
  - Data: specific information related to the message being exchanged, such as options being negotiated.
- Padding: used to ensure that the PPP frame size is an integer number of 32-bit words.
- CRC: also known as FCS, it consists of 16 or 32 redundant bits used for error detection. The PPP uses the same generator polynomials as those considered by the HDLC protocol (defined by (10.6)).

10.4.3 IEEE 802.2 Protocol

The Institute of Electrical and Electronics Engineering created the IEEE 802 committee, whose objective relied on the creation of LAN and MAN standards to assure interoperability among different existing technologies (Token Ring, Ethernet, etc.).

To achieve this goal, the IEEE 802 committee created several subcommittees. As can be seen from Figure 10.35, the IEEE 802.2 subcommittee was one of these, whose objective was the standardization of a protocol for the LLC sublayer. Such resulting protocol has the name of the subcommittee that created it, that is, [IEEE 802.2], having been adopted by ISO/IEC and renamed as [ISO/IEC 8802-2:1998]. Moreover, other subcommittees were created for the standardization of different MAC sublayer technologies used in LAN or MAN networks. As can be seen from Figure 10.35, important LAN/MAN sublayer standards include

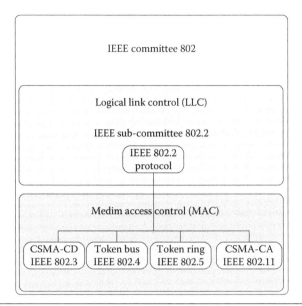

Figure 10.35   IEEE 802 committee and some relating data link layer subcommittees.

- IEEE 802.3—CSMA-CD
- IEEE 802.4—Token bus
- IEEE 802.5—Token ring
- IEEE 802.11—CSMA-CA

Note that the different LAN/MAN subcommittees were dependent on the LLC subcommittee (IEEE 802.2 subcommittee). Moreover, in addition to the subcommittees here described, others were created. Nevertheless, we only focus on those with interest for the current description.

An important note that is worth mentioning is the fact that all different IEEE LAN/MAN protocols (IEEE 802.3, 802.4, 802.5, 802.11, etc.) make use of the same LLC protocol, which is responsible for performing error control and flow control: the IEEE 802.2 protocol.

The LLC sublayer based on the IEEE 802.2 protocol data unit (PDU*) transports, in the payload data field, the packets generated by the network layer (upper layer). Moreover, the PDUs are transported in the frame payload data fields of the MAC sublayers (IEEE 802.3, 802.4, 802.5, 802.11, etc.). As can be seen from Figure 10.36, the IEEE 802.2 PDU is similar to that of the HDLC, being composed of four different fields:

- Destination service access point (DSAP): consist of a one octet address that identifies the LLC destination.
- Source service access point (SSAP): consist of a one octet address that identifies the LLC source.

---

* A PDU is the message format of the LLC sublayer.

DSAP	SSAP	Control	Payload data
1 octet	1	1–2	Variable (minimum 46 octets)

Figure 10.36   IEEE 802.2 protocol data unit format

- Control: consist of one or two octet field used to allow the handshaking associated to error control and flow control. It is worth noting that the redundant bits for error control using CRC are included as part of the MAC subayer (not part of the LLC PDU).
- Payload data: consists of a variable length field, carrying upper layer packets, with a minimum size of 46 octets. In case the packet size is lower than that, the payload data is zero padded to this value.

Depending on the type of handshaking performed by the IEEE 802.2 LLC, it may provide the following type of services to the network layer:

- Type 1: connectionless and nonconfirmed service. In this case, error control and flow control is not provided.
- Type 2: connection-oriented service. In this case, the protocol sliding window is used to implement the flow control, whereas the protocol go back N ARQ assures that the frames are delivered free of errors. Moreover, the duplication of frames is avoided and the correct sequence of frames is guaranteed.
- Type 3: connectionless but confirmed service. In this case, the protocol stop and wait is used to provide error control and flow control. Nevertheless, this protocol is not as efficient as the connection-oriented one (type 2).

10.5 Medium Access Control Sublayer

As previously described, this sublayer is responsible for managing and controlling the access of hosts into a common transmission medium, within a LAN or a MAN. Depending on the type of management, the access control can be of synchronous or asynchronous types.

Synchronous type of access control mechanisms uses rigid allocation of bandwidth to hosts, regardless these hosts use or not such bandwidth. Example of synchronous access control mechanisms includes the FDMA, the TDMA, the CDMA, and so on. GSM networks make use of TDMA access control mechanism. In this case, a frequency carrier is split into eight time slots, one for carrying data relating to a user's communication in one direction.

Contrarily, asynchronous type of access control mechanisms allows the bandwidth being allocated as a function of the users' need and of the network available capacity. The asynchronous allocation of resources can be performed using different methods:

- Demand assignment multiple access (DAMA): similar to synchronous mechanisms, it is also based on FDMA, TDMA, or CDMA. However, the number of frequency carriers, time slots, or code sequences is dynamically allocated, as a function of the users' need and of the network available capacity. This requires a centralized management of the resources.

- Round robin: this access control mechanism is based on a token that circulates in the network in a certain direction. When a host receives the token, it is allowed to send data to the shared medium for a certain maximum period of time. After such period, the host must stop the transmission and the token is forward to its adjacent host. In the next round, the host may continue its transmission of data. After having received the token, in case the host has no data to transmit, it passes the token immediately to the adjacent host, without consuming a rigid time period. The advantage of the round robin mechanism relating to the TDMA relies on the fact that if a host does not need to use the resources (has no data to transmit), it does not consume a fixed bandwidth. Example of protocol that makes use of this access control mechanism includes the token ring and the token bus.

- Contention: this distributed access control mechanism is based on the ability of any station to start transmitting in case the shared medium is idle (free). In case the channel is busy (being used by another station), it waits until the channel becomes idle, after which it starts transmitting. This mechanism is very simple and works well with low and medium traffic conditions. Since the probability of collision increases exponentially with the increase of the traffic level, this mechanism tends to collapse under high traffic conditions. Examples of contention access control mechanisms include the CSMA-CD* or the carrier sense multiple access with collision avoidance (CSMA-CA†).

10.6 Medium Access Control Protocols

At the beginning of computer networks, there were only proprietary protocols developed by companies, and the shared market was the metric of success. Among the existing important LAN/MAN protocols, there were the token ring developed by IBM and the Ethernet created by the DIX consortium (**D**igital, **I**ntel, and **X**erox). Since these protocols were not standardized and were incompatible, the IEEE 802 committee created several subcommittees (see Figure 10.35). Each subcommittee became responsible for the standardization of each relevant LAN/MAN protocol. This was the solution to accommodate all important proprietary protocols, each one with some advantages and disadvantages, instead of standardizing a single protocol, while leaving all the others out.

As can be seen from Figure 10.35, the IEEE created a group of subcommittees for the creation of different MAC sublayer standards. However, these standards are directly

---

* The CSMA-CD is the contention mechanism used by the IEEE 802.3 protocol.
† The CSMA-CA is the contention mechanism used by the IEEE 802.11 protocol.

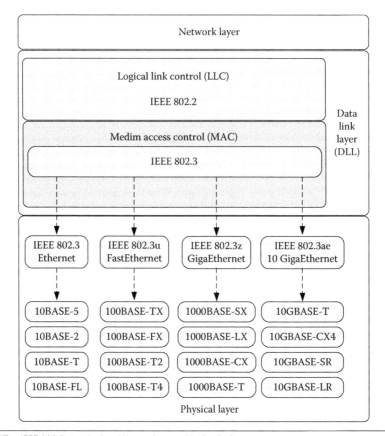

Figure 10.37 IEEE 802.3 standard and its versions and technologies.

related to the physical layer used to allow its implementation (10BASET, 10GBASEFX, etc.). An IEEE 802.3 network includes the definition of the access control to the shared medium, the frame format, and so on. Moreover, the corresponding physical layer includes the transmission rate, the type of cable, the maximum segment size, the type of line encoding technique (e.g., Manchester, non return to zero, etc.), the type of connectors, and so on. As can be seen from Figure 10.37, these different physical parameters vary depending on the IEEE 802.3 version. 10BASET and 10GBASEFX are two examples of physical layer technologies used to implement an IEEE 802.3 network.

The following subsections describe the most important LAN/MAN IEEE standards.

10.6.1 IEEE 802.3 Protocol

Ethernet technology, initially developed and commercialized by the DIX consortium, was the basis for the development of the IEEE 802.3 standard. Nevertheless, there are small differences between these two namely small variations of the content of the different frame fields. The current description focuses on the IEEE 802.3, which is the currently mostly implemented standard [IEEE 802.3].

Preamble	S F D	Destination address	Source address	Type Size	Payload data	CRC
7 octets	1	6	6	2	46–1500 (variable)	4

Figure 10.38 IEEE 802.3 frame format.

The IEEE 802.3 is a standard responsible for the access control to the shared medium based on a contention mechanism. In addition, the physical layer is also defined. It receives data from the upper sublayer (IEEE 802.2 LLC) and encapsulates it into the payload data field of the IEEE 802.3 MAC frame, followed by the addition of the corresponding frame header. Then, the frame is broadcasted in the network, which consists of a shared transmission medium.*

Figure 10.38 depicts the IEEE 802.3 frame format, with the indication of different fields and corresponding sizes. The frame includes the following fields:

- Preamble: it consists of a flag used to allow the synchronization of the receiver with the transmitter at both physical and data link layer (i.e., at bit and frame synchronization). It is composed of 56 bits (7 octets) consisting of alternating 0 and 1 logic state bits.
- Start of frame delimiter (SFD): it consists of eight bits (1 octet) with the logic state bits "10101011." It is used to signalize the start of a frame. Moreover, the SFD, together with the preamble, are used to allow the receiver to synchronize with the transmitter.
- Destination address: it consists of the physical address of the destination Network Interface Card (NIC). The physical address is also referred to as hardware address or as MAC address. It is composed of six octets (48 bits), with the following decomposition:
  - The initial 24 bits are called Organizationally Unique Identifier (OUI), being globally allocated by the IEEE to the organization or hardware manufacturer.
  - The final 24 bits are internally managed and allocated by the organization or manufacturer in a way to avoid duplications.
- Source address: it consists of the physical address of the source NIC with the structure defined for the destination address.
- Type/Size: it consists of a two octet field used to identify either the frame size or the frame type:
  - If the field content is a decimal number lower than 1500, such number represents the number of octets transported in the payload data field of the frame.
  - If the field content is a decimal number higher than 1536, the field is used to represent the type of layer 3 data transported in the payload data field of the frame.

---

* The IEEE 802.3 standard assumes a hub as the central node of the network. This device makes the transmission medium a shared resource.

- Payload data: it is used to transport LLC data (and the LLC data is used to transport layer 3 data, i.e., packets*). In case its content has a size lower than 46 octets, the upper sublayer (IEEE 802.2) performs zero padding to the PDU up to this value. Zero padding is the action of adding zeros to assure that a field presents a certain minimum size.

- CRC: also referred to as FCS, it consists of redundant bits transmitted with the data to allow the receiver to check the integrity of the frame using the cyclic redundancy check. The error verification is applicable to the whole frame except the preamble, SFD, and CRC. Note that the LLC protocol (upper sublayer) does not make use of any error control technique. It lets the MAC sublayer verify the errors. In case the MAC sublayer detects an error, it informs the LLC sublayer, which is then responsible for requesting the repetition of the PDU. The LLC sublayer implements the handshaking necessary to allow the flow control and error control (receive ready, reject, etc.). The generator polynomial used by the IEEE 802.3 standard is†

$$P(x) = x^{32} + x^{26} + x^{23} + x^{22} + x^{16} + x^{12} + x^{11}$$
$$+ x^{10} + x^{8} + x^{7} + x^{5} + x^{4} + x^{2} + x + 1 \tag{10.7}$$

The contention mechanism used by the IEEE 802.3 standard is the CSMA-CD, defined as follows (see Figure 10.39):

- A host that intends to initiate a frame transmission within the network starts by listening to the channel (carrier sense).

- If the shared medium is idle (i.e., if there is not any host transmitting), the host may initiate the transmission. Otherwise, if another host is transmitting (medium is busy), it waits for a back-off period of time. After this period, the host rechecks the channel availability to start the transmission.

- Although the channel could be idle and a host could have started the transmission, another host could have started the transmission in a very close moment. This causes a collision.‡ To assure that such collisions are detected, a host needs to keep listening to the medium (receiving) while transmitting. If a host detects a collision, the host must proceed with the transmission such that the minimum frame size (512 bit) is transmitted and such that the transmission is finalized

---

* In case of the IEEE 802.3 standard, the maximum transmission unit (MTU) is 1500 octets. This corresponds to the maximum payload size of a frame, which also corresponds to the maximum size of a packet (layer 3 data), added of the LLC overhead. As previously described, in case the packet size is higher than 1500 octets, fragmentation is implemented by a router at the border of an IEEE 802.3 network.

† This is the same generator polynomial used by the HDLC protocol in 32 bits mode.

‡ The reason for the hosts not to, initially, detect each other, results from the fact that the network has a certain length, which corresponds a certain period of time for the propagation of the electric signals. When a host starts the transmission, the generated electric signals need to propagate along the transmission line to achieve another space location within the physical network.

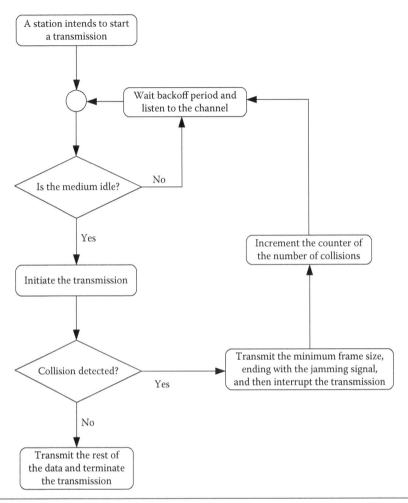

Figure 10.39 CSMA-CD contention mechanism.

with the jamming signal.* Then, the host must interrupt its transmission. This is plotted in Figure 10.40 for the case where a collision has been detected at the start of the frame (within the initial 44 payload octets). Otherwise, the host transmits immediately the jamming signal and then interrupts the transmission. This procedure ensures that all receivers detect a collision by checking the received corrupted CRC, including those located in the other physical extreme of the network. After the jamming signal, both hosts must interrupt their transmissions and try to restart after a listening strategy defined as follows:

- Nonpersistent: if the medium is idle, transmit immediately. If the medium is busy, wait a back-off period† and listen again to the medium. Note that

---

* The jamming signal consists of a number of random bits (six octets), transmitted at the end of the frame, instead of the four CRC octets. When a receiving host verifies the CRC, it realizes that an error has occurred and discards the frame.

† The back-off period is defined at the end of Section 10.6.1.

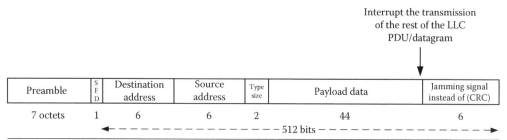

Figure 10.40   Transmitted frame in case a collision is detected within the initial 44 payload octets.

it may happen that the medium is already idle, but the host is still waiting a random period.

- 1-persistent: if the medium is idle, transmit immediately. If the medium is busy, keep waiting and listening to the medium. Then, when it becomes idle, transmit immediately. If more than a single host is waiting for the medium to be idle, the transmissions of these waiting hosts will certainly collide. This represents a disadvantage of the 1-persistent scheme.
- P-persistent: if the medium is idle, the transmit probability is p, while the probability to delay one time unit* is $(1 - p)$. If the medium is busy, wait for being idle and then use the previous procedure. If the transmission is delayed, use the previous procedure after the waiting time. The value of p varies, but typical values are between 0.5 and 0.75.

The back-off period used in IEEE 802.3 is referred to as binary exponential back-off. It consists of a random waiting period, whose mean value is doubled during the initial ten retransmission attempts. The mean value of the random waiting period is kept unchanged for more six additional retransmission attempts. In case 16 unsuccessful collisions occur, the host gives up attempting and sends an error to the upper layer. Then, the cycle restarts after a larger back-off (this value increases with the increase of the number of groups of 16 attempts).

10.6.1.1 Maximum Collision Domain Diameter   The maximum collision domain diameter (MCDD)[†] is defined as the distance between the two furthest network nodes. From this parameter we may calculate the total time it takes for the shortest frame (minimum frame size) to travel round trip between the two furthest network nodes in that domain. Note that this time takes into consideration the round trip time (double) to allow the collision (interfering) signal returning to the transmitting source.

---

* This value approximates the propagation delay.

[†] A collision domain consists is experienced in a LAN/MAN which uses a hub as a central node and a contention access mechanism (CSMA-CD or CSMA-CA).

The minimum frame size of an IEEE 802.3 network (at 10 and 100 Mbps) is 512 bit.[*] The period of time necessary to transmit this minimum frame is referred to as slot time. The slot time corresponds to the maximum period of time where a collision can be detected. Beyond such slot time, a collision never occurs, as there is enough time for a certain transmission to reach and being detected by all other network nodes. If the minimum frame size were shorter, the MAC sublayer could not detect a collision, and the collision/error would have to be handled by another layer (e.g., by the TCP) or the data would be corrupted. This is avoided by defining the maximum collision domain diameter.

The maximum collision domain diameter consists of the maximum length (diameter) of a domain, where a collision can be handled by the MAC sublayer. It is dimensioned for half of the slot time period (round trip), resulting in 256 bits duration. Nevertheless, to take into account additional delay in network devices (e.g., repeaters) the one-way calculations are performed for a total of 232 bits.

Considering a 10 Mbps IEEE 802.3 network, the time for transmitting 232 bits comes

$$T = \frac{\text{number of bits}}{\text{bit rate}} = \frac{232}{10^7} = 23,2 \ \mu s \tag{10.8}$$

From physics, the MCDD comes

$$MCDD = v \cdot T \tag{10.9}$$

where v is the propagation speed of electric signals in the transmission medium (coaxial cable, twisted pair, optical fiber) and T is the result of (10.8). Taking into account the approximate propagation speed in the coaxial cable of $v = 1,22 \cdot 10^8$ m/s (note that the light speed in the vacuom is $v = 3 \cdot 10^8$ m/s), from (10.9) the MCDD becomes 2800 meters.

Using the same principle, the time for the transmission of 232 bits in a 100 Mbps IEEE 802.3 network comes $T = 2,32 \ \mu s$, and the MCDD results in approximately 280 meters for a coaxial cable (205 meters for a twisted pair).

Normally, the maximum network size is constrained by the maximum segment size,[†] instead of the MCDD. This is the case of a 10BASET network, where the maximum segment size is limited to 100 meters, while the MCDD is 280 meters. Nevertheless, for speeds of 1 Gbps or above, the MCDD becomes the constraint. In this case, a mechanism called carrier extension is used to overcome such limitation. The carrier extension consists of a number of bits transmitted after and together with the regular frame, when its length is shorter than 4096 bits (see Figure 10.41). The objective is to assure that the resulting frame length (original frame plus carrier extension) is a minimum of 4096 bits long (used in the calculation of the 1 Gbps IEEE 802.3 slot

---

[*] This is calculated from Figure 10.38, taking into account the minimum number of bits after the start of frame delimiter.

[†] Note that the maximum segment size depends on the attenuation and distortion, among other factors.

Figure 10.41   Carrier extension.

Figure 10.42   Frame bursting.

time). Naturally, in case the frame length is longer than 4096 bits, the carrier extension is not added. Note that the minimum frame size is kept as 512 bits long, and thus, interoperability with 10 and 100 Mbps IEEE 802.3 networks is assured. The resulting extension of the slot time from 512 bits to 4096 bits extends the MCDD of the 1 Gbps IEEE 802.3 standard from around 20 meters up to around 200 meters.

An alternative solution that can be adopted by 1 Gbps or faster IEEE 802.3 networks is the frame bursting procedure. Instead of spending bandwidth at transmitting the carrier extension, if the transmitting host has several short frames to transmit, these frames can be sent together, linked with the interframe gap (IFG). The IFG consists of a predefined bit pattern. This improves the network performance, relating to the carrier extension use. Figure 10.42 depicts the frame bursting procedure. Note that the initial frame is always sent using the carrier extension, whereas the others can be transmitted using the frame bursting procedure. The maximum length of the linked short frames is limited to 56,536 bits. The use of the frame bursting results in a higher minimum frame size, which translates in a higher MCDD.

The above described CSMA-CD mechanism is applicable to a network that makes use of a linear hub, which translates in a shared transmission medium. The hub is a repeater, which repeats in every outputs the bits present at one of the inputs. Consequently, the medium becomes shared and the CSMA-CD is applicable. Nevertheless, in case the central node of the network is a switch (or a bridge), instead of a hub, since this device only switches the frames to the output where the host with the frame's destination MAC address is located, the effect of the CSMA-CD is neglected. In fact, the CSMA-CD is also applicable but, since the use of the switch makes the transmission medium not shared, the effect of the CSMA-CD is neglected.

In any case, note that the IEEE 802.3 is a standard defined for the worse case scenario, which is the case of a network with a hub as a central node. The use of a switch is a modification of the IEEE 802.3 standard, where the MAC contention mechanism based on the CSMA-CD is not followed. Moreover, in case a switch is used in a 1 Gbps or faster IEEE 802.3 network, the carrier extension and frame bursting do not need to be used.

The CSMA-CD facilitates the access to the shared medium as long as the traffic rate is below 40–50%. Beyond this threshold, the performance of the CSMA-CD degrades heavily.

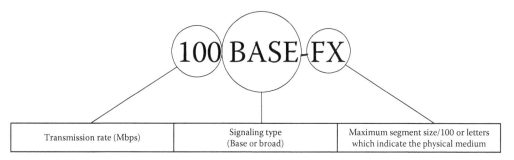

Figure 10.43 Identification of IEEE 802.3 technologies used in the physical layer.

Table 10.1 Physical Characteristics Used in IEEE 802.3 (10 Mbps)

	IEEE 802.3–10 MBIT/S (ETHERNET)			
	10BASE-5	10BASE-2	10BASE-T	10BASE-FX
Cabling	12 mm coaxial cable (RG8/RG11)	6 mm coaxial cable (RG58)	UTP Cat.3, 4, or 5 (4 pairs)	Multimode fiber (62.5/125 mm) (2 fibers for full duplex)
Physical topology	Bus	Bus	Star	Star
Encoding	Manchester	Manchester	Manchester	Manchester
MSS (meters)	500 (2500 maximum with up to 4 repeaters)	185 (300 if repeaters are not employed)	100	2000
Type of connectors	BNC	BNC	RJ45	ST or SC connectors

10.6.1.2 Physical Layer Used in IEEE 802.3 Networks As can be seen from Figure 10.37, there are several physical layer implementations of an IEEE 802.3 network. An example of a physical implementation is the 100BASE-FX, whose meaning is described in Figure 10.43. The first group of digits (100) refers to the transmission rate, while the second group (BASE or BROAD) refers to the signaling type (baseband or broadband). Finally, the last group of letters or figures refers to the maximum segment size divided by 100 (e.g., 5 from 10BASE-5 stands for the maximum segment size of 500 meters) or to the transmission medium (FX from 100BASE-FX stands for optical fiber).

Tables 10.1 through 10.4 present the physical characteristics of the IEEE 802.3 at 10 Mbps, 100 Mbps, 1 Gbps, and 10 Gbps, including parameters such as the type of cabling used, the physical topology, the line encoding technique, the maximum segment size (MSS), and the type of connectors used.

Naturally, increased speeds have been achieved with latest standards. Optical fiber cables have played an important contribution to achieving higher speeds and higher ranges without the need to use repeaters/regenerators (i.e., increased maximum segment size). From the two types of optical fiber cables, the single mode is the one that achieves the best performance. Another important parameter that allows achieving higher transmission rates is the use of more efficient line coding techniques, as well as the combination of the multiple pairs for transmission or receiving data.*

---

* As an example, the 1000BASE-T allows transmitting at a speed of 1 Gbps using four UTP Cat.5 pairs. These pairs are all used for transmit, all for receive, or some pairs for transmit and others for receive.

Table 10.2   Physical Characteristics Used in IEEE 802.3u (100 Mbps)

	IEEE 802.3U–100 MBIT/S (FASTETHERNET)			
	100BASE-T		100BASE-X	
DESIGNATION	100BASE-T2	100BASE-T4	100BASE-TX	100BASE-FX
Cabling	UTP Cat.3, 4, or 5 (2 pairs)	UTP Cat.3 (4 pairs: one for Tx, another for Rx, two pairs are negotiated)	STP or UTP Cat.5 (2 pairs)	Multimode fiber (62.5/125 mm) (2 fibers for full duplex)
Physical topology	Star	Star	Star	Star
Encoding	Manchester	8B/6T	4B/5B with scrambling	4B/5B
MSS (meters)	100	100	100	412
Type of connectors	RJ45	RJ45	RJ45	ST or SC connectors

Table 10.3   Physical Characteristics Used in IEEE 802.3z (1 Gbps)

	IEEE 802.3Z–1000 MBIT/S (GIGABITETHERNET)			
DESIGNATION	1000BASE-X			
	1000BASE-CX	1000BASE-LX	1000BASE-SX	1000BASE-T
Cabling	STP (two pairs)	Multimode fiber (62.5/125 mm) (2 fibers)	Single mode optical fiber (2 fibers)	UTP Cat.5 (four pairs negotiated for Tx or Rx)
Physical topology	Star	Star	Star	Star
Encoding	8B/10B	8B/10B	8B/10B	4D-PAM5 with scrambling and FEC
MSS (meters)	25	220	5000	100 (200 with repeater)
Type of connectors	DB9 or HSSDC	ST or SC connectors	ST connectors	RJ45

Table 10.4   Physical Characteristics Used in IEEE 802.3ae (10 Gbps)

	IEEE 802.3AE–10 GBIT/S		
	10GBASE-T	10GBASE-SR	10GBASE-ER
Cabling	UTP Cat.5 or better (four pairs)	Multimode fiber (62.5/125 mm) (2 fibers)	Single mode optical fiber (2 fibers)
Physical topology	Star	Star	Star
Encoding	4D-PAM10 with scrambling and FEC	64B/66B	64B/66B
MSS (meters)	100	300	40000
Type of connectors	RJ45	ST or SC	ST

It is important referring that the channel impairments tend to increase with the increase of the transmission rates. This tends to cause a degradation in the bit error rate (BER) performance. The use of error detection mechanisms makes the receiver to keep requesting repetitions of frames. To avoid the resulting increased overhead and delay, the high transmission rates made over transmission mediums of low performance (e.g., twisted pairs) is normally associated to FEC (error correction) and scrambling (to improve the signal quality).

Finally, it is worth noting that the IEEE 802.3ae (10 Gbps) can be used in both LAN and MAN networks. The 10GBASE-ER* can be used as a MAN, as an alternative to SDH, ATM, or even MPLS.

10.6.2 IEEE 802.5 Protocol

The token ring protocol, initially developed and commercialized by IBM, was the foundation for the standardization of the [IEEE 802.5], having been adopted by ISO/IEC and redesignated as [ISO/IEC 8802-5:1998]. Nevertheless, there are small differences between the token ring and the IEEE 802.5 standard, namely small variations on the content of the different frame fields. The current description focuses on the IEEE 802.5 standard, as more adopted in LAN than the token ring.

The IEEE 802.5 consists of a MAC sublayer standard, which defines a MAC mechanism based on the above described round robin, as well as the physical layer. This sublayer receives data from the upper sublayer (IEEE 802.2 LLC) and encapsulates it into the payload of the IEEE 802.5 MAC frame, followed by the addition of the corresponding frame header. Then, the frame is transmitted over the LAN.

The IEEE 802.5 standard considers the transmission speed of 4 Mbps using UTP cabling or 16 Mbps with STP cabling. Currently, the speed of 100 Mbps is already possible. The differential Manchester line encoding technique is used in IEEE 802.5 standard.

The logical topology comprised by the IEEE 802.5 standard is the ring, where a token circulates in a certain direction. When a host receives the token, it is allowed to send data to the common transmission medium for a certain maximum period of time. After such period, the host must interrupt the transmission and the token is forward to the adjacent host. In the next round, the host may continue its transmission. After having received the token, in case the host has no data to transmit, it immediately forwards the token to the following adjacent host, without consuming a rigid time period. This represents an advantage relating to the TDMA, where fixed time slots are allocated to different users, regardless they have or not data to transmit.

An important advantage of this standard relating to the IEEE 802.3 relies on its ability to assure a maximum delay for a host to transmit a frame. This maximum delay corresponds to the maximum time necessary for a token to circulate in the ring. This characteristic is especially important in case of high traffic load, where the successive collisions experienced in IEEE 802.3 networks tend to collapse it or to insert a delay higher than that acceptable for the provision of quality of service (e.g., for IP telephony or video streaming). In addition, IEEE 802.5 networks allow prioritization of traffic, which is another advantage as compared to IEEE 802.3 networks. On the other hand, The IEEE 802.3 is much simpler than the IEEE 802.5, without the need to have a

---

* ER stands for extended range.

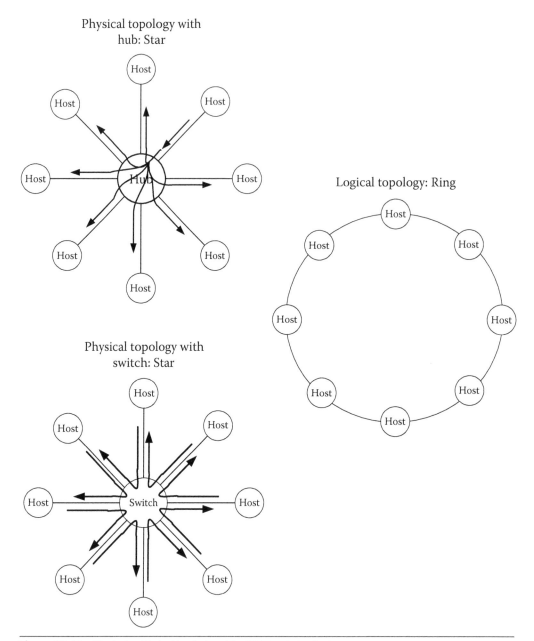

Figure 10.44   Physical and logical topologies of the IEEE 802.5 standard.

management station responsible for the control of the token. In any case, the IEEE 802.3 tends to be the preferable standard due to its simplicity and due to its good performance under low traffic load conditions.

Albeit the ring is the considered logical topology of the IEEE 802.5, the physical topology is the star, where a hub or a switch is used as a central node. This can be seen from Figure 10.44. In case the central node is a hub, the token and the frames are sent to all stations (common transmission medium) but only the next adjacent host processes the data. Note that it was previously described that a network that makes use of a hub as a central node has a logical bus topology. Nevertheless, since the IEEE 802.5 standard uses a token to control the access to the shared medium, the logical topology is transformed into a ring. In this case, although a token is used to implement the round robin access mechanism, the hub makes the medium a diffusion channel.* The delay inserted in the signal by a hub corresponds to one bit period. This is the time necessary to implement the regeneration of bits.

Contrarily, in case a switch is used, the transmission medium is not shared, and this device only forwards frames and tokens from a host to its adjacent, following a certain direction of the logical ring. In this case, the channel is not of diffusion type. The delay inserted by a switch in the signal depends on its type (store-and-forward, cut-through, etc.).

The operation of the MAC sublayer defined by the IEEE 802.5 standard is as follows:

- A token circulates in the network, from host to host, in a certain direction of the ring.
- If a certain host has data to transmit, when the token is received, the host keeps it and transmits data for a maximum period of time corresponding to the token holding period (THP). The default THP value is 10 ms[†] During this period, the host can transmit one or more frames. Note that a host is only allowed to transmit data, in case the priority of the frame is higher than the priority of the token.[‡] This priority management represents a great advantage of IEEE 802.5 relating to IEEE 802.3 networks.
- A host listens to the channel, accepting the frames whose destination MAC address is itself. Otherwise, the frames are forward to the adjacent host.
- The frames circulate from host to host, around the ring, being copied by the destination, and removed by the host that generated it, after one round.
- Once a host terminates the transmission of data (or in case it has no data to transmit), it immediately forwards the token to its adjacent host of the ring.

The IEEE 802.5 standard requires a host to act as a management station. Any host of the network may act as a management station. Its functions include the following:

- Frame removal, in case the host that generated the frame did not remove it after one round.

---

* Similarly to the IEEE 802.3 standard, with a hub as a central node.
† IEEE 802.5 networks working at 4 Mbps may transmit approximately one frame with the maximum load (4500 octets), or several lower length frames, in one THP period. At a speed of 16 Mbps, more frames are allowed in each THP period.
‡ The priority of the token is defined by the management station, based on the priority of the traffic that the different hosts have to transmit.

- Generation and insertion of the token into the ring, as well as its monitoring. Moreover, the management station is also responsible for defining the priority level of the token.
- Detection of lost tokens. Due to channel impairments, a token may lose its sense. In this case, a new token has to be inserted, while the old one must be removed.

Figure 10.45 depicts the format of an IEEE 802.5 frame, while Figure 10.46 depicts the format of an IEEE 802.5 token.

The content of the frame fields are described in the following:

- Start delimiter (SD): it delimits the start of the frame, using a violation of the differential Manchester line coding technique.
- Access control (AC): it contains priority bits to distingue between a frame and a token. Moreover, it also contains priority bits and other bits used for monitoring purposes.
- Frame control (FC): it indicates the type of data transported in the payload data field.
- Destination address (DA): it consists of the address of the destination host.
- Source address (SA): it consists of the address of the source host.
- Payload data: it transports IEEE 802.2 PDU. The maximum length of the frame payload data is 4500 octets.
- CRC: also referred to as FCS, it uses the same CRC function as the one defined for IEEE 802.3 (defined by (10.7)), applied to all frame fields, except the SD, CRC, ED, and FS.
- Ending delimiter (ED): it indicates whether a frame is the last one of a group of frames or an interim frame within such group. Moreover, it gives an indication about when a CRC error is detected in a frame.
- Frame status (FS): consists of a group of bits that are changed by the destination host when it is copied. It is based on these bits that either the transmitting host or, alternatively, the management station removes the frame from the ring.

S D	A C	F C	Destination address	Source address	Payload data	CRC	E D	F S
1	1	1	2–6 octets	2–6	46–4500 (variable)	4	1	1

Figure 10.45 IEEE 802.5 frame format.

	S D	A C	E D	
	1	1	1	octet

Figure 10.46 IEEE 802.5 token format.

← SFS →							◄EFS►	
PA	S D	F C	Destination address	Source address	Payload data	CRC	E D	FS
8 octets	1	1	6	6	0–4500 (variable)	4	1	1,5

Figure 10.47 Fiber distribution data interface frame format.

PA	S D	F C	E D
8 octets	1	1	1

Figure 10.48 Fiber distribution data interface token format.

Note that the three fields contained in the token are also present in the frame, presenting the same meanings.

10.6.3 Fiber Distribution Data Interface Protocol

The fiber distribution data interface (FDDI) is another MAC sublayer standard based on the round robin access mechanism, with MAN applications. The basic operation of the FDDI standard is very similar to that of the IEEE 802.5. While the IEEE 802.5 is an IEEE standard and used in LAN, the FDDI was standardized by American National Standards Institute (ANSI) [ANSI X3.139] and by ISO as [ISO 9314-2] for the MAC sublayer. The corresponding physical layer was standardized as [ANSI X3.148] and [ISO 9314-1].

The FDDI considers two rings based on optical fiber, an operational and a backup ring. The switching from one to another is automatically performed in case of failure. It allows data rates of 100 Mbps, covering an area of up to 100 kilometers and up to 500 network hosts. The regeneration distance is 2 km for multimode fiber and 60 km for single mode fiber. Moreover, the digital coding technique is the 4B/5B, which achieves a better efficiency than the differential Manchester used in the IEEE 802.5 standard.

Similar to the IEEE 802.5 standard, the FDDI was designed to work with the IEEE 802.2 LLC,* as an upper sublayer. It receives data from the upper sublayer and then encapsulates it into the payload data of the FDDI MAC frame, followed by the addition of the corresponding frame header, as depicted in Figure 10.47. Moreover, the content of the token is depicted in Figure 10.48, whose fields are the same as those described for the frame content.

The content of the frame fields are defined in the following:

- Start of frame sequence (SFS): it comprises these fields:
  - Preamble (PA): a sequence of predefined bits used for synchronism purposes.
  - SD: a sequence of predefined bits that delimits the start of a frame.
  - FC: it indicates the type of data transported in the payload data field.

---

* Although the FDDI standard is not an IEEE standard, it interfaces with the IEEE 802.2 LLC sublayer.

- DA: it consists of the address of the destination host.
- SA: it consists of the address of the source host.
- Payload data: it transports IEEE 802.2 PDU. The maximum length of the frame payload data is 4500 octets.
- CRC: it uses the same CRC function as the one defined for IEEE 802.3, as defined by (10.7), applied to the FC, DA, SA, and payload data.
- End of frame sequence (EFS): it comprises two fields:
  - ED: it consists of a group of bits used to signalize the end of a frame. Moreover, these bits give indication about whether a frame is the last one or an interim.
  - FS: consists of a group of bits that are changed by the destination host when it is copied. It is based on these bits that either the transmitting host or, alternatively, the management station removes the frame from the ring. Moreover, this field allows the signalization when an interim node detects an error in a frame.

10.6.4 IEEE 802.11 Protocol

It consists of a technology dedicated to interconnect and allow Internet connection of wireless devices in a LAN environment, using the IEEE 802.11 standard. Wi-Fi, often known as Wireless Fidelity, is a trademark of Wi-Fi Alliance for certified products based on IEEE 802.11 standard [IEEE 802.11]. This standard specifies the physical layer and the MAC sublayer, offering services to a common 802.2 LLC.

Originally developed for cable replacement in companies, Wi-Fi quickly became very popular in providing IP connectivity in environments such as offices, restaurants, airports, residential, campus, and so on, covering typically ranges of the order of 100 m outdoors and 30 m indoors [IEEE 802.11].

The initial version of Wi-Fi, invented in 1991 by NCR Corporation/AT&T, in The Netherlands, was standardized as IEEE 802.11 supporting 1 or 2 Mbps in the 2.4 GHz band, using either requency hopping spread spectrum (FHSS) or direct sequence spread spectrum (DSSS) [Geier 2002]. This version was upgraded by the following newer versions:

- IEEE 802.11a, as a standard version that consists of an extension to IEEE 802.11 that allows up to 54 Mbps in the 5 GHz band using OFDM transmission technique, instead of FHSS or DSSS.
- IEEE 802.11b, allowing a data rate of 11 Mbps in the 2.4 GHz band, using DSSS due to its relative immunity to interference.
- IEEE 802.11g, as an extension to IEEE 802.11b that allows up to 54 Mbps in the 2.4 GHz band, using either OFDM or DSSS transmission techniques.
- IEEE 802.11n, as an upgrade to allow over 100 Mbps in the 5 GHz band by using both OFDM transmission technique and the multistreaming MIMO scheme.

Wi-Fi uses the CSMA-CA algorithm as the MAC sublayer protocol. This medium is similar to CSMA-CD considered but using the request to send (RTS) and clear to send (CTS) messages, sent by the frame sender and destination, respectively. These additional messages allow the CSMA-CA reaching a performance improvement, as compared to the CSMA-CD, by alerting the other stations that a frame transmission is going to take place [Ohrtman 2003]. In addition, this solves the hidden terminal problem,* which can be experienced in ad-hoc mode (making use of using an access point) and in infrastructure centralized mode (with an access point). Just as in an Ethernet LAN, having more users results in a reduction of throughput (within the coverage area). Therefore, its efficiency is limited to a reduced number of users and/or reduced traffic.

Wi-Fi-based products require at least 20 MHz for each channel (22 MHz in the 2.4 GHz band for IEEE 802.11b) and have specified only the license exempt bands 2.4 GHz ISM (Industrial, Scientific, Medical), 5 GHz ISM and 5 GHz Unlicensed National Information Infrastructure for operation [Ferro 2005].

With the IEEE 802.11 family of standards, a wireless access point (WAP) connects a group of wireless devices into a single cable device (normally a router). A WAP is similar to a network cable switch, performing the frame switching based on the MAC address.

Besides allowing connectivity in infrastructure centralized mode (using a WAP), Wi-Fi also allows ad-hoc networks (peer-to-peer interconnection). This means that wireless devices can interconnect directly, without using an IEEE 802.11 WAP. In addition, by using two wireless bridges, an IEEE 802.11 link can be established to interconnect two cable LANs, as long as the two bridges are within Wi-Fi wireless range.

The IEEE 802.11 networks are formed by cells known as basic service set (BSS). The BSS are created using a WAP (in infrastructure centralized mode) or without a WAP (in ad-hoc mode). An extended service set consists of multiple interconnected WAP (via wireless or via cable), which allows the interconnection of different BSS. Each BSS has a service set identifier (SSID). The SSID is 32 octets long, being used to identify the name of the BSS.

The IEEE 802.11 frames can be of three different types: data, control, and management. The generic format of all IEEE 802.11 frames can be seen from Figure 10.49. The preamble consists of the synchronism signal (80 bits) composed of alternating 0 and 1 bits, followed by the SFD consisting of 16 bits with the following pattern: 0000 1100 1011 1101. The PLCP header is always transmitted at 1 Mbps and contains information about the PDU length, as well as information about the transmission rate that will be used to transmit the frame. The MAC data is described below, whereas the final CRC field consists of 16 CRC bits used for error detection of the frame header.

---

* The hidden terminal problem refers to the situation where a terminal's transmitting coverage area may not be within the receiving coverage area of another terminal that belongs to the same network, whereas an intermediate node (wireless access point or terminal working in ad-hoc mode) may be in both areas of coverage.

Preamble	PLCP Header	MAC data	CRC

Figure 10.49 IEEE 802.11 frame format.

←——————————————— MAC Header ———————————————→

Control	Duration ID	Address 1	Address 2	Address 3	Sequence	Address 4	Payload data	CRC
2 octets	2	6	6	6	2	6	0–2312 (variable)	4

Figure 10.50 IEEE 802.11 message authentication code data format.

Protocol version	Type	Sub type	To DS	Fm DS	More fragment	Retry	Pwr manag	More data	WEP	Order
2 bits	2	4	1	1	1	1	1	1	1	1

Figure 10.51 Decomposition of the control field (IEEE 802.11).

Figure 10.50 shows the format of the IEEE 802.11 MAC data, whose fields have the following meanings:

- Control: this is the initial field, whose breakdown of subfields is depicted in Figure 10.51. It is used for control of the MAC sublayer. This field includes the following subfields:
  - Protocol version: identifies the protocol version.
  - Type: contributes to identify the type of IEEE 802.11 frame (control, management, or data).
  - SubType: together with the type, identifies the type of IEEE 802.11 frame.
  - To DS: this bit is set to 1 when the frame is sent to the WAP (for forwarding).
  - From DS: this bit is set to 1 when the frame is received from the WAP.
  - More fragments: this bit is set to 1 in case the frame is carrying a fragment of a packet. The first fragment has this bit set to 0.
  - Retry: this bit is set to 1 in case the frame is a retransmission of a previously sent one.
  - Power management: indicates the power management of the sender after the transmission of the current frame.
  - More data: means that the station has more frames to transmit. This is useful to prevent receiving stations to enter power save mode.
  - WEP: this bit is modified when the frame has been processed.
  - Order: this bit is set to 1 when the frames are sent in sequence (frames and fragments are commonly not sent in sequence).
- Duration ID: may have different meanings. In power-save pool frames, this is the station ID. In other frames, this field identifies the duration used for network allocation vector (NAV) calculation.*
- Address 1: recipient MAC address.

---

* The NAV refers to the duration used by a transmitting station to send a frame and by the receiving station to send the corresponding ACK. This is measured either after the transmission of the RTS or of the CTS.

- Address 2: the MAC address of the station that transmitted the frame. It can be a wireless station or a WAP.
- Address 3: if the To DS subfield of the control field is set to 1, the address 3 is the original source MAC address. If the From DS subfield of the control field is set to 1, the address 3 is the destination MAC address.
- Sequence: it includes two subfields:
  - Fragment number field: identifies the sequence number of each frame (12 bits).
  - Sequence number field: identifies the number of each fragment (4 bits).
- Address 4: source MAC address used in ad-hoc mode.
- Payload data: contains the LLC PDU data received from the upper sublayer (IEEE 802.2 LLC sublayer). Its size varies from 0 up to a total of 2312 octets.
- CRC: also referred to as FCS, consists of a 32 redundant bits used for error detection. It uses the same CRC generator polynomial as the one defined for IEEE 802.3 (defined by (10.7)).

10.6.5 Digital Video Broadcast Standard

The digital video broadcast (DVB) is a suite of standards that define the physical layer and data link layer of digital video distribution [Watkinson 2001]. While the MPEG consists of a set of codec standards, defining the way the analog video is digitized and compressed, the DVB standards define the way the digital video is transmitted (e.g., modulation scheme, error correction type and associated code rate, frame structure, etc.). It is worth noting that the digital video transmitted using a DVB standard is encoded and compressed using a MPEG standard (DVB-MPEG). An exception is the DVB for handheld devices, entitled DVB-H, which comprises its own specific encoding and compression algorithm.

The DVB system includes a variety of standards for different transmission mediums, namely the following:

- Terrestrial television (VHF/UHF): DVB-T and DVB-T2
- Cable: DVB-C and DVB-C2
- Satellite: DVB-S, DVB-S2, and DVB-SH
- Digital terrestrial television for handheld terminals: DVB-H, DVB-SH

Different modulation schemes and error correction codes are used in different DVB standards. Transmission mediums more subject to channel impairments make use of lower order modulation schemes, as well as lower error correction code rates.

Note that the mobile TV, using the LTE infrastructure, is expected to be a competitor to DVB-H-based TV broadcast.

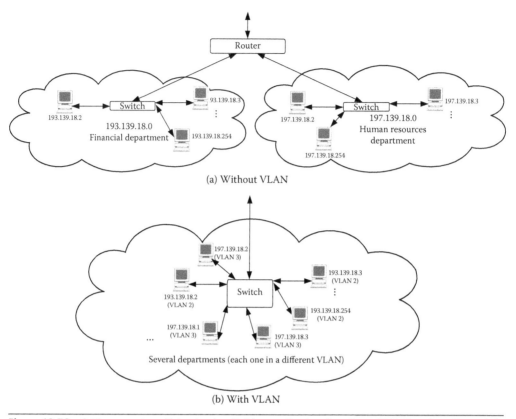

Figure 10.52   Getting access to multiple networks with (a) conventional configuration or (b) VLAN configured in a switch.

10.7 Virtual Local Area Networks

Virtual Local Area Networks (VLAN) can be created in switches allowing different logical networks with a single physical network. VLAN are used for different purposes, namely

- For providing access to different networks with a single switch
- For creating different broadcast domains with a single switch
- For achieving logical isolation among different hosts (security)

Providing access to different networks with a single switch: the access to different networks (or subnetworks) is normally granted by the use of a router. However, creating multiple VLANs in switches allows gaining access to multiple networks without having to make use of a router to separate the networks. In addition, a single switch may be used, instead of multiple switches for multiple networks, and thus, VLAN allows a reduction in the amount of hardware. This can be seen from Figure 10.52. Moreover, it is possible to use more than a single switch within a physical network to have access to multiple logical networks. This is advantageous to reduce the amount of cabling (a single cable is used between switches, instead of multiple cables from a single switch into multiple hosts) or when more switch ports are needed than those made available by a single switch. In this case, different switches must be connected in trunk mode and an encapsulation protocol must be configured. The most used encapsulation protocol is the IEEE 802.1q. An example of VLAN configuration in multiple switches is provided in Annex C.

Creating different broadcast domains with a single switch: normally, a broadcast domain corresponds to a whole organization. However, it is known that most of the broadcast traffic is limited to a department, while the rest of the organization's hosts are overloaded with such undesired traffic. The creation of different VLAN allows reducing the amount of undesired traffic that the hosts receive. This is achieved by creating different VLAN for different broadcast domains (e.g., different departments). This represents an improvement of network performance. Figure 10.52b depicts an example of two VLAN created in a switch, each one to be used by a different organization department (e.g., VLAN 2 for the financial department and VLAN 3 for the human resources department*). Consequently, each department will benefit of a different broadcast domain. It is also worth noting that with the use of a switch configured with multiple VLAN, the flexibility is greatly improved. Let us suppose the case where each department is physically located in different rooms, and a new employee who belongs to the human resources department is placed in the room that physically belongs to the financial department. With the use of VLAN, the network administrator only has to reconfigure the switch port that serves such new employee from the financial VLAN into the human resources VLAN.

Logical isolation among different hosts (security): Let us suppose that the employees of an organization need to have access to three different networks: Intranet, Extranet (or Internet), and IP telephony. Using the conventional approach and assuming that isolation needs to be assured among these different networks (for security purposes), three different physical networks are needed. Nevertheless, if three VLANs are created, a single physical network can be used, whereas three different logical networks are implemented over it. In this case, the traffic cannot be exchanged between different virtual networks (VLAN) (except if such capability is explicitly configured using a router). This can be seen from Figure 10.53b, where the network 193.139.18.0 (data) is isolated from the network 197.139.18.0 (IP telephony), using a common physical network infrastructure. The only way to enable the exchange of data among different networks is making use of a router. In this case, the switch port connecting the router is configured in trunk mode, whereas the router's interface needs to be configured using as many subinterfaces as the number of VLAN to be interfaced with and using an encapsulation protocol (e.g., IEEE 802.1q). Annex C provides an example of configuration in routers and switches, where interconnection between different VLANs is possible. Note that there can be multiple switches configured with multiple VLANs, where the interconnection between them is performed in trunk mode (using e.g., the IEEE 802.1q protocol), whereas the switches' interfaces used to provide access to computers are configured in access mode.

---

* Note that the VLAN 1 is the default.

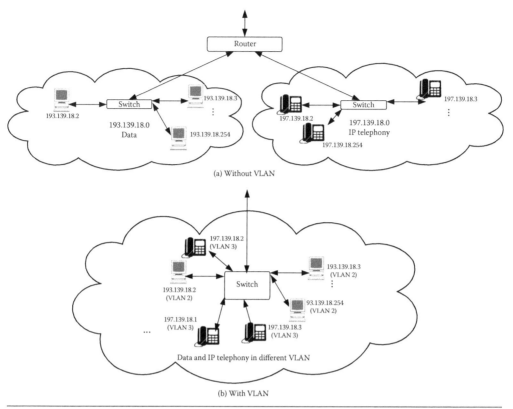

Figure 10.53   Isolating networks (a) without VLAN and (b) with VLAN using a single physical infrastructure.

The reader should refer to Section 11.4 for the configuration of Cisco Systems devices (routers and switches) using Cisco Internetwork Operating System (IOS). Moreover, an example of VLAN configuration in multiple switches is provided in Annex C.

End of Chapter Questions

1. What VLAN are used for?
2. What are the functionalities provided by the logical link control sublayer? What the mostly known LLC sublayer protocol is?
3. What does the MAC sublayer is used for? In which type of networks does the MAC sublayer exist?
4. What are the advantages and disadvantages of the IEEE 802.3 protocol relating to the IEEE 802.5?
5. Define the CSMA-CD mechanism.
6. What is the difference between the CSMA-CD and the CSMA-CA?
7. In which known LAN standard is the CSMA-CA mechanism used?
8. In which layer and sublayer is the error control capability implemented?

9. Consider the CRC generator polynomial $P(x) = x^2 + 1$ with the information source 11011. Calculate the sequence of transmitted bits.

10. Consider that the received bits sequence is 1101101 and that the CRC generator polynomial is $P(x) = x^2 + 1$. Verify whether the received bit sequence is free of errors.

11. Which types of flow control protocols do you know? How do they work?

12. Which types of error control protocols do you know? How do they work? How do they relate with flow control protocols?

13. What does a network control protocol is used for?

14. What is the difference between a network control protocol and a link control protocol?

15. What is the difference between the go back N ARQ and the selective reject ARQ mechanisms?

16. Having a code rate of ¾, how can we obtain a code rate of ½?

17. What does a link control protocol is used for?

18. In which scenarios is the use of error correction mechanisms preferable, instead of error detection?

19. What is the difference between the 10BASE-T and the 1000BASE-T technologies?

20. What does puncturing stands for? What are its advantages and disadvantages?

21. What does code interleaving stands for? What is its advantage and why is it used?

22. What is the relationship between the Hamming distance and the error detection capability?

23. What is the encapsulation protocol normally used in VLAN?

24. Which types of mechanisms can be used for signalizing the start and end of frames?

25. What is the difference between the 1000BASE-CX and the 1000BASE-SX technologies?

26. What is the relationship between the Hamming distance and the error correction capability?

27. Which physical layer mechanisms are used by 1000BASE-T to allow a throughput of 1000 bps?

28. What is the IEEE 802.1d protocol used for?

29. What does the maximum collision domain diameter stands for?

30. What does the carrier extension stands for and for which purpose is it used?

31. In CSMA-CD, what does back-off period stands for? Which values may it take?

32. What is the code puncturing used for?

33. What is the difference between the sliding window protocol and the go back N?

34. What is the maximum collision domain diameter of a IEEE 802.3 network, working at 100 Mbps?

35. Which kind of listening strategies can be used with the CSMA-CD mechanism?
36. What is the difference between a recursive and a nonrecursive error correction code?
37. Define the CSMA-CD mechanism.
38. What is the logical topology of a IEEE 802.5 (token ring) network?
39. What is the maximum throughput obtained with FDDI?
40. What is the difference between a bridge and a switch?
41. What is the difference between a hub and a switch?
42. What is the difference between a switch and a router?
43. For which purposes is the spanning tree protocol used for?
44. In which modes can the IEEE 802.11 operate?
45. What is the throughput made available by the different versions of IEEE 802.11?
46. Consider a convolutional code whose outputs are defined by $k_1 = n_1 \oplus n_0 \oplus n_{-1}$, $k_2 = n_1 \oplus n_0$, $k_3 = n_1 \oplus n_{-1}$. Considering that the input sequence is 10101, what is the output sequence?
47. How can we mitigate the effects of loops in a network?
48. Having a single physical network, how can we create several logical networks?
49. What does frame bursting stands for?
50. How does the sliding window protocol work?
51. Which type of forward error correction codes do you know? What are the differences among them?
52. What is the difference between a systematic and a nonsystematic error correction code?
53. What is the difference between the positive acknowledgement with retransmission (PAR) and the negative acknowledgement (NAK)? What are their relative advantages and disadvantages?
54. How is the code rate of an error code defined?
55. Define the differences between block codes and convolutional codes.
56. Which kind of LLC protocols can be implemented in a permanent circuit for data communications?
57. What is the control field of the HDLC protocol used for?
58. Which kind of HDLC frames do you know?
59. What are the S frames of the HDLC protocol used for?
60. What is the advantage of using piggybacking in HDLC? How is it implemented?
61. Which kind of links can the PPP use?
62. What is the protocol fields of the HDLC protocol used for?
63. By using HDLC frames for controlling links (e.g., NCP, LCP, CCP, ECP), which fields can be used in the payload data?
64. What is the difference between the PPP and the HDLC protocol?

11

NETWORK LAYER

The Internet protocol (IP) is the most important network protocol of transmission control protocol (TCP)/IP protocol stack. It provides a basic service to the delivery of datagrams, enabling a wide list of application and services. The basic functionalities of the IP are as follows:

- Definition of the address space in the Internet
- Routing of datagrams across the network nodes
- Fragmentation and reassembling of datagrams

To implement these functionalities, the IP makes use of datagrams, which consist of its elementary message format, used for end-to-end transmission* of data. To allow this, actions are to be taken by each node (router) of the network. A router is a device that makes network layer switching of datagrams. Normally, different routers exist between the source and the destination of an end-to-end connection. An end-to-end connection is composed of a concatenation of several layer 2 links (data link layer), and the router is the device that is between two successive layer 2 links (normally of different types). In the IP, each router has to decide about the best way to forward the datagrams and, to allow such routing, the datagram transports information about the destination address.

As previously described in Chapter 2, a datagram is composed of transport layer data (segment), to which an IP header is added. This is the basic concept of encapsulation, where, at the transmitter side, the upper layer data (segment) is transferred to an adjacent lower layer (datagram) and added with overhead (source and destination IP address, sequence numbers, etc.). The opposite operation, called de-encapsulation, is performed at the destination side, where a lower layer removes its layer's header (e.g., IP header) before the data (IP payload) is transferred to an upper layer (as a segment).

Since the IP is a connectionless protocol, there is no need to establish the connection before the data is exchanged. Moreover, since the IP is a datagram-based protocol (instead of virtual circuits), different datagrams may follow different paths, and the datagrams may arrive the destination out of order. In case data reliability is necessary for the provision of quality of service, the TCP is adopted. The TCP is responsible for ordering datagrams, correcting datagrams with errors, detecting and correcting lost datagrams, and discarding duplicated datagrams.

---

* End-to-end transmission can be viewed as the transmission between a client and a server or between two client computers.

IP version 4 (IPv4) is the most currently used version. The IP version 6 (IPv6), initially referred to as IP next generation, was standardized by [RFC 2460], and we face, currently, a transition phase for its full migration. Among the different advantages of the IPv6 relating to the IPv4, the former provides a much higher addressing space. Moreover, this protocol presents advanced security capabilities, the ability to provide improved quality of service (QoS), and the configuration of hosts in an IPv6 network is much more simplified (plug and play).

11.1 Internet Protocol Version 4

11.1.1 Internet Protocol Version 4 Classfull Addressing

As introduced in Chapter 2, an IPv4 address consists of a 32-bit address, being composed of four groups of eight binary numbers (10101101.00001111.11111100.00000011). For the sake of simplicity, an IPv4 address is normally displayed in four groups of decimal numbers, using the dotted-decimal notation (e.g., 173.15.252.3). The Internet Assigned Numbers Authority (IANA) is the entity responsible for the assignment of IP address ranges to regional network information centers, whereas regional network information centers are responsible for the assignment of IP addresses to domains, such as Internet service providers (ISP), companies, and so on.

There are two different possible ways of identifying the address class of an IP address. One way relies on the observation of the leftmost octet of an IP address in decimal notation. As can be seen from Table 11.1, the leftmost octet value determines the border between classes.

The second possibility requires the conversion of the leftmost octet from decimal to binary and the observation of the position of the leftmost zero. Class A has the leftmost zero in the most significant bit (MSB). Class B has the leftmost zero in the second position, and the MSB is 1. Class C has the leftmost zero in the third position, and the two left bits are 1.

Taking the IP address 173.15.252.3 as an example, entering the value 173 into Table 11.1, it is clear that this is a class B address. Using the second method, converting the value 173 into binary we obtain 10101101. Since the leftmost zero is in the second position, we conclude that this IP address belongs to class B. From Figure 11.1, being of class B, we know that the two leftmost octets identify the network, and the

Table 11.1 Mapping Between the Address Class and the Leftmost Octet Value

CLASS	DECIMAL RANGE	BINARY RANGE
Class A	0 … 127	0XXXXXXX
Class B	128 … 191	10XXXXXX
Class C	192 … 223	110XXXXX
Class D	224 … 239	1110XXXX
Class E	240 … 255	11110XXX

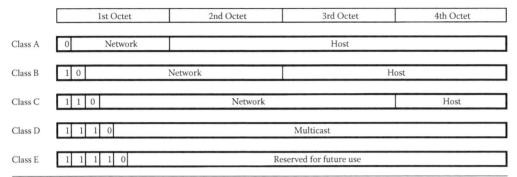

Figure 11.1   Classes of internet protocol version4 addressing.

Table 11.2   Reserved Internet Protocol Addresses

RESERVED IP ADDRESSES	
Own IP address	0.0.0.0
Network address	X.0.0.0 (class A)
	X.X.0.0 (class B)
	X.X.X.0 (class C)
Broadcast in own network	255.255.255.255
Broadcast in a specific network	X.255.255.255 (class A)
	X.X.255.255 (class B)
	X.X.X.255 (class C)
All networks	255.X.X.X (class A)
	255.255.X.X (class B)
	255.255.255.X (class C)
Loopback	127.X.X.X

two rightmost octets identify the host within a certain network. In fact, the network IP address is composed of the network part of the IP address with zero value in all bits of the host part of the IP address. Therefore, the network IP address of our example is 173.15.0.0.

There are some addresses that have specific applications, and for this reason, they cannot be used as the others. These addresses are listed in Table 11.2.

As can be seen from Table 11.2, the IP address with all zero value bits (0.0.0.0) stands for the own IP address. This is the default address before having being assigned an IP address to a host (e.g., by the dynamic host configuration protocol (DHCP) or manually by a network administrator). Note that the notation included in Table 11.2 is decimal, but an octet with zero decimal value corresponds to all zero value bits in the octet (00000000). Similarly, an octet with a decimal value 255 corresponds to all bits with one-value in the octet.

An IP address with all zero value bits in the position of the host part of the IP address (e.g., X.X.0.0 for class B*) represents the network address. The IP address

---

* Where X stands for any decimal number between 0 and 255.

with all one-value bits in the position of all octets (255.255.255.255) corresponds to the broadcast address in the own network. The IP address with all one-value bits in the position of the octets that represent the host part of the IP address corresponds to the broadcast address in a specific network (e.g., X.X.255.255 is the broadcast address within the network X.X.0.0 [class B]). The IP address with all one value bits in the position of the network part of the address (e.g., 255.255.X.X for class B) represents all hosts with the host part of the IP address X.X, in all class B addresses. Finally, the IP address with 127 decimal values in the leftmost octet represents the loopback address. This is used for testing purposes, when a host sends data to itself.

Taking into account the IP address classes defined in Figure 11.1 and the reserved IP addresses listed in Table 11.1, the number of networks and the number of hosts per IP address class can be quantified (see Table 11.3). The class A has a total of eight bits assigned to the network, but the leftmost bit is fixed to zero. Therefore, seven bits are used to identify networks ($2^7 = 128$), but the all zero value bits cannot be used (reserved for the identification of the own IP address) and the decimal value 127 is reserved for feedback, leaving a total of $2^7 - 2 = 126$ existing class A networks. Similarly, 24 bits are used to address hosts within each class A network, but the all zero value bits (own network) and the all one value bits (broadcast address) cannot be used, leaving a total of $2^{24} - 2 = 16,777,214$ hosts within each network. Following the same rational, the number of class B and C networks and hosts listed in Table 11.3 can easily be deducted.

In addition to the aforementioned reserved IP addresses, each address class has its own reserved address range, used as private addresses. Private addresses are addresses reserved by IANA to be used only inside local area networks (LANs). Private IP addresses are also referred to as site-local addresses. A network administrator may use these private addresses without any permission from the IANA. Nevertheless, the router should implement a network address translation (NAT) or port address translation (PAT),* such that the private IP addresses are not used in the Internet world. The private IP address ranges are listed in Table 11.4, whereas Table 11.5 lists the link local addresses. Link local addresses are known as automatic private Internet protocol addressing (APIPA) in Windows operating systems. Similar to private addresses, APIPA addresses are only to be used inside private networks. Moreover, a host using an APIPA address, which needs to exchange data with the Internet world, needs to use NAT/PAT address translation.

The link local addresses (APIPA) [RFC 3927] may be used as an alternative to the Dynamic Host Configuration Protocol (DHCP) protocol, consisting of a way that a network may use to allow hosts getting IP addresses. In case a host contacts the DHCP server for obtaining an IP address without any response and if the APIPA is configured in the network interface card (NIC), the operating system automatically assigns an IP address within the APIPA range as listed in Table 11.5. The APIPA protocol assures that this address is unique within the LAN. This procedure allows

---

* The NAT/PAT are defined later in this section.

Table 11.3 Quantification of the Number of Networks and Hosts per Address Class

CLASS	NUMBER OF NETWORKS	NUMBER OF HOSTS
A	$2^7 - 2 = 126$	$2^{24} - 2 = 16,777,214$
B	$2^{14} = 16384$	$2^{16} - 2 = 65,534$
C	$2^{21} = 2,097,152$	$2^8 - 2 = 254$

Table 11.4 Private (Site Local) Internet Protocol Address Ranges

	PRIVATE IP ADDRESS RANGES
Class A	10.0.0.0 – 10.255.255.255
Class B	172.16.0.0 – 172.31.255.255
Class C	192.168.0.0 – 192.168.255.255

Table 11.5 Link Local (APIPA) Address Ranges

LINK LOCAL (APIPA) ADDRESS RANGES	
ADDRESSES	MASK
169.254.0.1 – 169.254.255.254	255.255.0.0

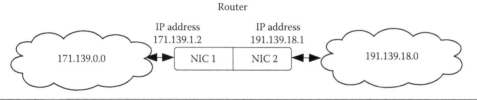

Figure 11.2 Example of a router with two network interface cards.

obtaining an IP address without making use of the DHCP server or without having to manually configuring it. This procedure is very useful in small size networks.

A router is a network layer device that is used to interconnect two or more networks or wide area network (WAN) segments. In addition, these two or more interfaces are normally of different type, that is, the data link layer protocols in each of the interfaces is different. Note that a router has an IP address per NIC installed and configured. As can be seen from Figure 11.2, NIC1 is connected to a class B network and has a class B IP address (171.139.1.2), whereas NIC2 is connected to a class C network and has a class C IP address (191.139.18.1). From this example, it is seen that a router does not have an IP address. A router has as much IP addresses as the number of interfaces (NIC). The configuration of IP addresses in Cisco 2600 series routers is described in Section 11.4, and an example is included in Annex C.

11.1.2 Internet Protocol Version 4 Classless Addressing

The IP address space has been described in classfull mode. This mode is very inefficient, since the rigid allocation of bits to host or network part of the IP address leaves many IP addresses unused. Therefore, due to this reason and due to the rapid growth of the number of Internet hosts, the available addressing space is depleted. The measure that can mitigate this limitation is making use of the IP address is classless mode.*

In the classless mode, although bits are left preassigned to classes, we add another level of flexibility through the use of a subnet mask. In classless mode, the border between the network and the host identification is determined by the subnet mask. Mask of a network is an address with one-value bits in the part that belongs to the network part of an IP address, and zero value bits in the part that belongs to the host part of the IP address. Using the classless mode, the border between network and host part of the IP address is no longer defined by the complete octets, that is, by 8-bit segments. Classless mode can use any number of bits to address network and host parts of addresses, and the address range is not divided into classes. The IP range previously defined for any of the classes (e.g., class A) can now be utilized without any restriction between network and host assignments.

A different notation for specifying the network prefix consists of counting the number of bits with value one in the subnet mask and appending that number to the address with a slash (/) separator. This is referred to as bit-count notation of the subnet mask. In the classless mode, some bits normally used to address hosts (in classfull mode) are now used to address sub-networks within a certain network. The remaining bits are then used to address hosts within each sub-network.

Completing the logic AND operation between the binary version of an IP address and the binary version of the subnet mask lead us to the network and sub-network parts of an IP address. Similarly, the host part of an IP address can be obtained by performing the logic NAND operation between an IP address and the subnet mask.

Let us examine a host with an IP address 10.20.230.140 and with subnet mask 255.255.192.0 (see Figure 11.3). The subnet mask in binary representation is 11111111. 11111111.11000000.00000000. We may refer to this IP address and subnet mask pair as 10.20.230.140/18, since the subnet mask has a total of 18 bits with value one bits (bit count notation).

In the example of Figure 11.3, performing the AND operation between the IP address and the subnet mask, we conclude that the host is in the network address 10.20.192.0. Similarly, performing the NAND operation, we conclude that the host part of the IP address is 0.0.38.140. This can be seen from Figure 11.4.

Since the IP address is to be written and read at network layer, computers (network end stations) and routers (network nodes) are the devices that need to have access to it. Routers are the devices responsible for routing the datagrams between the source and the destination, using the datagrams' destination IP address for this purpose,

---

* Other mechanisms such as private addresses, DHCP, NAT/PAT also contribute to optimize the reduced IPv4 address space.

IP address - Decimal	10	20	230	140
IP address - Binary	00001010	00010100	11100110	10001100
Subnet mask - Decimal	255	255	192	0
Subnet mask - Binary	11111111	11111111	11000000	00000000
Result of AND operation - Binary	00001010	00010100	11000000	00000000
Net/subnet address is the result of AND operation - Decimal	10	20	192	0

Figure 11.3 Calculation of the net or subnet part of the IP address.

IP address - Decimal	10	20	230	140
IP address - Binary	00001010	00010100	11100110	10001100
Subnet mask - Decimal	255	255	192	0
Subnet mask - Binary	11111111	11111111	11000000	00000000
Result of NAND operation - Binary	00000000	00000000	00100110	10001100
Host part of the IP address is the result of NAND operation - Decimal	0	0	38	140

Figure 11.4 Calculation of the host part of the IP address.

in conjunction with routing tables. Routers may perform their function only based on the destination address or using the subnet mask as well, in case classless mode is being considered and the router is compatible with. Traffic routing between IP networks only based on IP address is called classfull routing. On the other hand, the methodology of routing taking subnet mask into consideration is referred to as classless inter-domain routing (CIDR) [RFC 1519]. While in the classfull mode, routers forward datagrams based on the network part of the destination's IP address, using the CIDR routing, these devices read both the network part and the sub-network part of the destination's IP address and use these two components to decide about which output interface to use to forward the datagram.

Because of lack of IPv4 addresses, making use of classless mode, sub-networks can be created with a single IPv4 network address. In this case, only a sub-network IP address can be assigned to an organization, instead of a whole network IP address. This results in a much more efficient use of the limited IPv4 address space.

Let us consider an example of an organization with the need to have six IP addresses.* If the organization is given a class C network IP address in classfull mode, from Table 11.3, we know that it has a total of 254 full IP addresses (host IP addresses), while using only six. Consequently, the organization is leaving 248

---

* This can be the number of simultaneous IP addresses used dynamically by a higher number of hosts, to have access to the Internet using the NAT.

unused host IP addresses, and these unused IP addresses cannot be used by anyone else. With classless mode, the organization can be given, for example, a class C network IP address, but with a subnet mask 255.255.255.248 (this corresponds to /29 in the bit-count notation). In this case, since the default (classfull) class C subnet mask is 255.255.255.0 (/24) and the classless subnet mask is 255.255.255.248 (/29), a total of $29 - 24 = 5$ bits are used to create sub-networks (a total of $2^5 - 2 = 30$ sub-networks), while the remaining 3 bits are left to address hosts within each sub-network (a total of $2^{32-29} - 2 = 2^3 - 2 = 6$ hosts). The subtraction of the number two in the computation of the number of sub-networks and hosts results from the fact that the all zero value bits (represent the network/sub-network IP address) and the all one value bits cannot be used (represent the network/sub-network broadcast IP address). The sub-networks that can be created with five bits are 00001, 00010, 00011, … , 11110. The sub-network with all zero value bits (00000) and all one value bits (11111) are not used, as they are used by the classfull network address and by the classfull broadcast address within the classfull network. Note that some routers already allow the assignment of the first and last sub-networks. In such case, there is no need to perform the subtraction by two in the computation of the number of sub-networks. Nevertheless, in this book, we consider that these two sub-networks cannot be used. Therefore, we consider that the sub-network with all zero value bits (000) and all one-value bits (111) are not used, as they are used to identify the sub-network address and the broadcast address within the sub-network. The six hosts that can be addressed with three bits are 001, 010, 011, 100, 110, and 110.

Tables 11.6, 11.7, and 11.8 show the subnet masks that can be used with different address classes.

For high number of bits used to address sub-networks (taken from the bits used to address hosts in the classfull mode), the calculation of subnetting becomes too complex to be performed using the binary notation. There is a rule that allows the calculation of subnetting using the decimal notation, which becomes much simpler than using the binary calculation. Such subnetting calculation using the decimal notation is summarized in Table 11.9. Moreover, Tables 11.10, 11.11, and 11.12 present examples of classless subnetting calculation for class C, B, and A IP addresses, respectively.

Table 11.6 List of Possible Class C Subnet Masks (in Decimal and Bit Count Notation)

CLASS C SUBNET MASKS	
255.255.255.0 (default)	/24
255.255.255.192	/26
255.255.255.224	/27
255.255.255.240	/28
255.255.255.248	/29
255.255.255.252	/30

Table 11.7   List of Possible Class B Subnet Masks
(in Decimal and Bit Count Notation)

CLASS B SUBNET MASKS	
255.255.0.0 (default)	/16
255.255.192.0	/18
255.255.224.0	/19
255.255.240.0	/20
255.255.248.0	/21
255.255.252.0	/22
255.255.254.0	/23
255.255.255.0	/24
255.255.255.192	/26
255.255.255.224	/27
255.255.255.240	/28
255.255.255.248	/29
255.255.255.252	/30

Table 11.8   List of Possible Class A Subnet Masks
(in Decimal and Bit Count Notation)

CLASS A SUBNET MASKS	
255.0.0.0 (default)	/8
255.192.0.0	/10
255.224.0.0	/11
255.240.0.0	/12
255.248.0.0	/13
255.252.0.0	/14
255.254.0.0	/15
255.255.0.0	/16
255.255.192.0	/18
255.255.224.0	/19
255.255.240.0	/20
255.255.248.0	/21
255.255.252.0	/22
255.255.254.0	/23
255.255.255.0	/24
255.255.255.192	/26
255.255.255.224	/27
255.255.255.240	/28
255.255.255.248	/29
255.255.255.252	/30

According to the rules defined in Table 11.9, the number of sub-networks for a class C network address with the subnet mask 255.255.255.128 would be 0. This results from the fact that, with such subnet mask, only one bit is being used to address sub-networks, which would result in $NSN = 2^1 - 2 = 0$. Consequently, the subnet mask 255.255.255.128 is not included in the subnet mask list for class C addresses.

Table 11.9   Subnet Calculation

SUBNET CALCULATION	
Number of sub-networks (NSN)	NSN = $2^x - 2$, where x is the number of one value bits used to address sub-networks
Number of hosts per sub-network (NHPN)	NHPN = $2^y - 2$, where y is the number of zero value bits in the subnet mask
Sub-network addresses	Multiple of the base number. Base number = 256 − subnet number. Subnet number is the decimal value of the octet that identifies the sub-network in the subnet mask. The resulting sub-network address is composed of the network part of the original IP address, with the multiples of the base number as the sub-network part and with all zero value in the bits that identify the hosts within each sub-network
Broadcast in the sub-network address	Corresponds to the addresses that precede the following sub-network addresses
Valid host address within each sub-network	All valid IP addresses between two sub-network addresses, except the sub-network addresses and their broadcast addresses

Table 11.10   Example of Subnet Calculation for Class C (Network IP Address 195.1.2.0/28)

SUBNET CALCULATION (IP ADDRESS = 195.1.2.0; SUBNET MASK = 255.255.255.240)	
Number of sub-networks	NSN = $2^4 - 2 = 14$
Number of hosts per sub-network	NHPN = $2^4 - 2 = 14$
Sub-network addresses	Base number = $256 - 240 = 16 \rightarrow$ sub-networks are 195.1.2.16, 195.1.2.32,...,195.1.2.224
Broadcast in the sub-network address	Broadcast addresses are 195.1.2.31, 195.1.2.47,...,195.1.2.239
Valid host addresses within each sub-network	Address range for sub-network 195.1.2.16 is 195.1.2.17 → 195.1.2.30; address range for sub-network 195.1.2.32 is 195.1.2.33 → 195.1.2.46,...

Table 11.11   Example of Subnet Calculation for Class B (Network IP Address 128.1.0.0/18)

SUBNET CALCULATION (IP ADDRESS = 128.1.0.0; SUBNET MASK = 255.255.192.0)	
Number of sub-networks	NSN = $2^2 - 2 = 2$
Number of hosts per sub-network	NHPN = $2^{14} - 2 = 16,382$
Sub-network addresses	Base number = $256 - 192 = 64 \rightarrow$ sub-networks are 128.1.64.0 and 128.1.128.0
Broadcast in the sub-network address	Broadcast addresses are 128.1.127.255 and 128.1.191.255, respectively, for the first and second sub-networks
Valid host addresses within each sub-network	Address range for sub-network 128.1.64.0 is 128.1.64.1 → 128.1.127.254; address range for sub-network 128.1.128.0 is 128.1.128.1 → 128.1.191.254

Similarly, the subnet mask 255.255.255.254 is also not used, as the number of hosts would be zero (NHPN = $2^1 - 2 = 0$). Note that these subnet masks are not listed in Table 11.6.

As can be seen from Table 11.7, the subnet mask 255.255.255.128 is not considered in class B addresses.[*] This results from the fact that, according to the rules defined in Table 11.9, the base number would be 128 and the sub-networks would be $x.y.0.128$, $x.y.0.256$, $x.y.1.128$, $x.y.1.256$, $x.y.2.128$,... Nevertheless, the number $x.y.z.256$ cannot

---

[*] The same for class A.

Table 11.12   Example of Subnet Calculation for Class A (Network IP Address 11.0.0.0/18)

SUBNET CALCULATION (IP ADDRESS = 11.0.0.0; SUBNET MASK = 255.255.192.0)	
Number of sub-networks	NSN = $2^{10} - 2 = 1,022$
Number of hosts per sub-network	NHPN = $2^{14} - 2 = 16,382$
Sub-network addresses	Base number = $256 - 192 = 32 \rightarrow$ sub-networks are 11.0.32.0, 11.0.64.0, 11.0.96.0, 11.0.128.0, 11.0.160.0, 11.0.192.0, 11.0.224.0, 11.1.0.0, 11.1.32.0,…,11.255.224.0
Broadcast in the sub-network address	Broadcast addresses are 11.0.63.255, 11.0.95.255, 11.0.127.255, 11.0.159.255, 11.0.191.255, 11.0.223.255, 11.0.255.255,…,11.255.223.255, respectively
Valid host addresses within each sub-network	Address range for sub-network 11.0.32.0 is 11.0.32.1 $\rightarrow$ 11.0.63.254; address range for sub-network 11.0.64.0 is 11.0.64.1 $\rightarrow$ 11.0.95.254,…

be used (the range of eight binary digits are only between 0 and 255). This originates that the number of sub-networks would only be half of those normally considered using the calculations defined in Table 11.9. Moreover, this would lead to an inefficient use of the address space, as would leave many addresses unused. Consequently, the best solution is to avoid using this subnet mask.

11.1.3 Network and Port Address Translation

Due to the rapid depletion of the IPv4 address space, several mechanisms had to be developed to optimize the unallocated address space. The classless mode of IPv4 address scheme was one of these mechanisms. Other mechanisms include the NAT, the PAT, and the DHCP.

Table 11.4 shows the private address ranges for different IPv4 address classes. These addresses can freely be used within LANs, without any permission from the Network Information Center, as long as these IP addresses are not used in the Internet world. A computer, located in a LAN with Internet access, may have either a public IP address or a private IP address as long as the router performs the address translation from a private into a public (for outgoing datagrams).*

This address translation can be implemented using one of two following methodologies:

- NAT: the router that interconnects the LAN with a metropolitan area network (MAN) or WAN for Internet access is running the NAT application, which is responsible for mapping private into public addresses. Depending on the number of simultaneous computers that need to have access to the Internet, a single IP address or multiple public IP addresses are used by the NAT. The NAT changes the source IP address in every outgoing datagrams, whereas the destination IP address are changed in every incoming datagrams. The NAT procedure is depicted in Figure 11.5.

---

* And from a public into a private (for incoming datagrams).

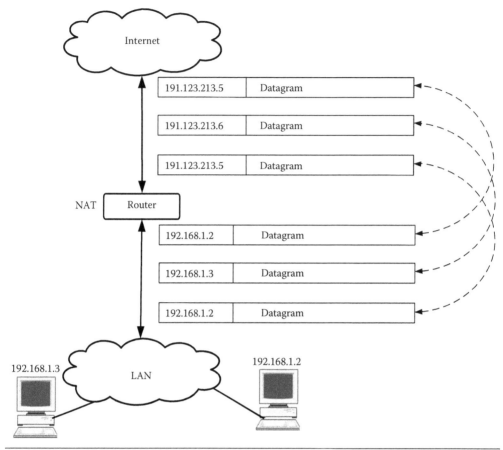

Figure 11.5 Network address translation.

- PAT: the router that interconnects the LAN with a MAN or WAN for Internet access is running the PAT application, which is responsible for mapping private IP addresses into a single public IP address. The PAT changes the source IP address of a datagram from a private address into a unique public IP address. Moreover, the PAT changes the outgoing segment's source port into one out of a range of ports used by the PAT router in the Internet world. Consequently, although the outgoing datagrams from different computers present the same source IP address, they are differentiated by different source ports. The opposite operation is performed by the PAT for incoming datagrams. Consequently, for incoming traffic, the PAT converts different destination's ports in segments (and a single public IP address) into the different private IP addresses used by different LAN computers.* The PAT procedure is depicted in Figure 11.6.

---

* After this conversion, since the LAN works at layer 2, the router consults the ARP table to find the MAC address, which corresponds to the destination's private address. In case such MAC address is not part of the ARP table, a broadcast ARP packet is sent to the LAN to identify the MAC address that corresponds to a certain IP address (this procedure is known as ARP protocol). In this case, such new entry is added to the ARP table.

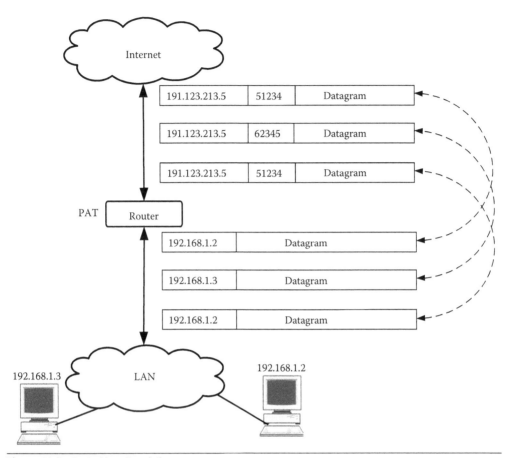

Figure 11.6 Port address translation.

It is worth noting that, very often, routers run simultaneously the NAT and the PAT protocols. In this case, instead of having a single public IPv4 address, the router has two or more addresses, as well as a range of port addresses to use in the Internet world.

Although allowing a better address space usage, these protocols also prevent the IP addresses from being known from the Internet world.* This represents an added value from the security point of view.

11.1.4 Dynamic Host Configuration Protocol

The DHCP [RFC 2131] is another example of a protocol that contributes to achieve a better usage of the depleted IPv4 address space. The DHCP assigns dynamically IP addresses to hosts. This way, when hosts are not connected or are switched off, they do not have any IP address assigned. Most of the home routers run the DHCP server. This allows a router assigning private IP addresses to different LAN computers. The DHCP is also normally used by ISP, such that every time a client switches on a

---

* This is normally used to avoid a server's IP address (e.g., a web server) from being known from the Internet.

router, it receives a new IP address. Note that, similar to the NAT and PAT, the dynamic modification of IP addresses performed by the DHCP also represents an added value from the security point of view.

11.1.5 Internet Protocol Version 4 Datagram

As previously described, at the transmitting host, a segment is encapsulated into the payload field of an IP datagram, being de-encapsulated at the receiving host. Similarly, an IP datagram is successively encapsulated and de-encapsulated into the payload of different data link layer frames. At intermediate nodes (routers), an IP datagram is taken from the payload frame, the destination IP address is read, and the decision about the output interface to use to forward it is taken, based on the routing table. Afterwards, re-encapsulation of the IP datagram into the payload of the new data link layer frame takes place. To allow the routing of an IPv4 datagram, hosts* need to have access to a myriad of datagram fields. Figure 11.7 depicts the different fields of an IPv4 datagram header.

The different IPv4 header fields are described in the following:

- Version: as can be seen from Figure 11.7, it is a 4-bit length field, presenting the value 4 for the IPv4.
- Internet header length (IHL): it is a 4-bit length field, composed of a number that expresses the header length (including options and padding), in groups of 32 bits.
- Differentiated services code point (DSCP). It is used in differentiated services (DiffServ) [RFC 2474] for the provision of QoS by intermediate routers. These requirements include delay, jitter, packet loss, or throughput.† This field is identified by a number between 0 and 15. Using the DiffServ option, datagrams with lower DSCP numbers have priority over those with higher DSCP numbers. In the past, this field was named type of service (ToS).
- ECN: it stands for explicit congestion notification [RFC 3168], being an optional feature. It enables network congestion notification.
- Total length: it represents the total datagram length, expressed in octets. The minimum datagram length is 20 octets (only the header, without options or payload), and the maximum datagram length is 65,535 octets. In case the maximum data field length‡ of the frame (maximum transmission unit [MTU]) is shorter than the datagram length, fragmentation is carried out.

---

* Computers, routers, gateways, and so on.

† The audio or video streaming are delay- and jitter-sensitive services, whereas the file transfer is a packet loss/bit error rate sensitive service.

‡ Note that the payload length of the data link layer has normally a variable length, depending on the amount of data (packet size) to transport among two adjacent nodes.

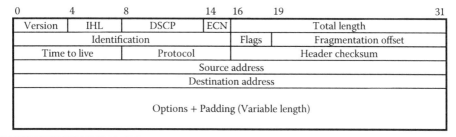

Figure 11.7   IPv4 header.

- Identification: it consists of a sequence number used for identifying fragments of an original IP datagram.
- Flags: it consists of three bits as follows:
  - Bit 0: reserved (must be zero value)
  - Bit 1: do not fragment
  - Bit 2: more fragments
- Fragmentation offset: it gives information about which part of the fragmentation does a certain datagram belongs.
- Time to live: it specifies the number of hops that a datagram may be subject to. Each router decrements this field by one. When this field reaches zero, the datagram is discarded.
- Protocol: it gives information about the upper layer protocol encapsulated in a datagram [RFC 790] (e.g., 6 for TCP and 17 for UDP).
- Header checksum: error detection redundant bits applied to the datagram header. In case an error is detected in a hop, the datagram is discarded.
- Source address: 32-bit address of the source host. In case NAT is in use, this source IP address may not be the original one.
- Destination address: 32-bit address of the destination host. In case NAT is in use, this destination IP address may not be the original one.
- Options: because of its rare use, this field is optional, including some prior routing instructions such as the following:
  - Source routing: a list of router addresses through where the datagram should follow. The implementation of source routing partially transforms the IP switching from datagram method into virtual circuits method.
  - Route recording: it records information about the routers which the datagram passed through.
  - Stream identification: used for streaming services (e.g., video streaming or voice over IP).
- Padding: used to ensure that the header length is an integral number of 32-bit words.

11.2 Internet Protocol Version 6

The massification of the Internet, in combination with the inefficient use of the limited classfull IPv4 address space, drove the need for a new Internet address scheme. The depletion of the IPv4 address space was the most important motivation for the development of the IPv6 [RFC 2460] [RFC 1752]. The IPv6 makes use of a 128-bit address, instead of the 32-bit address considered by the IPv4. This makes available a total of $2^{128} = 3,40 \times 10^{38}$ addresses, which allows accommodating the future terminals that result from the massification of the Internet and from the machine-to-machine communications. Mechanisms used to optimize the limited IPv4 address space, such as the NAT, PAT, and DHCP, are not required in IPv6.

Besides the wider address space made available by the IPv6, there were other Internet specifications that were intended to be improved with the new IP version. The IPv6 includes the following improved features:

- Improved QoS mechanisms: with the integration of the different services into a common telecommunications infrastructure, there was the need to make routing faster and smarter, such that a more efficient and effective way of implementing packets* differentiation. This is a characteristic of the IPv6, which allows a better handling of packets belonging to different services. Some mechanisms were improved to reduce the resources consumed by intermediate routers. An example is the lower number of the fields of an IPv6 header, as compared to the number of IPv4 fields (despite the longer mandatory IPv6 header, namely due to longer IPv6 addresses). This translates to lower resources consumption by intermediate nodes, as well as faster routing. Another example is the IPv6 source routing (optional mechanism), which presents several improvements relating to the IPv4 source routing. This is a mechanism that allows specifying some of the routers that will be visited by packets belonging to a certain flow of data. In addition, the flow label field (mandatory header field) allows prioritized handling by intermediate routers. These mechanisms allow improving the QoS for different types of services.
- Improved security functions: while the security architecture for IP (IPsec) is an IPv4 option, the exchange of packets in IPv6 networks is always carried out using the IPsec framework. This mechanism prevents attacks against the confidentiality, integrity, and authenticity. The reader should refer to Chapter 14 for a detailed description of the IPsec.
- Support for higher user mobility: the host (interface) identification of an IPv6 address is unique and can be automatically generated from the message authentication code (MAC) address. Consequently, the movement of a user across different networks gives him the possibility to keep connectivity

---

* Note that the IPv6 uses the term packet, instead of datagram (as considered by IPv4).

(using the same interface part of the IPv6 address) without administrative intervention.* With such automatic MAC/host address conversion, the ARP is no longer needed. Moreover, the Internet control message protocol version 6 (ICMPv6) used by IPv6 improves the auto-configuration of hosts, allowing an easier and quicker discovery of neighboring nodes.

- Faster and easier routing: the IPv6 address is hierarchical, consisting of a global routing prefix, subnet identification, and interface identification. The global routing prefix is normally assigned to sites by geographical regions. Since the IPv4 addresses were not organized by regions, inter-domain routers had to consult enormous routing tables to allow the selection of the output interface to forward a datagram. Conversely, since most of the IPv6 routing is performed inside a certain hierarchical level, this results in shorter routing tables, which allows a more efficient use of the routers resources (CPU, memory, bandwidth, etc.), enabling a faster routing.

IPv6 considers same type of dynamic routing protocols as IPv4 to generate routing tables, but only with the minimal modifications to accommodate the changes of the new address space.

11.2.1 Internet Protocol Version 6 Addressing

Similar to the IPv4, the entity responsible for the assignment of groups of IP addresses among different regions is the IANA. As can be seen from Figure 11.8, an IPv6 address [RFC 2373] is composed of 128 bits (16 octets).† For the sake of simplicity, an IPv6 address is displayed in eight groups of four hexadecimal digits,‡ separated by colons. Each group of four hexadecimal digits is within the range 0000–FFFF.

Within an IPv6 address, zeros can be omitted. As an example, the address 3D0F:00F3:0000:0000:0000:0000:0000:5A3C may simply be referred to as 3D0F:00F3::5A3C (contiguous zeros may be omitted).§ The position of the omitted zeros is marked with double colons. Nevertheless, this abbreviation can only be used once in an IPv6 address. Otherwise, it would be impossible to know how many zeros had been replaced by double colons. Moreover, within each group of four hexadecimal digits, the leftmost zeros can be omitted. Therefore, the address 3D0F:00F3::5A3C is equivalent to 3D0F:F3::5A3C (i.e., F3 instead of 00F3)

---

* Only the network/sub-network part of the IPv6 address changes. This can be discovered by establishing contact with neighboring nodes (without administrative intervention).

† The theoretical limit of the IPv4 address space is 2^{32}, corresponding to 4 294 967 296. Due to the rapid growth of the Internet, the available IPv4 address space is depleted. IPv6 solves this problem, as its address is composed of 128 bits, which makes a wide address space available for the Internet world (2^{128} addresses).

‡ Note that an hexadecimal digit may encode a total of four bits.

§ Another example can be the address 3D0F:00F3:0000:0000:0000:0000:0000:0000, which is equivalent to 3D0F:00F3::

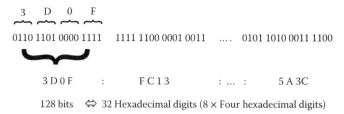

Figure 11.8 An example of an IPv6 address in both binary and hexadecimal notation.

Table 11.13 Generic Composition of a Global Unicast IPv6 Address

GLOBAL ROUTING PREFIX	SUBNET ID	INTERFACE ID
64-n bits	n bits	64 bits

or the address 3D0F:00F3:0000:FE01:0000:01C3:0003:5A3C is equivalent to 3D0F:F3:0:FE01:0:1C3:3:5A3C.

A global unicast IPv6 address presents the generic composition described in Table 11.13. Note that the interface identification (ID) is always composed of 64 bits. Moreover, the 64-n bits allocated to the global routing prefix plus the n bits allocated to the subnet ID makes a total of 64 bits. The value of n depends on the address type.

The global routing prefix identifies the network, whereas the subnet ID identifies the sub-network (these two items identify where the host is connected to). The interface ID is the part of the address that uniquely identifies an interface (this item identifies who you are). This address composition is similar to the composition of an IPv4 address in classfull mode. A host that moves across different networks or sub-networks keeps the same interface ID, but changes the global routing prefix and/or subnet ID. Consequently, it can be stated that a host only has an interface ID, not a whole IPv6 address.

To understand the IPv6 address composition, it is worth introducing the notion of prefix.

A prefix may be used for two different purposes:

- To identify an address range for any purpose.
- To separate the network and sub-network part of an address from the interface part of an address.

In the scope of the identification of an address range, the notation X::/Y stands for the address X:: with the prefix /Y. This consists of an address range starting by the address X000…000 with Y fixed binary digits within the address X, up to ZFFF…FFF. In case the address is in binary notation, since Y stands for the number of fixed binary digits, the fixed bits Z equals the fixed bits X. In case the address is in hexadecimal notation, since the hexadecimal digit X is composed of four bits, if the number of fixed binary digits corresponds to a number lower than four, the hexadecimal Z may differ from X by varying the remaining 4-Y bits. Let us consider the example of the global unicast address range, which is defined by 2000::/3 (the first three binary digits are fixed). This represents the address range starting in the address 2000:: (0010 0000 0000 …) up to 3FFF:FFFF:FFFF:FFFF:FFFF:FFFF:FFFF:FFFF (0011 1111 1111…).

Using the prefix to separate the network and sub-network part of an address from the interface part of an address, the concept is the same as used in IPv4 addresses using the bit count notation.

As an example, from the address 31CF:0123:4567:89AB:CDEF:0123:4567:89AB/64, we may conclude that the network and sub-network part of the address has a total of 64 bits (with the address 31CF:0123:4567:89AB:0000:0000:0000:0000) and that the remaining 64 bits are used to identify the interface within the network (CDEF:0123:4567:89AB).

The IPv6 considers three types of addresses:

- Unicast, which identifies a single network interface [RFC 2374]. As described earlier, the global unicast addresses are located within the address range 2000::/3 (the three leftmost bits of the IPv6 address are fixed to 001).
- Anycast, which identifies any address within a certain address group. For example, an anycast address may refer to the closer node within a certain unicast address range. Anycast addresses use the same address space as unicast, but for different purposes.
- Multicast, which identifies all addresses within a certain a group of interfaces. Note that there are no broadcast addresses in IPv6. The broadcast address is a special case of the multicast address. The multicast addresses are located within the address range FF00::/8 (the eight leftmost bits of the IPv6 address are fixed to 11111111).

Table 11.14 summarizes the reserved IPv6 address ranges, pre-allocated for different purposes. As can be seen from Table 11.14, there are three different types of unicast addresses.

Table 11.14 Reserved IPv6 Address Ranges

ADDRESS TYPE	BINARY PREFIX	IPV6 NOTATION
Unspecified	000…0	::/128 or 0:0:0:0:0:0:0:0
Loopback	000…1	::1/128 or 0:0:0:0:0:0:0:1
Link-local unicast	1111111010	FE80::/10
Site-local unicast	1111111011	FEC0::/10
Global unicast	001	2000::/3
Multicast	11111111	FF00:/8

- The global unicast address has a global scope, globally utilized in the Internet.
- The site-local unicast address corresponds to the IPv4 private address. The site-local unicast address can be used within a certain site (these addresses are unique within such site).
- The link-local unicast address corresponds to the IPv4 APIPA address. The link-local unicast addresses can only be used within a specific site. The link-local configuration is used by the operating system in the absence of having another type of IPv6 address. These addresses are used between on-link neighbors and for neighbor discovery on the same link. An IPv6 router never forwards link-local traffic beyond the link. These addresses are local and not advertised in the Internet world. The procedure assures that these addresses are unique within the site.

The composition of a global unicast address is summarized in Table 11.15.

As can be seen from Table 11.15 and Figure 11.9, IPv6 addresses have several hierarchical layers (TLA, NLA, SLA). These aggregation levels are allocated by geographical regions. This results in shorter routing tables, as most of the routing is performed within a certain level. The upper or lower level prefixes are only used if the address is not within its level. Moreover, very often a fixed node/address is used to

Table 11.15 Generic Composition of a Global Unicast Address

001	TLA ID	RES	NLA ID	SLA ID	INTERFACE ID
					UNIQUELY IDENTIFIES AN INTERFACE
		PUBLIC DOMAIN		SITE DOMAIN	
3 bits	13 bits	8 bits	24 bits	16 bits	64 bits
It means that the IPv6 address is of global unicast type	Top level aggregation identification (TLA ID): Identify long haul providers (the highest level in the IPv6 routing hierarchy) Similar to the network addresses, groups of TLAs are allocated by IANA to regional network information centers, which then allocates them to global ISP. There are a total of 213 = 8192 TLA ID.	For future TLA or NLA ID expansion.	Next level aggregation (NLA) ID: Identifies a regional site within a global ISP (a TLA ID).	Site level aggregation ID: Allows the creation of subnets within sites.	Uniquely identifies a network host. It uses the EUI-64 notation and is automatically generated from the MAC address.

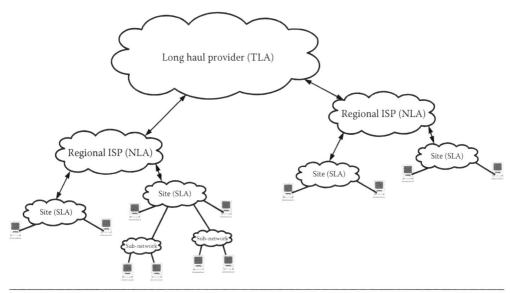

Figure 11.9 Hierarchical structure of IPv6 address space and routing.

reach different layers. This also gives a contribution for shortening the routing tables. This organization of the address space results in shorter routing tables, which translates in faster routing of the packets along the network, which facilitates the QoS.

The interface ID of a unicast address is 64 bit long and uses the notation EUI-64. It is directly generated from the NIC MAC address. As previously described, a MAC address is composed of 48 bits, grouped into six octets separated by colons, where each octet is denoted by two hexadecimal digits (e.g., 01:23:45:67:89:AB). The resulting EUI-64 address consists of the MAC address subject to the hexadecimal sequence "FFFE" added in the middle of the MAC address. Taking the previous MAC address as an example, the resulting EUI-64 address becomes 0123:45FF:FE67:89AB. Note that the interface ID groups four hexadecimal digits, instead of two, as considered by the MAC address. Using this conversion, a host automatically obtains an IPv6 address from the MAC address (which is allocated in factory), without the need to obtain an address from a DHCP server.* The host needs to assure that this automatically generated interface ID is not in use by any other host in the network. This is achieved through broadcast queries.

This interface ID of the IPv6 address is kept unchanged, regardless the network and sub-network where the host is. An IPv6 address within a site is identified by a total of 80 bits (64 bits for interface ID plus 16 bits for the site domain). Moreover, since the complete IPv6 address consists of the concatenation of the network ID, subnet ID, and interface ID, all that a host needs to know to find out its complete IPv6 address is to discover the network/sub-network ID. This can be obtained through advertisement with the neighboring routers. Consequently, a user/host who moves across different networks/sub-networks only changes the corresponding part of the IPv6

---

* In some special situations, such auto-configuration is not suitable. In such cases, the DHCPv6 (DHCP version 6) protocol may be used to configure the hosts.

Table 11.16 Generic Composition of a Multicast Addresses

1111 1111	FLAGS	SCOPE	GROUP ID
8 bits	4 bits	4 bits	112 bits
It means that the IPv6 address is of multicast type	Identifies whether the multicast address is permanently or transiently assigned	Identifies the multicast address scope: interface-local, site-local, organization local or global	Identifies a multicast group

address, while keeping the interface ID unchanged. Moreover, since an IPv6 address is automatically obtained from the MAC address and from information obtained from neighboring routers, it is easily concluded that the IPv6 protocol does not need to use the ARP protocol.

Table 11.16 shows the generic composition of multicast addresses. As can be seen, the leftmost bits consists of a fixed group of eight one value bits (ff in hexadecimal), followed by four bits (flags) which gives an indication about whether the multicast address is permanently or transiently assigned. The following four bits identify the scope, namely if the multicast address is only within a sub-network, a network, a local ISP, a long haul provider, or global. Finally, the 112 rightmost bits are used to create multicast groups. Note that a multicast group may coincide with the broadcast of a network, sub-network, site, ISP, or global, but may also consist of a group of individual and specific hosts distributed along different networks, sub-networks, or ISPs.

The transition from IPv4 into IPv6 is normally implemented in several steps. Therefore, there is the need to allow a smooth migration, which includes the possibility to find a mapping between IPv4 and IPv6 addresses.

When the traffic is mainly IPv6, whereas part of the network is still IPv4, a common procedure is the establishment of an implicit IPv6 tunnel in the IPv4 network to allow the exchange of IPv6 packets. In this case, the IPv6 protocol makes use of the IPv4 network as a data link layer. This requires the communication between IPv4 and IPv6 nodes, as well as the use of 6to4 unicast address mapping. The 6to4 unicast address uses the 16 leftmost bits fixed to 2002 (i.e., 2002::/16) of type 2002:XXYY:ZZWW::/48, defined as follows:

- XX is the hexadecimal notation of the leftmost IPv4 octet.
- YY is the hexadecimal notation of the second IPv4 octet.
- ZZ is the hexadecimal notation of the third IPv4 octet.
- WW is the hexadecimal notation of the rightmost IPv4 octet [Blanchet 2006] [Davies 2008].

As an example, the IPv4 address 193.123.213.25 can be translated into the following IPv6 address: 2002:C17B:D519::/48. Note that the 32 bits of the IPv4 address use the fields Res and NLA ID, as described in Table 11.15. Following this approach, IPv4 nodes may be used to allow the exchange of IPv6 packets. An IPv4 router of the tunnel reads the destination IPv6 address to find the output interface to forward the packet to, but only considers a part of the address (bit 17 to 48).

Conversely, when nodes of a network are configured for IPv6, whereas the packets are still IPv4, encapsulation of IPv4 datagrams into IPv6 packets are normally carried out. Another possibility is a hybrid configuration, whereas some nodes are IPv4 and others are IPv6. In these cases, the approach consists of encapsulating an IPv4 address into an IPv6 address. This is normally referred to as IPv4-mapped IPv6 address, being defined as 0:0:0:0:FFFF:<ipv4 address> [Blanchet 2006].* In this case, only the rightmost 32 bits out of the 64 bits allocated to the interface ID are used (as described in Table 11.15).

Finally, it is worth noting that a URL that makes use of the IPv6 protocol uses a notation that consists of including the IPv6 address inside brackets. As an example, the URL address http://[2123:0123:4567::89AB:CDEF]:/test.html corresponds to the IPv6 address 2123:0123:4567::89AB:CDEF.

11.2.2 *Internet Protocol Version 6 Packet*

As previously described, to reduce the required processing by routers, which translates in faster routing, the number of fields included in the IPv6 packet header was reduced, as compared to the IPv4 datagram header. Example of fields existing in the IPv4 that do not exist in the IPv6 mandatory fields include IHL, ID, flags, fragment offset and header checksum. Nevertheless, the mandatory part of the IPv6 header is longer than that of the IPv4. This is due to the longer IPv6 addresses (128 bits), as opposed to the 32-bit size IPv4 addresses.

Figure 11.10 depicts the IPv6 header, with the mandatory part and the extension part. The mandatory header has a fixed length of 40 octets, being depicted in Figure 11.11, whereas the extension header is composed of fields necessary for handling special functionalities. The field lengths are depicted in Figure 11.10, expressed in number of octets. Note that some fields were moved from the (fixed part of the) IPv4 header into the extension part of the IPv6 header. This way, the mandatory part contains only the fields that are to be read by all the intermediate nodes. The aim was to reduce the overhead, simplify the processing, and speeding up the routing. The extension header fields consist of the following [RFC 4260]:

- Hop-by-hop options: it is used to carry special information required to be processed by intermediate routers, such as resource reservation protocol (RSVP) for IPv6 or to implement the jumbo payload. The payload has a size up to 64,000 octets without special options. A larger payload can be implemented using the jumbo payload option, when properly signalized in the hop-by-hop options field.
- Routing: this field is used to implement source routing. The implementation of source routing partially transforms the IP switching from datagram

---

* Naturally, this is equivalent to ::FFFF:<ipv4 address>.

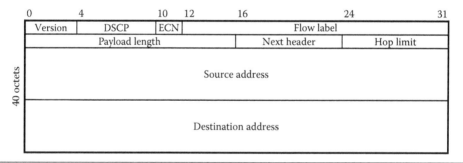

Figure 11.10   Generic decomposition of the IPv6 header.

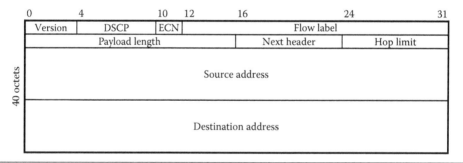

Figure 11.11   Mandatory part of IPv6 header.

method into virtual circuits method. The addresses of the intermediate nodes that compose the fixed part of the route are included in this optional field.

- Fragment: this field is used to implement fragmentation and reassembling of packets.
- Authentication: it consists of a mechanism that provides protection from attacks against the authenticity and integrity of packets and from anti-replay attacks. This is an option of the IPsec framework, as detailed in Chapter 14. Note that while the IPsec was optional in IPv4, the exchange of IPv6 data between routers is always implemented using the IPsec mechanisms.
- Encapsulating security payload (ESP): it consists of a mechanism that provides protection from attacks against the confidentiality, authenticity, integrity, and

anti-replay attacks. This is an option of the IPsec framework, as detailed in Chapter 14.

- Destination options: it consists of optional information to be handled by the destination router.

Note that the IPv6 fragmentation is only implemented by end stations of an end-to-end connection, whereas IPv4 fragmentation is carried out by intermediate nodes (routers). In the case of the IPv6, intermediate routers never implement fragmentation. The exchange of IPv6 packets must be preceded by a path MTU discover, which dictates the maximum fragment size transmitted. Consequently, as opposed to the IPv4, the fragment offset is not included in the mandatory part of the IPv6 header.

As can be seen from Figure 11.11, the mandatory part of the IPv6 header contains the following fields:

- Version: identifies an IPv6 packet.
- DSCP: being used for the provision of QoS by intermediate routers. These requirements include delay, jitter, packet loss, or throughput. Using the DiffServ option, datagrams with lower DSCP numbers have priority over those with higher DSCP numbers.
- ECN: being used to explicitly notify nodes of congestion situations. The fields DSCP and ECN are also jointly referred to as traffic class.
- Flow label: used to signalize the packets belonging to a specific flow of data that requires special attention by intermediate routers. This is used to preserve the required QoS. The flow label field allows prioritized handling by intermediate routers (e.g., for voice, audio, or video streaming packets).
- Payload length: consists of a 16-bit size field and corresponds to the length of the extension headers plus the payload. This field varies between 0 up to 65,535 octets. Note that the maximum payload is 64,000 octets without special options. The extension header hop-by-hop field allows implementing a larger payload entitled jumbo payload. In case the data field length of the frame (MTU) is shorter than the maximum packet length, fragmentation is carried out, which needs to be properly signalized in the fragment field of the extension header.
- Next header: this is a pointer to the header of the upper layer protocol (TCP, UDP, ICMP) carried in the packet payload. In case an extension header is used, this field points toward the first extension header. Note that the next header field also exists in the extension headers. In this case, this field points towards the following extension header or to the upper layer protocol (next header field of the last extension header);
- Hop limit: this field corresponds to the time to live field of an IPv4 header. It specifies the number of hops that a datagram may be subject to. Each router decrements this field by one. When this field reaches zero, the datagram is discarded.
- Source address: 128-bit address of the source host.
- Destination address: 128-bit address of the destination host.

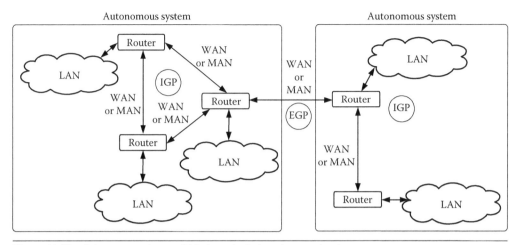

Figure 11.12 Autonomous systems and interior and exterior gateway protocols.

11.3 Routing

As previously described, routing is used to make datagrams arrive the destination after passing through several intermediate nodes (routers). These devices perform layer 3 switching (selection of the output interface to forward a datagram) based on the destination IP address and based on information contained in routing tables. A routing table gives information about the output interface to use in order to forward datagrams to a certain destination network.

The Internet is organized in clouds, which are interconnected. As can be seen from Figure 11.12, an autonomous system can be viewed as cloud. An autonomous system is composed of several LAN, MAN, and WAN, which belong to an organization (e.g., an ISP or a system of corporate LANs).

Within an autonomous system, routers forward datagrams between different LAN using interior gateway protocols (IGP). IGPs consist of a group of protocols used to generate entries into routing tables to allow forwarding datagrams along the organization's network (autonomous system). Similarly, exterior gateway protocols (EGP)* consists of a group of protocols used to generate entries into routing tables to allow forwarding datagrams between different autonomous systems. A router without connection with the exterior of an autonomous system only uses one or more IGPs. A router that connects with other routers of the same autonomous system and with routers of different autonomous systems (these routers are referred to as border routers) need to use both IGP and EGP.

The interconnection between autonomous systems has similarities with the interconnection of different LANs. Both can be viewed as clouds. Nevertheless, while LANs within an organization are interconnected using IGP-type protocols, the EGP-type protocols are used to interconnect different autonomous systems. Moreover, while switches are used within LANs to switch traffic between different hosts (layer2 switching),

---

* The border gateway protocol (BGP) is an example of an EGP.

interior routers (without connection with the exterior of an autonomous system) are used to switch traffic between different interior LANs (layer 3 switching).

Very often, in order to make a datagram arrive a certain destination, there is the need to cross several different autonomous systems. Therefore, different autonomous systems need to communicate, such that they update their routing table entries.

11.3.1 Routing Algorithms and Protocols

Different routing algorithms are used by different routing protocols to implement their function: compute a routing table that allows calculating the router's output interface to use in order to make the packet arrive a certain destination network. Routing protocols are used by routers to exchange knowledge information about the network. This information is then used to calculate the entries into the routing tables, that is, to calculate paths. Although the exchange of data generated by routing protocols to allow the creation of routing tables is carried out over UDP segments, these protocols belong to the network layer (layer 3). This can be seen from Figure 11.13. Consequently, there is a lower layer protocol (routing protocol) invoking a higher layer

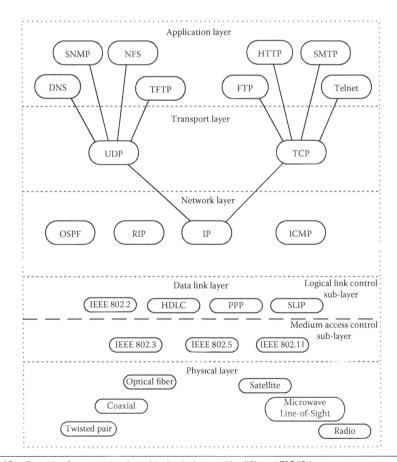

Figure 11.13 Example of some protocols and technologies used by different TCP/IP layers.

O   173.168.235.0/24 [110/782] via 174.168.235.1, 00:00:00, Serial1/0
C   174.168.235.0/24 is directly connected, Serial1/0
C   175.168.235.0/24 is directly connected, FastEthernet0/0
C   176.168.235.0/24 is directly connected, Serial1/1
O   177.168.235.0/24 [110/1562] via 176.168.235.2, 00:00:00, Serial1/1

Figure 11.14   An example of an IPv4 routing table.

protocol (transport protocol—UDP). This is possible in TCP/IP protocol stack, while the OSI reference model only allows an upper layer to make use of the services provided by a lower layer, and not the reverse.

Figure 11.14 shows an example of an IPv4 routing table. The different IP addresses shown in Figure 11.14 are class B, but the prefix /24 corresponds to class C. Consequently, we may conclude that the device (a host or a router) which has this routing table is using the classless addressing mode. From the second line, we see that the network 174.168.235.0, with subnet mask 255.255.255.0 (prefix /24), is directly connected (C on the left) through a serial interface with the device. The network 175.168.235.0/24 is directly connected through a Fast Ethernet interface. The network 176.168.235.0/24 is directly connected through a serial interface. From the first line, it is shown that the network 173.168.235.0/24 can be reached via other device (with the IP address 174.168.235.1, which belongs to the network 174.168.235.0 and which, from the second line, is directly connected to the router that has this routing table). The first line also gives us information that the remote device with IP address 174.168.235.1 can be reached through a serial interface (Serial1/0). Therefore, we conclude that this is a remote network (not directly connected). Moreover, the letter O means that the routing protocol used to generate this entry of the routing table was the open shortest path first (OSPF). Routing protocols are used to allow devices creating routing table. Finally, from the last line, it is shown that the remote network 177.168.235.0/24 can be reached via other device (176.168.235.2) and that the protocol OSPF is being used to generate this routing table entry. The information inside brackets [110/1562] is as follows: 110 is the administrative distance and corresponds to the OSPF protocol; 1562 is the metric distance (or cost), which corresponds to a value used by routers to make decisions about the path to use to make the datagrams arrive the destination. In case there is more than one path to reach a certain destination, the router selects the route with the smaller metric distance.

Routing algorithms are normally grouped into two main categories:

- Nonadaptive algorithms
- Adaptive algorithms

Two types of nonadaptive algorithms exists: the static and the flooding algorithm.

Using nonadaptive algorithms, the entries into the routing tables are not calculated using network topologies, traffic, or bandwidth metrics. The selection of the path to utilize is previously computed and preloaded into the router.

A static entry into the routing table consists of an action manually performed by the network administrator. Although different paths may exist for datagrams to arrive a certain destination, using a static entry, the network administrator forces the traffic to follow a certain fixed path. One disadvantage results from the fact that in case such path is suddenly interrupted (or subject to bad conditions), the traffic is not automatically switched over into the other path. Such re-routing of the traffic can only be performed through human intervention, that is, changing the required entry into the routing table. In this case, the traffic is interrupted during a certain time period.

Flooding is another nonadaptive algorithm. Using the flooding algorithm, a router forwards an input datagram into all different outputs. Consequently, the datagrams will arrive the destination with redundancy (the same datagram will be repeated) and through all possible paths. It is assured that one of the paths is the optimum one. An important disadvantage results from the fact that the network becomes overloaded with many repeated datagrams, which may degrade its performance.

On the other hand, adaptive algorithms support decision making of routers, which translates in routing table entries, taking into account one or a combination of metrics, such as the network topology, the traffic load, the bandwidth, and so on. In addition, different adaptive algorithms use different mechanisms to obtain information about the used metric. Since these metrics suffer variations in time, the routing table entries are permanently being updated and changing. As a result of this metrics variation, a certain traffic flow (e.g., a download) may initially follow a certain path (a sequence of routers) but, after a certain time period, the remaining datagrams may be routed through different intermediate nodes. Moreover, adaptive protocols compute the entries into routing tables without human intervention.

There are two families of routing protocols that make use of adaptive algorithms:

- Distance vector
- Link state

11.3.1.1 Distance Vector Protocols   Distance vector routing protocols compute the best path based on a metric to each destination node within an autonomous system (for IGP-type protocols) or in remote autonomous systems (for EGP-type protocols). The used metric* can be the number of hops, the bandwidth, the traffic, the delay, and so on. Some distance vector protocols make use of a combination of several of these elementary metrics. The computation of the shortest distance (best path) to a remote node using a distance vector protocol is as follows:

- Set the cost for itself as 0, and a certain value for adjacent nodes (1 when the metric is the number of hops).
- The preliminary routing table is sent to adjacent nodes with this cost information.

---

* The metric is also referred to as cost or metric distance.

- Each adjacent node saves the most recent received routing table and use this information to compute its own routing table by adding its cost/metric information to those included in the received routing table. Then the lowest distance to each destination is the one included in the new routing table. Finally, the node resends this new routing table to other neighbors (neighbors of neighbors).

From a routing table, each node knows which output interface should be used to forward packets to another network. Nevertheless, distance vector routing protocols do not compute the overall network topology. Distance vector protocols just compute the distances to other nodes. The network nodes send periodically the whole routing tables to the neighboring nodes. This is the way nodes use this protocol family to update their routing tables. In case such received routing table is not accurate, such inaccuracy is propagated along the whole network. Moreover, the amount of overhead generated in high dimension networks is high and the convergence time* is also high.

The mostly known distance vector routing protocols are the routing information protocol (RIP) and the interior gateway routing protocol (IGRP). The metric used by the RIP protocol is the number of hops, being exchanged between neighbors every 30 seconds. A maximum of 15 hops is allowed for a datagram to be delivered to the destination. A higher number of hops results in the datagram's discard.

Two important versions of the vector distance protocol exists: the RIP version 1 (RIPv1) [RFC 1058] and the RIP version 2 (RIPv2) [RFC 1723]. On the other hand, the IGRP consists of a Cisco Systems proprietary protocol.

IPv6 considers same type of routing protocols as IPv4, but with minimal modifications to accommodate the changes. The IPv6 uses an improved version of the RIPv2, entitled RIPng (RIP of the next generation).

11.3.1.2 Link State Protocols   Link state routing protocols compute a topologic database in each router, with the overall map of network connectivity. The map has the form of a graph, showing all network nodes and their interconnection. Link state protocols may use configured bandwidth in the interfaces or delay (among other parameters), as metric parameters. This map of topology is built as follows:

- Each node finds the neighboring nodes by sending a "hello" packet. A node which receives a "hello" packet respondes with a "hello" packet as well. Using this procedure each node measures the metric to its neighbors.
- Then, each node sends to the neighboring nodes link-state advertisement packets. These packets are sent periodically or when connectivity changes. Such packets include information about adjacent neighboring nodes as well as the corresponding metrics. Then the neighboring nodes resend the link-state advertisement packet to the neighbors of neighbors, following the flooding

---

* Time necessary to built in the routing table.

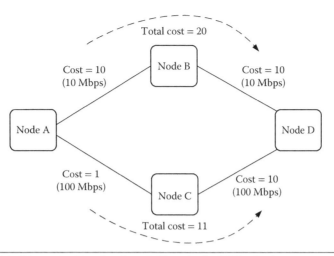

Figure 11.15 Metric distances (costs) and routing priorities using the OSPF protocol.

procedure. From the received link-state advertisement packets, the nodes can then build the network topology, including the indication of the cost between adjacent nodes.

Once such a network map is built, each node computes the shortest (best) path to each destination node within its autonomous system (for IGP-type protocols) or to remote autonomous systems (for EGP-type protocols). Such shortest path computation is performed using the Dijkstra algorithm [Dijkstra 1959], as described below. The entries into the routing tables consist of the calculated shortest paths using this algorithm. The shortest path consists of the path between a reference router and each destination node, which has the lowest metric distance.

Contrary to the distance vector protocols family where the whole routing tables are propagated, nodes using link state protocols only send to the neighboring nodes changes in the topology,* and this changes are only sent whenever a change in the topology is detected (instead of periodically). The mostly known distance vector routing protocols are the OSPF and the intermediate system to intermediate system (IS-IS). Note that the metric used by the OSPF protocol is $metric = 10^8/BW$, where BW stands for the bandwidth configured in the router's interface. Figure 11.15 depicts an example with bandwidths and resulting metric distances. Comparing the OSPF against other routing protocols such as the RIP, the following advantages are presented:

1. The OSPF allows ToS
2. The OSPF allows load balancing, that is, optimized distribution of the traffic in accordance with the different possible paths
3. The OSPF allows subdivision of an autonomous system in sub-groups, through the use of different areas

---

* For example, new nodes that enter the network, nodes that leave the network, links that become out of order, a cost/metric to another node that suffers a variation, and so on.

The IPv6 uses the OSPFv6, which is the OSPFv3 properly modified to accommodate the IPv6 address space [RFC 2740], as well as to integrate the link-local address of the IPv6.

The configuration of static and dynamic routing protocols in Cisco 2600 series routers is described in Section 11.4 and an example is included in Annex C.

11.3.1.3 Dijkstra's Algorithm This algorithm is used to compute the shortest path (minimum distance) between any two nodes. Consequently, this is the tool utilized by link state protocols for the calculation of routing tables.

This procedure needs to be implemented independently for all nodes. Each node is considered, at a time, as the reference node (initial node), and the minimum distance is calculated to all other network nodes. Then, another node is used as the initial node, and the procedure is repeated until all network nodes have been used as initial nodes. We assume that the distance between adjacent nodes is already known (calculated using the "hello" packets and propagated using the LSP packets). Note that the distances can be different in different directions. The Dijkstra's algorithm considers the following steps:

1. Assign an initial value zero to the initial node (node A, in example of Figure 11.16) and infinity to all other nodes. The distance of a node is the distance between the initial node and such remote node.
2. Set all nodes as unvisited and the initial node as the reference one.
3. Calculate the distance from the initial node to each one of the adjacent nodes through the several different paths (routes). Assume the selected path between the initial node and each adjacent node as the path that presents the lowest distance (shortest path). Note that the lowest distance may have several hops, instead of a single one. Mark these adjacent nodes as visited and register the nodes' distance as lowest computed distance (instead of infinity). Note that the path to the visited nodes will not be computed again (the best path has already been selected).
4. Repeat the distance calculation to adjacent of adjacent nodes, using the procedure previously described for the adjacent nodes. Increase the distance from the initial node successively by one, until the distance between the initial node and all network nodes have been calculated.

Let us consider the example of the network of Figure 11.16 to describe the Dijkstra's algorithm. Using the node A as the initial node, we set the distance to itself as 0 and infinity to all others. According to the procedure 3 above, the minimum distance to node B is 10 (direct path) and the minimum distance to node C is 15 (two hop path, through node B).* Set these the node's distances. Then, according to the procedure 4 above, the minimum distance to node D is to be computed. The distance through node B has a distance of 25, while the distance through node

---

* Note that the direct path between nodes A and C does not correspond to the lower distance.

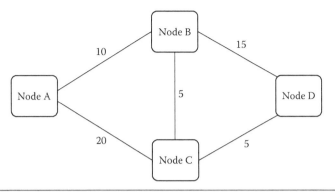

Figure 11.16 Example of a network whose shortest path is calculated with the Dijkstra's algorithm.

Table 11.17 List of Administrative Distances

PROTOCOL	ADMINISTRATIVE DISTANCE
Directly connected Interface	0
Static route	1
IGRP	100
OSPF	110
IS-IS	115
RIP	120
EGP	140
Unknown source	255

C is 20.* Consequently, the node D distance is set to 20, and the path between node A and node D is selected through node B and node C. As a conclusion, the calculated distances are the following:

- Node A: 0
- Node B: 10 (direct path)
- Node C: 15 (through node B)
- Node D: 20 (through node B and node C)

From this data, the routing table of node A can be constructed with the minimum distances to all remote nodes. Consequently, the initial node (node A) has knowledge that it has to send all the packets to node B (the interface to use).

11.3.1.4 Administrative and Metric Distances   In case there is more than one path to a certain destination, the router selects the path (by this order):

1. With the smaller administrative distance[†] (listed in Table 11.17)
2. With the smaller metric distance (in case there is more than one route with the same administrative distance, i.e., calculated with the same protocol)

---

* Note that the distance to node C was already set as 15 (mark as visited). Consequently, the route between node A and D through node C was not recomputed.

[†] A smaller administrative distance corresponds to a more reliable method used to calculate the route.

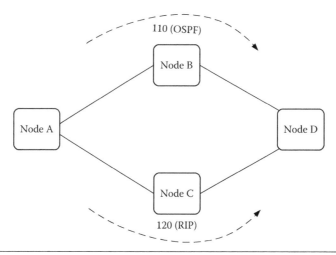

Figure 11.17   Administrative distances and routing priorities.

Figure 11.17 depicts an example with two paths between the node A and node D. There is one path through node B, whose calculation is performed with the OSPF protocol, and another path through node C, whose calculation is performed with the RIP protocol. According to the rule (1), the node A forwards the data through the node B because the OSPF protocol presents a lower administrative distance (110) than the RIP (120).

In the example of Figure 11.15, both paths are calculated with the OSPF protocol, and the route through node B presents a cost (or metric distance) of 20, whereas the route through node C presents a cost of 11. Since, according to rule (2), the selected route is the one that presents lowest metric distance, the route through C is the selected one.

11.3.2 Internet Control Message Protocol

The ICMP is a protocol used by operating systems to allow the exchange of error messages, control messages, for diagnostic purposes, or for flow control purposes of a network. This protocol is used by computers and routers.

As can be seen from Figure 11.13, the ICMP* belongs to the network layer,[†] being defined by [RFC 792]. Nevertheless, the ICMP protocol is implemented over the IP protocol (which is also a network layer protocol). In fact, an ICMP message is directly encapsulated into an IP datagram, which provides a connectionless and nonconfirmed service.

---

* The ICMP used with the IPv4 is also referred to as ICMPv4, whereas the ICMP protocol used with IPv6 is called ICMPv6.
† Note that, in the TCP/IP protocol stack, the network layer is also referred to as Internet layer.

Examples of ICMP messages include the following:

- Time to live exceeded message: this message is sent back to a sender of a datagram when it is discarded.
- Destination unreachable message: this message is sent back to a sender of a datagram when the destination's IP address could not be found.
- Echo request/reply message: the echo request message is sent to a certain destination's IP address to check whether or not it exists. The response to the echo request message is the reply message. These two messages are used by the ping command.
- Redirect message: this message is used when a path of a datagram is different from that considered by previous datagrams.

11.3.3 Fragmentation and Reassembling

A datagram is encapsulated into a frame at the transmitter side and de-encapsulated at the receiver side. Therefore, the maximum datagram/packet* size depends on the frame type used to encapsulate it.

A router is a device that is responsible for performing switching of packets between different networks or between a network and a network segment. The router normally make use of different access technologies (data link layer) in different interfaces, which translates in different frame types and maximum payload sizes. The MTU is the largest number of bytes that can be carried in a frame payload.† Note that the payload length of the data link layer frame is normally variable, depending on the amount of data to transport between two adjacent nodes.

If a router is switching a packet from a data link layer with a longer MTU into a data link layer with a shorter MTU and if the datagram length is longer than the shorter MTU, fragmentation needs to be implemented (i.e., each packet needs to be decomposed into a group of smaller pieces). Conversely, when a router is switching a packet from a data link layer with a shorter MTU into a longer MTU, reassembling may occur (i.e., the small pieces are grouped so as to recreate the packet with the original size).

Contrary to the IPv4 where fragmentation is carried out by intermediate nodes (routers), IPv6 fragmentation is implemented only by end stations of an end-to-end connection. In this case, intermediate routers never implement fragmentation. Consequently, the IPv6 protocol must discover the path MTU, in advance, before data exchange is initiated.

Table 11.18 shows the MTU of different data link layer technologies used in LAN.

---

* The datagram is the term used by IPv4, whereas the IPv6 uses the packet term. Since many of the functions described in this chapter are applicable to both IPv4 and IPv6, these terms are used interchangeably.

† Note that frame header is not included.

Table 11.18 Default MTU for Different Data Link Layer Technologies

DATA LINK LAYER	MTU (BYTES)
PPP	296
SLIP	296
Ethernet	1500
ISDN	1500
HDLC	1600
IEEE802.11	2312
IEEE802.5	4464

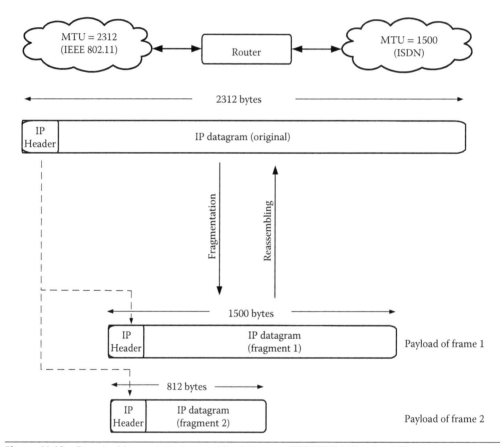

Figure 11.18 Example of fragmentation/reassembling between an IEEE 802.11 LAN and an ISDN segment.

It is worth noting that different packet fragments transport the maximum load (allowed by the MTU), whereas the last packet fragment transports the rest of the fragmented packet. Figure 11.18 depicts a router responsible for interconnecting a IEEE 802.11 LAN with an ISDN WAN segment. Since the LAN has a MTU = 2312 bytes and since the WAN has a MTU = 1500 bytes, the router needs to implement fragmentation in one direction (LAN to WAN) and reassembling in the opposite direction (WAN to LAN). As can be seen, the first fragment transports the maximum load (1500 bytes), whereas the second fragment transports the rest of the packet (2312 − 1500 = 812 bytes).

11.4 Cisco Internetwork Operating System

Most of the Cisco Systems routers and switches make use of the Cisco Internetwork Operating System (Cisco IOS) to implement their functions and to allow the configuration of different parameters. Depending on the device version, there are different versions of Cisco IOS, with some differences among them. The current description considers the Cisco IOS included in Cisco 2600 series routers and in Cisco Catalyst 2900/2950 series switches.

These network devices have typically the following memories:

- RAM (random access memory), which is responsible for temporarily storing frames or packets (layer 2 or layer 3 switching, respectively), storing the Cisco IOS, and storing the running configuration. The running configuration corresponds to the active configuration where the changes in the configuration of the device are stored (e.g., the configuration of an interface with an IP address, the name of the device, etc.).
- ROM (read only memory), which is responsible for storing the power-on-self-test (POST). The POST is used for testing the hardware. Old versions of Cisco IOS store the POST in the ROM memory.
- Flash memory, which consists of an electrically erasable programmable read-only memory (EEPROM), being responsible for saving Cisco IOS images.
- Nonvolatile RAM (NVRAM), which stores the startup configuration.

When a router or a switch is started up, it executes the following sequence of actions:

- It reads the POST from the ROM and executes the corresponding test.
- It reads the Cisco IOS from the EEPROM and loads it into the RAM memory.
- It reads the startup configuration from the NVRAM and loads it into the RAM memory in the form of running configuration.

11.4.1 Introduction to Cisco Internetwork Operating System

The access and configuration of Cisco Systems devices is performed using the command line interface (CLI) or setup configuration, through one of the following lines:

- Console port (CON)
- Auxiliary port (AUX)
- Telnet

The connectivity of a PC with a device through the CON or AUX needs to be configured with the following parameters:*

- Bit rate 9600 bps
- 8 data bits
- No parity
- 1 stop bit, no flow

The setup configuration only allows basic access and configuration of Cisco Systems devices. On the other hand, the CLI† consists of the most important interface through which the operator, administrator, and manager may fully interact with Cisco systems devices. Nevertheless, the user must be familiar with the language used in configuration. This language is covered in the following subsections. Depending on the type of interaction with Cisco Systems devices, the CLI allows three basic command modes:

- User mode: this is the default mode after having logged in into the device. It allows the user to execute only basic commands relating to device user utilities (not administration utilities). This mode is identified by the symbol > in the prompt, following the router's name.
- Privileged mode: this is the mode that allows viewing the details of the running configuration. Moreover, the access into the configuration mode is performed from the privileged mode. This mode is associated to the ability to execute device administration utilities. After having entered into the user mode (default mode), the user must type the "enable" command to enter the privileged mode. Once in the privileged mode, the symbol # appears in the prompt, following the router's name. To restrict the access to the privileged mode, which gives full access to the device, a password protection may be configured. The privileged mode can be accessed through either the CON line or through telnet, whereas the auxiliary line only gives access to the user mode. The privileged mode may be left by typing the "disable" command. Moreover, the CLI may be left by typing the "logout" command.
- Configuration mode: this mode allows modifying the running configuration. To enter this mode, the user has to first enter the privile`ged mode, after what the user must type the "configure terminal" command. This mode is identified by the characters (config)# in the prompt, following the router's name. To save the inserted parameters into the running configuration, the user needs

---

* Many communication applications may be used to allow communicating with a Cisco systems device. An important application used to allow communicating with the Cisco Systems device is the Windows HyperTerminal.
† To enter the CLI, the user should respond "no" to the question: "Enter the initial configuration dialog?"

to enter the sequence of characters CTRL-Z or to type the "exit" command. The configuration mode includes different sub-modes, namely the following:

- Interface configuration: it allows configuring the interface, namely the interface IP address, interface description, and so on. This submode is identified by the characters (config-if)# following the router's name.
- Line configuration: it allows configuring the lines used to gain access to the device (CON, AUX, telnet). The configuration of an access password necessary to allow using a certain line is an example of a line configuration parameter. This submode is identified by the characters (config-line)# following the router's name.

It is worth noting that the help command is always available from the CLI, when in the privileged or configuration modes, just by typing "?" in the prompt. Alternatively, a user may type a sequence of characters that corresponds to the initial letters of a command followed by the "?." The prompt lists the commands initiated by the written characters.* Moreover, the command TAB may be used to complete commands (initiated by a sequence of characters).

11.4.2 Basic Configuration of Routers and Switches

As previously described, the configuration of the Cisco systems devices involves modifying the running configuration. The modification of the running configuration is possible using the configuration mode, which is accessed by typing "configure terminal"† from the privileged mode.

11.4.2.1 Configuration Mode Once in the configuration mode, before entering any of the sub-modes, basic parameters can be configured, namely the following:

- Setting the time and date: the time and date can be adjusted by typing "clock set HH:MM:SS DD MON YR."
- Configuring the secret password: the secret password is an access password to the privileged mode. This is configured by typing "enable secret x," where x stands for the password.
- Configuring the router's name: this is possible by typing "hostname y," where y stands for the router's name.

11.4.2.2 Line Configuration Submode One may enter the line configuration submode, by typing "line console, 0" "line aux 0," or "line vty 0 4," to enter the console line, auxiliary, or telnet, respectively. The most configured item in the line submode is the access

---

* For example, Typing "cl?" returns the following list of commands: "clear" and "clock."
† Note that the Cisco IOS allows using abbreviated commands. In this context, instead of having to type "configure terminal," the user may just type "conf t" to enter the configuration mode from the privileged mode.

password. To configure an access password to enter the line (e.g., CON), the user must type "login" (to request a login password) and then "password y," where y stands for the password. To not request a login password, the user must type "no login."

11.4.2.3 Interface Configuration Submode   One may enter the interface configuration submode by typing "interface int_typeslot/port",* where int_type is the interface type (such as Ethernet, Fast Ethernet, serial, etc.), and slot/port corresponds to the slot and port where the interface is installed in the router. Some of the parameters that can be configured in the interface submode include the following:

- Configuring an interface description: a description of an interface may be configured by typing "description z," where z is the description itself.[†]
- Configuring an interface IP address[‡]: the interface IP address is configured by typing "IP address int_IP_address subnet_mask,"[§] where int_IP_address stands for the IP address of the interface and subnet_mask stands for its subnet mask. This command should be followed by the "no shutdown" command, which is necessary to activate the interface. It is worth noting that, when a serial cable[¶] is used to interconnect two routers, the configuration must consider one router acting as DTE and the other acting as a DCE. For the communication to be possible, the router acting as DCE needs to send to synchronism clock to the router acting as DTE. This is achieved by typing "clock rate clock_rate,"[**] where clock_rate stands for the clock rate expressed in bit/s (bps). Finally, the interface configuration can be checked by typing one of the following commands: "show interface," "show interface etherhet0/0," or "show IP interface brief".

11.4.3 Network Configuration in Routers

Besides the configuration of basic parameters, as covered in the previous subsection, there is the need to configure static or dynamic routing protocols in routers. This is necessary for a router to be able to create and update routing tables and, consequently, to be able to forward packets. Moreover, computers and switches also require some level of configuration.

The configuration of a computer is relatively simple. It normally consists of configuring the IP address (which can be static or using a DHCP server), as well as the IP address of the default gateway.[††]

---

* For example, interface ethernet0/1.
† A description may consist of, for example, "interface to Amsterdam."
‡ This configuration does not apply to switches.
§ For example, IP address 191.123.213.10 255.255.0.0.
¶ The HDLC protocol is normally used as the data link layer protocol over serial cables to interconnect two routers.
** For example, clock rate 32000.
†† The default gateway consists of a router to which all the packets generated in the computer should be forward to.

With regard to the switch, which is normally the central node of a LAN, a configuration is required in case virtual local area networks (VLAN) are to be created. This topic is covered in the following subsection. Otherwise, only basic parameters such as device's name and description are configured, if desired. The configuration of routes are performed as described in the following subsections.

11.4.3.1 Static Route Configuration   The configuration of a static route is performed by typing "ip route net_address subnet_mask next_router_address,"* where net_address is the IP address of the destination network to be reached, subnet_mask is the corresponding subnet mask and next_router_address is the IP address of the interface of the next router to which the packets should be forward to, in order to reach the destination network. Naturally, these commands must be inserted as many times as the number of remote networks to be statically configured. An entry into a routing table can be removed by typing "no ip route net_address subnet_mask next_router _address." In case a router should forward all the packets to a single other router, the command to use consists of "ip route 0.0.0.0 0.0.0.0 next_router_address,"† followed by the command "ip classless" (which indicates that the router must not take into account the subnet mask when forwarding packets). Note that the first group of 0.0.0.0 stands for all networks, whereas the second group of 0.0.0.0 stands for all subnet masks. This command is known as the default route, being normally used when a router is only connected to another router. A common situation is when the router only has two interfaces: one connected to the LAN and another interface connected to another router. In this case, all the packets received from the LAN should be forward to the other router (default route).

11.4.3.2 Routing Information Protocol Configuration   The configuration of the RIP protocol is straightforward. After having entered into the configuration mode, the user only has to type "router rip," followed by as much commands of type "network net_IP_address"‡ as the number of interfaces that a router has.

11.4.3.3 Open Shortest Path First Configuration   After having entered into the configuration mode, the user must type "router ospf proc_number,"§ where proc_number stands for a random process number to be considered in the router's configuration,¶ followed by as much commands of type "network net_IP_address wild_card area area_number"** as the number of interfaces that a router has. Note that wild_card stands for the wild card, which is used instead of subnet mask (approximately the

---

* Ex: ip router 191.123.213.0 255.255.255.0 192.125.215.4.
† Ex: ip router 0.0.0.0 0.0.0.0 192.125.215.4.
‡ Ex: network 10.0.0.0.
§ For example, router ospf 100.
¶ The process number is to be processed when a VPN is established using the MPLS protocol.
** For example, network 191.123.213.0 0.0.0.255 area 0.

opposite of it) and area_number stands for the area number to which this router belongs.* As described for the OSPF protocol, an advantage of this protocol results from the fact that an autonomous system may be split into different areas, facilitating the management and simplifying routing tables.

After having configured all the routes, using the static procedure or dynamic protocols, the user may view the routing tables by typing "show ip route." Moreover, the command "traceroute dest_IP_address" may be used in the router to follow the sequence of nodes of a packet until they reach a certain destination IP address (dest_IP_address).† An example of an IPv4 routing table is depicted in Figure 11.14. Naturally, the inserted commands are only saved into the running configuration by inserting the sequence of characters CTRL-Z. The annex C contains an example of a network configuration, including the configuration of the different interfaces.

11.4.4 Configuration of Virtual Local Area Networks

In case multiple VLANs are to be created, this configuration needs to be programmed in switches. Different VLANs are used to reduce the broadcast domain of a network, for security reasons, or to improve the management flexibility level of a network.

Traditionally, a LAN corresponds to a single physical and logical network, which is associated to a single network or sub-network IP address.‡ A user who intends to send data to all the hosts within the LAN uses the corresponding broadcast IP address. Nevertheless, it is known that an organization (such as a company) is typically composed of different departments,§ and most of the broadcasts are restricted to the department. If we create different VLANs and associate each department to a different VLAN, the broadcast traffic can then be sent only to the own department, without having to overload the rest of the LAN with traffic. Naturally, each VLAN should be split into different sub-networks, each one with a certain broadcast address. Without this capability, a router would have to be included to create different LANs¶ within the same organization. In addition, this use of VLANs brings an additional level of flexibility, as a computer physically located in a room traditionally used by a certain department may be used by a person belonging to another department. The only action that is required is to assign the interface of the switch from one into another VLAN or to move the patch cord from one into another interface of the switch.

---

* The first area must be 0.
† Similarly, the network administrator may use the commands "ping dest_IP_address" and "tracert dest_IP_address" in computers to check the connectivity or to follow the sequence of nodes of a packet until they reach a certain destination IP address. Moreover, the command "ipconfig" allows the visualization of the IP configuration in host interfaces.
‡ In case NAT/PAT is considered, a LAN may use a single external IP address, whereas internally, different private or public addresses are considered.
§ Example: financial department, human resources department, and so on.
¶ One LAN per department.

An important characteristic of VLANs implementation relies on the inability of a switch to allow the exchange of data between different VLANs. This can only be achieved by using a router properly configured to allow this. This characteristic may be used to logically isolate different networks or sub-networks used for different purposes within the same physical network infrastructure. Consequently, a common network infrastructure may be used for different purposes: a VLAN may be used, for example, to provide Internet access, another VLAN to give access to the voice over IP, another VLAN to provide access to the Intranet, and so on. Depending on the service that a terminal host intends to gain access, such host is configured with the corresponding network or sub-network IP address. This represents great savings in terms of cabling and network devices, while gaining an additional level of flexibility.

The configuration of switches typically follows two different steps: (1) the creation of VLANs, and (2) the assignment of interfaces to VLANs, as follows:

- Creation of a VLAN: this is configured in the privileged mode (not in the configuration mode). First, the user must type "vlan database," followed by "vlan vlan_number name vlan_name," where vlan_number* is the number assigned to the VLAN and vlan_name stands for the name assigned to the VLAN.† The created VLANs can be saved into the running configuration by typing the "exit" command.‡

- Assignment of an interface to a VLAN: this is configured using the running configuration.§ Then, the user configures each switch interface by typing the command "interface int_typeslot/port,"¶ where int_type is the interface type (such as Ethernet, Fast Ethernet, etc.), and slot/port corresponds to the slot and port where the interface is installed in the switch. This command must be followed by "switchport access vlan vlan_number," which corresponds to the assignment of the previously selected switch interface to a certain VLAN number. These pair of commands needs to be inserted for each interface of the switch. Otherwise, the interfaces remain in the default VLAN (VLAN1). This configuration can be saved into the running configuration by inserting the sequence of characters CTRL-Z. Finally, the user may check the assignment of the interfaces to different VLANs by typing the command "show vlan," "show vlan brief," or "show running-configuration."

As previously described, an important characteristic of VLAN is the inability to exchange traffic between different VLANs. For this exchange to become possible, a router needs to be connected to the switch, while the switch interface connected to

---

* Note that the default VLAN is the VLAN1. This VLAN is not configurable.
† For example, human resources department.
‡ Note that the VLAN database is saved by using the "exit" command, instead of the CTRL-Z sequence of characters (which is still valid for saving the changes when in the configuration mode).
§ As previously described, this is accessed from the privileged mode by typing "configure terminal."
¶ For example, interface ethernet0/1 switchport access vlan 2.

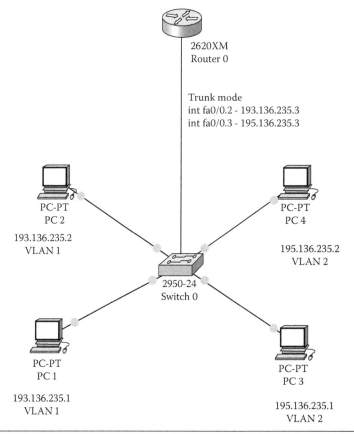

Figure 11.19 Example of a LAN with two VLANs configured and a router to allow the exchange of data between different VLANs.

the router needs to be configured in the trunk mode (instead of the above described access mode). This is achieved by typing "interface int_typeslot/port," followed by the command "switchport mode trunk" (Figure 11.19).

Moreover, the router's interface needs also to be properly configured to allow the exchange of traffic between different VLANs. This is performed by using the following sequence of commands in the configuration mode:

a. "interface int_typeslot/port" (the identification of the router's interface connected to the switch)
b. "no ip address" (do not assign any IP address to this router's interface*)
c. "no shutdown" (to activate this router's interface†)
d. exit (to exit without saving)
e. "interface int_typeslot/port.sub_int"‡ (the identification of the sub-interface to configure)

---

* Only the sub-interfaces will have IP addresses.
† Only the interface is activated, not the sub-interfaces.
‡ For example, int fa0/0.2, where 2 is the sub-interface to configure, which will be associated to a certain VLAN.

f. "encapsulation dot1q vlan_number" (uses the encapsulation protocol IEEE 802.1q for this sub-interface to connect to the vlan_number)

g. "IP address int_IP_address subnet_mask" (the IP address and subnet mask of the sub-interface)

The commands e., f., and g. above are to be typed as many times as the number of sub-interfaces to create and configure. Note that the IEEE 802.1q* consists of a protocol that manages multiple independent logical networks (VLANs) within a single physical IEEE 802.3 LAN network. The annex C contains an example of a network configuration where two switches are configured with VLANs, including a router which allows the exchange of data between different VLANs.

End of Chapter Questions

1. What is the metric distance considered by the RIP protocol?
2. Consider a router with the following routing table:
   > O 173.168.235.0/24 [110/782] via 174.168.235.1, 00:00:00, Serial1/0
   > C 174.168.235.0/24 is directly connected, Serial1/0
   > C 175.168.235.0/24 is directly connected, FastEthernet0/0
   > C 176.168.235.0/24 is directly connected, Serial1/1
   > R 173.168.235.0/24 [120/2] via 176.168.235.2, 00:00:00, Serial1/1

   a. Draw the network diagram corresponding to the information obtained from the routing table.
   b. According to the routing rules, what is the address of the next router to which the datagrams should be forward in order to reach a host with the destination IP address 173.168.235.3?
3. What is the broadcast address of the sub-network where the host with the IP address 10.96.2.1/12 is?
4. What does an APIPA address stands for? In which cases can it be used?
5. Why are IPv6 routing tables shorter than IPv4?
6. What is the PAT/NAT used for?
7. What is the DHCP used for?
8. What are the advantages of the IPv4 address space in classless mode relating to the classfull mode?
9. What are the mechanisms that can be used to maximize the limited address space made available by the IPv4 protocol?
10. Given the network address 10.0.0.0, with subnet mask 255.255.192.0, present the sub-network addresses, the first and last host addresses of each sub-network, and the broadcast addresses.
11. Enumerate the reasons which motivated the evolution of the Internet Protocol to IPv6. What are the advantages of the IPv6 relating to the IPv4?

---

* The IEEE 802.1q protocol is also referred to as VLAN tagging.

12. Consider a router with the following routing table:

    R 173.168.235.0/24 [120/3] via 174.168.235.1, 00:00:00, Serial1/0
    C 174.168.235.0/24 is directly connected, Serial1/0
    C 175.168.235.0/24 is directly connected, FastEthernet0/0
    C 176.168.235.0/24 is directly connected, Serial1/1
    R 173.168.235.0/24 [120/2] via 176.168.235.2, 00:00:00, Serial1/1

    a. Draw the network diagram corresponding to the information obtained from the routing table.
    b. According to the routing rules, what is the address of the next router to which the datagrams should be forward to reach a host with the destination IP address 173.168.235.3?

13. Given the network address 193.123.10.0/27, present the sub-network addresses, the first and last host addresses of each sub-network, and the broadcast addresses.

14. Given the network address 10.0.0.0, with subnet mask 255.255.255.192, present the sub-network addresses, the first and last host addresses of each sub-network, and the broadcast addresses.

15. Given the network address 10.0.0.0, with subnet mask 255.192.0.0, present the sub-network addresses, the first and last host addresses of each sub-network, and the broadcast addresses.

16. How many hosts can be allocated in a classfull class B network address?

17. How many class A network addresses exist?

18. What are the subnet masks possible to be used with a class B network address?

19. Why does the subnet mask 255.128.0.0 is normally not utilized with class C network addresses?

20. Consider the class C network with the address 193.136.235.0/27. Choose the true answer:
    a. Eight sub-networks can be created.
    b. The address of the first sub-network is 193.136.235.32.
    c. The number of hosts per sub-network is 62.
    d. All previous responses are correct.

21. Given the EUI-64 address (interface ID) 01:23:45:ff:fe:67:89:ab, of an IPv6 address, what is the corresponding MAC address?

22. What are the functionalities of the network layer?

23. Consider the network depicted below. Assuming a subnet mask 255.255.255.0, write the IOS commands that would be necessary to configure:
    a. In the router 2: a permanent static route to the remote networks
    b. In the router 1: the RIP protocol and the configuration of interfaces and sub-interfaces
    c. In the router 3: the OSPF protocol
    d. In the router 3: a default route to the remote networks
    e. In the switch 0: the creation of VLANs and the configuration of interfaces
    f. In the switch 1: the creation of VLANs and configuration of the interfaces

24. What does MTU stands for? And what is the relationship between MTU and frame size?

25. What is the ARP used for?

26. What does an autonomous system stands for?

27. Which kind of routing protocols do you know?

28. What are the differences between a link state routing protocol and a distance vector routing protocol?

29. What is the default maximum number of hops allowed by the RIP protocol before a datagram is discarded?

30. What does metric distance (or cost) stand for?

31. What is the sub-network address to which the host with the IP address 10.88.1.2/11 belongs?

32. Consider an host with the MAC address 01:23:45:67:89:ab. What is the corresponding interface ID of the IPv6 address?

33. What is the relationship between the MTU and the datagram length? What measures can be implemented to adjust these two parameters?

34. How many hosts can be allocated in a sub-network where the host with the IP address 10.76.1.2/11 is located?

35. How many bits compose an IPv6 address?

36. Why is IPv6 routing faster than IPv4?

37. Given an IPv4 address, how do you create the corresponding IPv4-mapped IPv6 address?

38. What are the two basic rules used by routers to decide about the output interface to use to forward datagrams to a certain remote network?

39. How in an IPv6 address composed?

40. What does fragmentation stands for?

41. What does administrative distance stands for?

42. Consider a router with the following routing table:
    R 173.168.235.0/24 [120/3] via 174.168.235.1, 00:00:00, Serial1/0
    C 174.168.235.0/24 is directly connected, Serial1/0
    C 175.168.235.0/24 is directly connected, FastEthernet0/0
    C 176.168.235.0/24 is directly connected, Serial1/1
    R 173.168.235.0/24 [120/2] via 176.168.235.2, 00:00:00, Serial1/1
    a. Draw the network diagram corresponding to the information obtained from the routing table.
    b. According to the routing rules, what is the address of the next router to which the datagrams should be forwarded to reach a host with the destination IP address 173.168.235.3?

43. What is the DHCP protocol used for?

44. What is the metric distance considered by the OSPF protocol?

12
TRANSPORT LAYER

The transport layer of the transmission control protocol/Internet protocol (TCP/IP) protocol stack is an end-to-end layer that aims to provide, as much as possible, the required quality-of-service (QoS) to the application layer, at the lowest cost. To assure this functionality, the transport layer aims to achieve a trade-off between the QoS parameters requested by the application layer and the QoS available at the network layer (which is a function of the instantaneous traffic).

The QoS parameters dealt with and negotiated by the transport layer includes the following:

- Requested bit rate in each direction
- Bit error rate
- Maximum end-to-end delay
- Maximum end-to-end jitter
- Priority rules
- Probability of failure in the connection establishment
- Maximum time to establish A transport connection
- Maximum time to terminate A transport connection
- Cost
- Security protection mechanisms

Depending on the service and application layer protocol, the four most important basic service requirements are as follows:

- Bit rate requirements*
- Sensitivity to loss of data
- Delay sensitivity
- Variable or constant bit rate service

As an example, the IP telephony (VoIP) service presents low requirements in terms of bit rate and low sensitivity to loss of data (typically, the voice service accepts a bit error rate [BER] better than 10^{-3}). Nevertheless, the IP telephony QoS requires a low delay, while presenting a high sensitivity to delay variations (jitter). The IP telephony data exchange is performed over user datagram protocol (UDP), whereas the RTP (upper layer) assures the correct sequence of the exchanged segments. Another

---

* Bit rate is commonly referred to as bandwidth. Nevertheless, Chapter 3 describes the relationship between these two factors.

example is a download, which makes use of the FTP. This requires an approximately high bit rate and a very high sensitivity to loss of data. On the other hand, this service presents a reduced sensitivity to delay and to jitter.

To assure that the required bit rate is made available to the application layer, the transport layer may group multiple network layer connections into a single application layer connection. Let us consider the case where the network layer has a 512 kbps connection established, and the application layer is requesting a 1024 kbps connection. In this situation, the transport layer may establish a second 512 kbps network connections and, in a transparent manner, offer a 1024 kbps connection to the application layer. Similar function may be offered when the application layer is requesting a datagram size higher than the maximum transmission unit (MTU). In this situation, the transport layer performs the segmentation (at the transmitting side) and reassembling (at the receiving side) of data, such that it fits within the MTU.

The address of the TCP/IP transport protocol is the port number, composed of a 16 bit address (see Figure 12.1). Port numbers between 0 and 1023 are assigned by the Internet Assigned Numbers Authority and are called well-known ports. These port numbers are to be used by servers and cannot be used by hosts (see Tables 12.1 and 12.2). Port numbers between 1024 and 49,151 are also registered ports but are used for less known services. Finally, port numbers between 49,152 and 65,535 are port numbers to be used by hosts.

A TCP/IP connection is identified by a source and a destination socket. A socket consists of the concatenation of an IP address with a port number. An example of a socket is 173.22.83.10:80. The first group of 32 bits is the IP address (173.22.83.10), whereas the second group of 16 bit corresponds to a port number (80). In this case, since 80 is a well-known port number, this transport layer address corresponds to a HTTP service provided by a server.

The application layer generates a continuous flow of data, which is segmented and transmitted by the transport layer. The message format of the TCP/IP protocol is called a segment and may be transmitted using one out of two different protocols:

- UDP
- TCP

These protocols are defined in the following subsections.

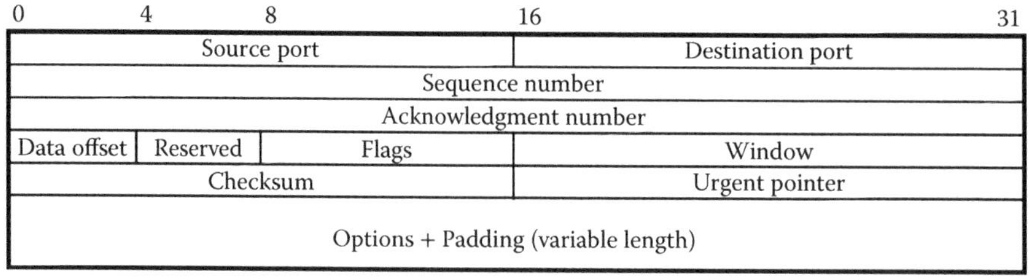

0 4 8	16 31
Source port	Destination port
Sequence number	
Acknowledgment number	
Data offset	Reserved	Flags	Window
Checksum	Urgent pointer
Options + Padding (variable length)	

Figure 12.1 Transmission control protocol header.

Table 12.1 A List of Some Well-Known TCP Port Numbers

TCP PORT NUMBER	SERVICE
21	FTP
23	Telnet
25	SMTP
80	HTTP
110	POP3
143	IMAP
443	HTTPS
161	SNMP

Table 12.2 A List of Some Well-Known UDP Port Numbers

UDP PORT NUMBER	SERVICE
53	DNS
69	TFTP
161	SNMP

12.1 Transmission Control Protocol

The TCP is a transport protocol used to provide the application layer with a connection-oriented service between two end entities. Since the TCP is connection-oriented, it requires the previous connection establishment before the data is exchanged. Moreover, since the TCP provides reliability,* it assures that the exchanged data, delivered by the transport data to the application layer, presents the following properties:

- Data is delivered with reliability: by using the positive acknowledgment with retransmission (PAR) procedure, the TCP protocol keeps the segment errors at an acceptable level.
- Packets are delivered in the correct sequence: as a result of instantaneous traffic conditions, routing tables are permanently being updated. Consequently, the routing of a connection suffers variations in time, which translates in a change of the route being used to forward packets between the two end entities. Therefore, a segment with a higher sequence order may arrive before a segment with a lower sequence order. The TCP assures the reordering or segments before delivery to the application layer.
- Packet losses are detected and corrected: as previously described, one of the various datagram's fields is the time to live. This field consists of the maximum number of nodes that a datagram may cross. If a datagram crosses a higher number of nodes, it is discarded. Consequently, by using the TCP protocol, such discard is detected by the transport layer, being requested the repetition of the segment transported in the discarded datagram.

---

* A connection-oriented service is always a confirmed service.

- The duplication of packets is avoided: duplication of packets occurs as a result of the flooding routing protocol. Since the TCP segments are numbered, the receiver assures that the segments are delivered to the application layer in the correct order.

The error detection is provided using the PAR procedure as follows:

- When a certain amount of data is sent, the transmitting entity activates a timer and wait for its acknowledgment (ACK) within a certain time period.
- If the amount of data is correctly received by the receiving entity, its ACK is sent back to the transmitter (service confirmation). In case an error is detected, the receiving entity does not send back the ACK. The error detection technique is implemented by the TCP using CRC codes.* Note that although the data link layer also makes use of error control mechanisms,† some residual‡ error exists. Moreover, packets can be discarded by intermediate routers and packets can be duplicated or can arrive out of order. Consequently, the TCP does not assume the service provided by the lower layers as with quality. Consequently, the TCP has mechanisms to deal with such impairments.
- In case the transmitting entity does not receive the ACK within the expected time period, it assumes that the amount of data did not arrive the destination correctly and repeats its transmission.
- Then the transmitter proceeds with the transmission of the rest of the data.

Note that this procedure allows error control and flow control. When the receiver informs the transmitter that a certain amount of data was correctly received (error control), it is also informing that it is ready to receive more data (flow control).

To make an efficient use of the available transmission medium, while providing flow control, the TCP makes use of the sliding window protocol.§ Therefore, instead of sending a segment at a time (as in the case of the stop and wait protocol), a group of segments are sent together. Moreover, signaling data (e.g., acknowledgments) are often exchanged using piggybacking, where a segment is used simultaneously to send data and to acknowledge data previously received from the opposite direction. The sliding window protocol is implemented using number of octets as a reference. Before the data is exchanged, the receiver informs the transmitter about the number of octets that can be sent without acknowledgments (the window size is expressed in number of octets). Such number of octets (window size) is encapsulated into several segments. Moreover, the maximum number of octets that can be transported in a single segment is known as maximum segment size (MSS).

---

* The reader should refer to Chapter 10 for a detailed description of the CRC codes.
† With some exceptions, such as the SLIP.
‡ Error control mechanisms do not assure 100% error-free frames.
§ The sliding window protocol is detailed in Chapter 10.

The TCP is used to support the following application protocols: FTP, Telnet, SMTP, HTTP, and so on. Figure 12.1 depicts the different fields of a TCP header, with the corresponding indication of each field's size expressed in number of bits. The TCP segment header is composed of two parts: (1) a fixed length part, with a total size of 20 octets, which includes 10 mandatory fields between the source port and the urgent pointer (each line of Figure 12.1 has a total of 32 bits, i.e., four octets); and (2) a variable length part, which corresponds to the options plus the padding.

The content of the several segment's header fields are listed below:

- Source port: it consists of a 16-bit transport layer address, which identifies the source of a transport layer connection.
- Destination port: it consists of a 16-bit transport layer address, which identifies the destination of a transport layer connection.
- Sequence number: it consists of a 32-bit field used to identify the sequence of the first octet within a segment.
- Acknowledgment number: it consists of a 32-bit field, which indicates the sequence number of the next expected data octet (to be sent within the next segment). In case the previous data was properly received by the receiving entity, this number corresponds to the sequence number immediately after the last octet contained in the last segment. Alternatively, in case of error, this number corresponds to the first octet of the segment received with error, which is meant to be retransmitted.
- Offset: it consists of a 4-bit number, representing the header length expressed in groups of 32 bits. This corresponds to the shift between the beginning of a segment and its payload. In fact this equals the options length (expressed in group of 32 bits) plus five (fixed length header).
- Reserved: it is a reserved 4-bit field.
- The flags field consists of eight individual flags, each with one bit (CWR, ECE, URG, ACK, PSH, RST, SYN, and FIN). When a flag is activated, it means that a certain function is in use as follows:
  - The URG flag activated means that the urgent pointer field is to be considered by the receiving entity of the segment.
  - The ACK flag activated means that the acknowledgment field is to be considered by the receiving entity of the segment.
  - The RST flag is used to terminate a confused connection or to refuse a new connection establishment.
  - The SYN flag is used in the establishment of a new connection.
  - The FIN flag is used to terminate a previously established connection.
- Window: it is a 16-bit size number, used by a receiving entity to inform the transmitter about the maximum number of octets that can be sent by the transmitting entity, using the sliding window protocol (flow control). It allows the transmission of several segments without acknowledgment, in total carrying a number of octets lower than the window field.

- Checksum: it consists of 16 redundant bits, used to perform error detection of the entire segment, using CRC codes.
- Urgent pointer: it consists of a 16-bit number. Adding this number to the sequence number, we obtain the last sequence number of the urgent data.
- Options: because of its rare use, this field is optional, including some instructions such as the following:
  - MSS: it consists of a 32-bit number, representing the largest amount of data, specified in number of octets that can be carried in a single segment. This number is negotiated during the TCP connection establishment. To avoid segmentation, the MSS should be smaller than the MTU minus the IP header length (the MTU depends on the data link layer used).
  - Window scaling: it consists of a 24-bit number, used to introduce a scale factor to the number expressed in the window field (which is limited to 65,535 octets). This is used in high data rate transmission to increase the size of the window.
  - Selective acknowledgment permitted: it is a 16-bit number used to implement selective repeat, instead of the pure sliding window protocol. This allows an efficiency gain in terms of bandwidth.
- Padding: used to ensure that the header length is an integral number of 32-bit words.

Note that the total segment size (LEN*) is not explicitly defined in any field. This can be calculated from the total length field of the IP datagram subtracting the Internet Header Length of the IP datagram and subtracting the segment offset. Naturally, before this subtraction is completed, all numbers need to be expressed in the same units (e.g., in octets or in bits).

Figure 12.2 depicts the handshaking used in a TCP connection establishment, where SP stands for source port and DP stands for destination port.

Similarly, Figure 12.3 depicts an example of TCP data exchange. As previously described, the TCP uses the sliding window protocol. From the first message depicted in Figure 12.3, it is seen that the receiving entity is authorizing the transmitting entity to send 200 octets using the sliding window protocol (WIN = 200), and that the maximum segment size is 100 (MSS = 100). Since the transmitting entity (server) has a total of 280 octets to send, this is split into several segments of 100 octets. The server sends two segments, which totalize 200 octets and waits for the acknowledgment from the receiver (client). Since an error occurred in the second segment, the acknowledgment number corresponds to the sequence number of the first octet of the second segment, which needs to be repeated. Then, the server retransmits the segment received with errors and the last segment (with only

---

* See Figure 12.3.

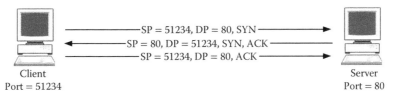

Figure 12.2 Transmission control protocol connection establishment.

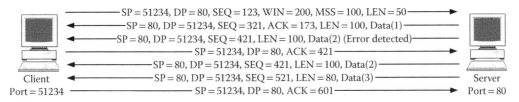

Figure 12.3 Example of transmission control protocol data exchange.

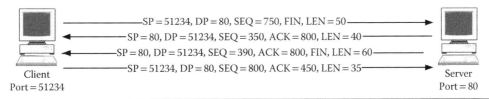

Figure 12.4 Ending a transmission control protocol connection.

80 octets). The data exchange is finalized with the acknowledgment sent by the receiving entity (client).

For the case of a segment received free of errors, the acknowledgment number corresponds to the sequence number of the first octet of the previously received segment summed with the number of octets sent in such segment (LEN). In case of error detection, the acknowledgment number corresponds to the sequence number of the first octet of the corrupted segment in the previous window of segments.

Figure 12.4 depicts the procedure to terminate a TCP connection. As can be seen, the client sends a segment to the server with the FIN flag activated, and this entity responds with ACK and FIN flags activated (in two different stages) This corresponds to the confirmation from the server. Finally, the client acknowledges the FIN message received from the server.

As previously described, the TCP requires that the receiving entity keeps sending successive acknowledgments to the transmitting entity and re-transmissions are made whenever acknowledgments are not received by the transmitting entity. This handshaking represents additional overhead in the network and additional signal delays. This is the reason why delay-sensitive services (e.g., voice and video streaming, etc.) do not make use of the TCP. These services are normally implemented using the UDP protocol, as defined in the following subsection. Contrarily, services that require data reliability, without presenting delay sensitivity, are normally implemented over the TCP.

12.2 User Datagram Protocol

The UDP is connectionless and provides an unconfirmed service to the application layer. As a connectionless service, it does not require the previous connection establishment, allowing the rapid exchange of data between entities.

Nevertheless, the data is delivered based on the best effort, that is, without assuring reliability of the exchanged data and without any kind of flow control. Since the UDP does not introduce delays (in opposition to the TCP), it is normally used by delay-sensitive services (which are not demanding in terms of data reliability) or by services whose data is periodically and redundantly retransmitted (such that an error is rapidly corrected by its update). Example of delay-sensitive but not error-sensitive service is the audio, the video streaming, or the IP telephony. Example of services that may be supported on the UDP due to its data redundancy is the SNMP, as well as the DNS protocol. The UDP only provides a minimum service to its upper layer, acting as a simple interface between the application layer and the network layer.

Figure 12.5 depicts the several fields of an UDP segment header, with the corresponding indication of each field's size expressed in number of bits. As in the case of the TCP segment header, the two initial fields are the source and destination addresses. The length field consists of a 16-bit number containing information about the whole segment length (header and data), expressed in octets. Note that this field does not exist in the TCP segment header, having to be calculated from the packet length and from the TCP segment header length (offset). Finally, the checksum is a group of 16 bits used to allow error detection of the UDP header. This field is optional. If not used, its value is set to zero. When an error is detected in the UDP segment header, the segment is just discarded. Note that, contrarily to the TCP error detection, the UDP error detection is never applied to the payload.

12.3 Integrated and Differentiated Service

As previously described, there are a number of QoS requirements that need to be followed to ensure that a service achieves the expectations. These requirements include the following parameters:

- Delay: services such as audio or video streaming can accommodate a low level of delay (are delay-sensitive)
- Jitter: this corresponds to a variation in delay. While a small amount of delay can still be supported by audio or video streaming, its variation rapidly degrades these type of services.

0	16	31
Source port	Destination port	
Length	Checksum	

Figure 12.5 User datagram protocol header.

- Throughput: the bit rate that is required to support a specific service. This can be fixed or variable. While the video streaming requires a high and fixed bit rate, the web browsing is not very demanding in terms of throughput, but is a service that requires a variable bit rate.
- Packet loss: the increase in the number of lost packets corresponds to a degradation of the BER. There are other factors that may degrade the BER, such as the errors generated in the physical and data link layer. Nevertheless, errors generated in physical and data link layer do not rely in the scope of the provision of QoS by transport layer. While the voice telephony can perform well with a BER up to 10^{-3}, a file transfer typically requires a BER better (lower) than 10^{-6}.

The need to achieve such QoS requirements is the result of the integration of different services into a common network infrastructure.* In order to provide QoS, there are two important models that can be implemented in TCP/IP networks. These models were normalized by the IETF, being named as follows [RFC 1633]:

- Integrated services (IntServ)
- Differentiated services (DiffServ)

12.3.1 Integrated Services

The IntServ [RFC 2211] [RFC 2212] [RFC 2213] [RFC 2215] aims to provide QoS in networks taking into account the different service requirements. The IntServ is a model that implements a group of mechanisms on the flow-by-flow basis. The provision of QoS is achieved by applying individual reservations defined by flow specifications† combined with the resource reservation protocol (RSVP) [RFC 2205] to a group of routers in a network. The packets belonging to an end-to-end connection, which pass through a group of routers benefit from a resource reservation, which may avoid latency or packets discard. Consequently, packets belonging to a service such as audio or video streaming may benefit from higher priority in intermediate routers. To allow the provision of resources, IntServ classify the different traffic into a wide range of levels of granularity.

The exchange of traffic is preceded by a connection establishment phase, which requests a certain level of QoS from the network (between two end entities), along a certain route. The intermediate routers receive such request and, based on the previously allocated resources, on the available resources and on the type of data, decide whether it is possible to assign the required resources (buffers, bandwidth, etc.) to the new connection. Such resource reservation is implemented using the RSVP protocol.

---

* This is also referred to as the convergence of the telecommunications.
† The flow specification depends on the type of traffic being exchanged (e.g., file transfer, web browsing, video streaming, etc.).

The drawback of the IntServ is that, since it rigidly assigns resources to connections and since the resources are limited, new connections may be refused or existing connections may be dropped, due to lack of resources. These effects of congestion are more visible in high dimension networks. Moreover, the amount of signaling needed to handle the QoS requirements in networks that implement IntServ tends to be too high. Consequently, IntServ is only normally adopted in low dimension networks.

12.3.2 Differentiated Services

DiffServ [RFC 2475] is an alternative model that can be implemented to tentatively allow the provision of QoS in high dimension networks, while reducing the required resources (overhead, bandwidth, processing, etc.). While the IntServ operates over individual flows of data, DiffServ is applicable to high volumes of data, reducing the required signaling, reducing the granularity and, consequently, the complexity. DiffServ handles different connections of the same type using the same principles.* This implies the prior negotiation of resources reservation for the traffic generated by a certain customer. Based on the customer's payment, an Internet service provider (ISP) offers a certain Service Level Agreement (SLA), which defines the maximum amount of data for each class of traffic (IP telephony, video streaming, file transfer, web browsing, etc.), as well as the type of warranties that are offered to each traffic class. Different traffic classes are identified using the differentiated services code point (DSCP) field of the datagrams headers. When a packet enters the ISP cloud, intermediate routers read the DSCP field, and the packet is treated according to the negotiated SLA. If the amount of traffic exceeds that negotiated in the SLA, packets may be dropped or delayed (or an additional fee may be charged to the customer). Moreover, in case of congestion, for the same traffic class (identified by the DSCP field), packets belonging to a customer with higher SLA have precedence.

DiffServ is the mechanism currently mostly implemented in the Internet for the provision of QoS.

End of Chapter Questions

1. What are the functionalities of the transport layer?
2. What is the difference between the IntServ and the DiffServ protocols?
3. What are the differences between the TCP and the UDP protocols?
4. Give examples of services and application layer protocols supported on TCP and UDP protocols.
5. Which kind of channel impairments is dealt with by the TCP protocol?
6. How is the QoS provided by the IntServ protocol?

---

* For example, different IP telephony connections or different file transfer connections belonging to customers with the same service level agreement are handled by the network in the same manner.

7. What is the reason why the exchange of IP telephony packets considers the UDP instead of the TCP protocol?

8. What are the phases of a TCP connection establishment?

9. A client intends to terminate a FTP connection with a server. Describe the messages' exchange necessary to terminate such connection.

10. How is the error control and flow control implemented by the TCP protocol?

11. Why do the TCP segments include a sequence number?

12. How is the QoS provided by the DiffServ protocol?

13. What are the disadvantages that may result from the use of the TCP protocol relating to the UDP protocol?

14. What is a well-known port?

15. What is the difference between jitter and delay?

16. What are the basic QoS requirements?

17. Knowing that the data link layer already normally makes use of error control mechanisms, why does the TCP protocol also implement these mechanisms?

18. What is the service level agreement used for by the DiffServ protocol?

19. A client intends to establish an HTTP connection with a server. Describe the messages' exchange necessary to establish the connection.

13

SERVICES AND APPLICATIONS

A data service is something that is provided to the end user for a specific function, making use of a specific program (application program). Examples of data services include the Internet protocol (IP) telephony, web browsing, e-mail, file transfer, and so on. These services can be provided making use of a variety of application programs. For example, Microsoft Outlook, Microsoft Outlook Express, and Groupwise are three examples of specific application programs that can be used to have access to e-mail (service). Web browsing is also a service that can be accessed making use of Internet Explorer, Mozilla Firefox, Google Chrome, and so on.

The application program described above has a different meaning than the application layer. As can be seen from Figure 13.1, the application layer is the layer of the open system interconnection (OSI) or transmission control protocol (TCP)/IP stack (network architecture), which provides some functionalities to the application program (user's program, external to the network architecture). Both are then used to provide a data service to the end user.

The tasks common to the different types of data services are typically included in other than the application layer. Those tasks were split along the other OSI or TCP/IP layers, as a function of their level of abstraction. Contrarily, the application layer is responsible for implementing tasks that are specific of the data service to be provided to the end user. The application layer comprises different protocols, one or more per data service. In addition, these protocols are different for different reference models (e.g., OSI or TCP/IP). While the file transfer access and management (FTAM) is the protocol of the OSI reference model (OSI-RM) which allows the file transfer, the TCP/IP defines the file transfer protocol (FTP) and the trivial FTP (TFTP).

As previously described, the upper layers of a network architecture are extreme-to-extreme layers, whereas the lower layers are point-to-point layers. The application layer defines functionalities in the interaction between ending hosts (e.g., between a client and a server).

Depending on the protocol stack used to implement the network architecture, there are different functionalities dealt with by the application layer. While the application layer of the OSI-RM is limited to the application layer itself, the TCP/IP model includes the functionalities that the OSI-RM assigns to the application layer, presentation layer, and session layer.

The session layer comprises a group of functionalities, which involves the management of a session. Such functionalities can be split as follows:

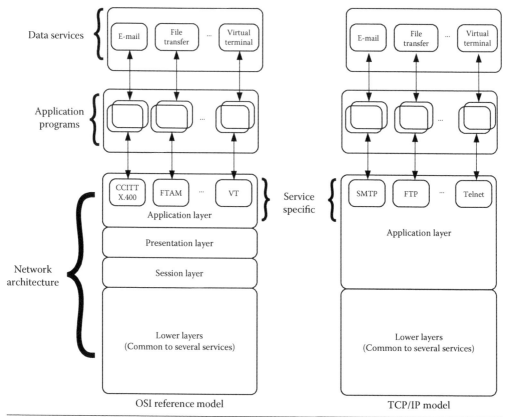

Figure 13.1   Difference between application layer (OSI and TCP/IP reference models), application programs, and services.

- Setup of sessions
- Synchronization
- Dialog management
- Activities management

Setup of sessions: a session consists of a logical link that is established between end entities. It has a meaning different from a physical link, which is established and managed by the transport layer. In case a physical link that supports a certain logical link fails, a new physical link is established and the session (logical link) continues with its exchange of data session protocol data unit (SPDU).*

Synchronization: it considers the periodic transmission of synchronization points that are properly acknowledged by the receiving entity. In case the physical link fails, the session data does not need to be retransmitted from the beginning, but from the last synchronization point correctly acknowledged. Note that, from the session layer perspective, an acknowledgement means that the received data has been properly processed.†

---

* The SPDU is defined in Chapter 2.

† The acknowledgement of good reception of data is performed at transport layer, whereas the acknowledgement of good processing of data is performed at session layer.

Dialog management: although the physical link may support full-duplex operation, some user processes only support, from the logical link perspective, half-duplex operation. The management of which entity is allowed to send data is managed at this layer, using a token that is transferred from one entity into another;

Activities management: the application data is typically split into a set of elementary activities. Each activity is properly managed and signalized using marks* to register that it has initiated or finalized. These marks are important in order to allow the receiving entity to identify the borders of an activity. An example of a sequence of two elementary activities can be a copy of a file followed by its deletion. The data cannot be deleted (second activity) without being sure that the copy (first activity) has been successfully performed. The ISO 8327/CCITT X.225 is an example of a standard that defines the session layer.

The presentation layer relates to the way the data is represented and exchanged in the interaction among different ending entities. Moreover, it provides means for the establishment and use of abstract syntax, which allows the exchange of data. An important syntax commonly used among different entities is the abstract syntax notation 1 (ASN.1). The task of this layer can be viewed as the definition of a language (standard) to use in the communication among different hosts. Instead of allowing the communication being performed with the internal language of a host (e.g., computer), a standardized language is established, and all hosts must use such language in the interaction with the others. This avoids the need to have successive translators (gateways) among all different possible host languages. This way, it is only needed to perform the translation from the host's internal language into a common standardized language. Note that the presentation layer also deals with encryption and compression of data, as these two functionalities change the language format used in communications. Such functionalities need to be properly negotiated and accepted before the exchange of data takes place. The ISO 8823/CCITT X.226 is an example of a presentation layer standard.

From the OSI-RM perspective, the application layer deals with the tasks that are specific of the application program, whereas the session and presentation layers take care of the tasks that are common to different application programs. From the TCP/IP model perspective, the application layer takes care of all tasks defined for the application layer, presentation layer, and session layer of the OSI-RM.

The following sections describe different data services with a description of the corresponding TCP/IP application layer protocols used to allow the access to such data service.

13.1 Web Browsing

Web browsing is the most important Internet service. It is a service that allows a user (client) to access multimedia information stored in a server. The server can be located in an Intranet, Extranet, or Internet. The access of information through the

---

* Exchanged through end entities.

Internet is referred to as web browsing. Moreover, web browsing is performed in a space called World Wide Web (WWW). It consists of a client/server service, where a user retrieves information from a web page. A web page may provide a myriad of media, such as text, tables, graphics, voice, images, video, and so on.

13.1.1 Hypertext Transfer Protocol

The hypertext transfer protocol (HTTP) is the TCP/IP application layer protocol that allows a client to use the WWW. The HTTP is a client/server protocol supported in the TCP, and therefore, a certain level of data reliability is assured. The different types of media are encoded in text. Since the data exchange is limited to text, the HTTP is simple. Moreover, since the HTTP it is supported by the TCP transport protocol, the access to each web page (transaction) follows three elementary different phases: connection setup, data exchange, and termination. This results in an independent treatment of each transaction (web page access). Once the client has successfully received the desired web page, the TCP connection is terminated.

A web page can be viewed as a file that is stored in a server. A client intending to retrieve a web page sends a GET message to the server with the requested URL (uniform resource locator), and the server responds with the corresponding web page (HTTP response). This can be viewed as a download of an HTML file.

In the data exchange between a client and a server, the HTTP may use either ASCII format or the multipurpose Internet mail extension (MIME) format. The MIME consists of a message format that allows other non-ASCII characters, as well as binary formats containing images, sounds, and other files. There are different versions of HTTP, namely the HTTP 0.9, 1.0, and 1.1 [RFC 2616], with small differences in the message formats.

The exchange of HTTP data between a client and a server can be established in two basic ways: direct mode or indirect mode. As can be seen from Figure 13.2, the direct mode comprises the establishment of a direct connection between a client and a server. Contrarily, the indirect mode comprises an intermediate node, where two or more successive TCP connections are established to establish the end-to-end connection and to get access to the web page. As depicted in Figure 13.3, the indirect mode may have to be employed due to the use of a proxy server or a gateway.

A proxy is a server that is located at the edge of an internal network.* All connections, established between an internal host (host located in the Intranet) and an external host (e.g., a server located in the Internet), are broken down into two successive connections for the purpose of avoiding direct connectivity and for the purpose of traffic filtering (i.e., network security). From the client perspective, the proxy acts as a server. Contrarily, from the server perspective, the proxy acts as a client. This represents a protective measure from certain types of network attacks.

---

* Typically in a demilitarized zone, as defined in Chapter 14.

Figure 13.2 Direct HTTP connection.

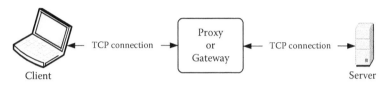

Figure 13.3 Indirect HTTP connection.

Another device that may break down a TCP connection is the use of a gateway. A gateway is typically a network device that changes a protocol into another. As an example, a host located in a Microsoft network, which needs to have access to a server located in an IP network, needs a gateway to perform the translation between the two network protocols. As described in Chapter 14, a gateway may also be used in the implementation of the security architecture for IP (IPsec) cryptographic protocol. In this case, the hosts from outside of an organization connect to an IPsec gateway* located in the interior of an organization through the Internet using the IPsec cryptographic protocol, while the gateway connects to a server in plaintext (in clear). Contrarily, the connections among internal hosts (Intranet) are performed in plaintext.

Note that, in some cases, the access to a web page can be performed without having to establish a connection with a server. A cache is a memory that stores previously accessed web pages for a certain period of time. Both hosts and proxies have cache memories. In the later case, when an internal network host request an access to a previously accessed web page (which could have been performed by another network host), the proxy sends the web page to the client without establishing the TCP connection with the server. This results in a more effective use of the external bandwidth and lower delays.

13.2 Electronic Mail

E-mail consists of a service that allows the exchange of messages among different end users. When a message is sent by a client, such message is uploaded to the message recipient's server and then forward to the message destination's server. Finally, the message is downloaded from the destination's server into the destination's host. Depending on the application layer protocol used to provide this network service, the message can be either only composed of text or contain other media files such as video, audio, and so on.

---

* An IPsec gateway is typically a firewall or a router that runs the IPsec protocol.

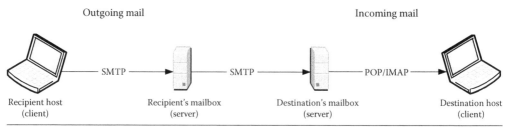

Figure 13.4   SMTP protocol.

13.2.1 Simple Mail Transfer Protocol

The Simple Mail Transfer Protocol (SMTP) [RFC 821] is the TCP/IP application layer protocol which allows a client uploading an e-mail to an e-mail server (user mailbox) or exchanging an email among e-mail servers, through an IP-based network, such as the Internet (see Figure 13.4). In case of e-mail forwarding, an e-mail may go through one or more intermediate servers, between the recipient's server and the destination's server. It is worth noting that the SMTP protocol does not support authentication.

The SMTP protocol comprises two elementary agents: user agents and message transfer agents. User agents allow users sending messages using Domain Name Servers (DNS) names of type mailbox@location. On the other hand, message transfer agents are responsible for forwarding messages to the destination server.

The SMTP protocol is supported by the TCP protocol, and therefore, a certain level of data reliability is assured. Nevertheless, in case a message as a whole (not its TCP segments) fails to be properly delivered to the destination's host, the mail's recipient is neither informed about the failure nor the message recovery is performed. As previously described in chapter 12, the default 2 SMTP server (TCP) port is 25.

The SMTP message format comprises text type data, consisting of a string of 7-bit ACSII characters. In case a message contains other types of media (pictures, videos, executable files, or others than 7-bit ASCII characters code*), the MIME format is utilized. Note that when a message contains annexes, the message format is automatically switched to MIME format. The mapping between MIME and SMTP messages is performed using e-mail clients or servers. The SMTP protocol does not allow a user agent (client) obtaining messages from a mailbox (server). This functionality is performed using one of the following protocols:

- Post Office Protocol (POP) [RFC 1225]: allows the transfer of e-mails between a server and a client, using authentication;
- Internet Message Access Protocol (IMAP) [RFC 2045]: presents some additional features than those available with POP, namely the ability to read

---

* Such as national language characters.

e-mails located in a server (through downloading or webmail), while leaving a copy in the mail server. It also allows the share of folders among different users, as well as the search of e-mails in a server.

13.3 File Transfer

The file transfer consists of a service that allows a client to download or upload files from or to a server. It consists of a service based on client–server concept.

13.3.1 File Transfer Protocol

The FTP is an IP-based application layer protocol [RFC 114] [RFC 959]. Since the file transfer is a service sensitive to errors, the FTP is supported on the TCP transport protocol. It uses two separate connections, respectively, for control and for data. The default TCP port used by the FTP server for connection control is 21. The control is used for clients' identification and authentication.

The TFTP consists of an FTP version with lower capabilities. While the FTP is connection-oriented (TCP-based), the TFTP is connectionless and supported on the user datagram protocol (UDP) transport protocol. Consequently, the TFTP results in a nonreliable FTP. Nevertheless, it is typically used for the transfer of small files, when data transmission is periodically repeated or in cases where the rapidity of the data transfer is an important requirement.

13.4 IP Telephony and IP Videoteleconference

Some of the reasons that motivated the implementation of the IP telephony and IP videoteleconference (VTC) are the possibility to integrate voice and data services into a common network infrastructure, the low cost of the equipments (as compared to cir-cuit switching), and the high availability and the high available bandwidth of modern deployed packet-switching networks.

The IP telephony or IP VTC comprises audio and/or video* acquisition and encoding using a codec (see Chapter 6). The obtained constant bit rate data is then encapsulated into packets. These packets are then transmitted through a packet-switching network (e.g., IP-based network, such as the Internet), using routers as intermediate nodes, instead of PABX.† At the destination, the transported data is then dc-encapsulated and the extracted digital data is then converted into analog data,‡ for human presentation.

---

* Note that the original audio/video consist of analog information.
† As considered in conventional circuit-switching networks.
‡ Using the same codec as the one used for the data encoding at the transmitter side.

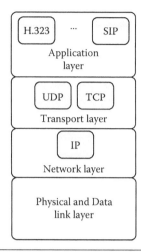

Figure 13.5   IP multimedia protocols and the TCP/IP stack.

As previously described in Chapter 1, both IP telephony and VTC are delay intolerant real-time traffic. This means that their performance degrade heavily when the media transfer is subject to a certain amount of delay (or delay variation, i.e., jitter). A typical maximum acceptable end-to-end delay in signals is 200 ms, which allows an acceptable performance.

Chapter 6 described analog and digital encoding standards and techniques used in both audio (telephony) and video (television). The following subsections focus on two protocols, which are commonly used for the exchange of both telephony, VTC, and application data, through packet-switching networks such as the Internet. Their insertion into the TCP/IP stack is depicted in Figure 13.5.

13.4.1 H.323

The ITU-T defines a set of recommendations for voice and video over different network types. The [ITU-T H.323] is an ITU-T recommendation, which consists of an application layer protocol and specifies the procedures and protocols for the exchange of voice, video, and data conferencing for packet-switching networks such as IP-based or IPX-based, for both point-to-point and point-to-multipoint multimedia communications.

Similarly, the [ITU-T H.320] defines similar services for ISDN networks, [ITU-T H.324] for PSTN, and [ITU-T H.321] [ITU-T H.310] specifies it for ATM networks.

To implement the above described services, the [ITU-T H.323] comprises four different network component types (see Figure 13.6) [Liu 2000]:

- Terminals: consist of the equipments (hardware and software) used for human interfacing. This can be a personal computer or a personal digital assistant (PDA) running a softphone or a packet-switching-based telephone, and so on.

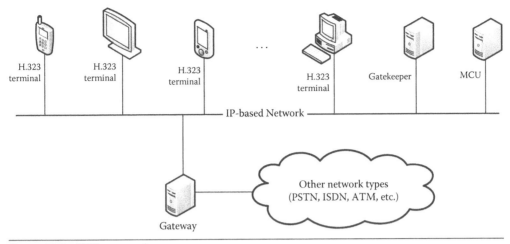

Figure 13.6 H.323 components.

- Gatekeeper: it is responsible for performing address resolution, namely for mapping a telephone number or username/e-mail address into a transport address.* It performs a task similar to the DNS mapping protocol.[†] In case the terminals have knowledge about the transport addresses of destination endpoints, a gatekeeper may not be necessary. Such mode of operation is referred to as "direct mode." Nevertheless, a gatekeeper also performs additional functionalities such as admission control[‡] and bandwidth management. Therefore, the use of a gatekeeper is essential to avoid congestion, especially in case of medium to large networks;
- Gateway: in case interoperability is required among different network types,[§] a gateway is used. Such network device assures voice and video codec translation, data format translation, and signaling translation. This network component is needed only when translation is necessary;
- Multipoint control unit (MCU): this component allows H.323 point-to-multipoint communications (conferencing). Without the MCU, only point-to-point communications are possible.

Figure 13.7 depicts the H.323 protocol stack used in IP-based networks.

As can be seen, depending on the functions that are to be accomplished, different protocols are used by the ITU-T H.323, as follows [Thom 1996] [Minoli 2002]:

- Real-time protocol (RTP): it is responsible for controlling a session.[¶] Since H.323 voice and video traffic is transferred using the UDP transport protocol,

---

* Concatenation of an IP address with a port number.
[†] Nevertheless, instead of converting an alphanumeric name into an IP address (DNS), it translates a telephone number or username/e-mail address into a transport address (gatekeeper).
[‡] A gatekeeper must authorize a call.
[§] For example, between an IP-based and a PSTN-based or an ISDN-based network.
[¶] Nevertheless, the TCP/IP protocol stack does not comprise the session layer, whose functionalities are performed by the application layer. Therefore, it can be stated that the RTP is an application layer protocol.

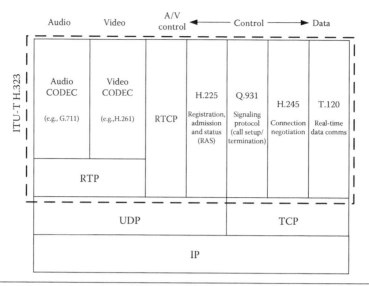

Figure 13.7   H.323 protocol stack for Internet protocol-based networks.

the RTP is an intermediate protocol that adds a header, with a number of fields. This complements the UDP service by adding some level of reliability* to the exchanged data. The RTP fields are:

- Sequencing of packets: it is implemented through the addition of a sequence number field into the RTP header. This allows the delivery of packets in the correct sequence, the detection of lost packets,[†] and the avoidance of duplication of packets.
- Payload identification: this field allows the identification of the audio and/or video codec, which can be used for different purposes, such as bandwidth management.
- Frame identification: used to signalize the start and end of frames.
- Source identification: used to identify the originator of packets. This is relevant in case of point-to-multipoint communications.
- Intramedia synchronization: used to implement buffering and avoid a certain level of jitter.

- Real-time control protocol (RTCP): it consists of a control protocol, which allows the RTP performing its tasks. It indicates the identity of a caller and provides a variety of network information which can be useful for keeping the network performing well (e.g., average delay, jitter, lost packets, etc.).
- H.225 (RAS): it is used by terminals to register with the gatekeeper and to request it a call.

---

* Note that corrupted packets are not detected and/or corrected by the RTP.
† Such information is only used for statistics purposes.

- Q.931 signaling protocol: used for call setup and termination between two or more terminals.
- H.245: used in a communication for the negotiation of audio and/or video codec.
- T.120: it consists of a protocol that manages the real-time data conferencing such as instant messaging, file transfer, application sharing, and so on. Note that these types of data communications are performed using the TCP, instead of the UDP transport protocol (used in voice and video communications).

Figure 13.8 plots the different phases of a voice and/or video call, with the indication of the corresponding protocols.

13.4.2 Session Initiation Protocol

Contrarily to the ITU-T H.323, which defines a set of standards for each specific functionality, the session initiation protocol (SIP) [RFC 326] [RFC 2543] [RFC 3261] is an application layer protocol, specified by the Internet engineering task force, being limited to a signaling protocol, used for setting up, configuring, and terminating multimedia sessions [Minoli 2002]. This is performed using an HTTP-like client–server request/response model, using text-type messages.* Similar to the HTTP, the multimedia calls performed with the SIP can make use of authentication and encryption protocols such as IPsec, SSL or TLS.† Moreover, similar to the ITU-T H.323, the SIP is used for voice, video, and/or data‡ point-to-point and point-to-multipoint communications over IP-based networks. Nevertheless, the exchanged traffic is transparent to the SIP [Schulzrinne 2000], using other protocols such as the RTP, RTCP,§ or the session description protocol (SDP). The SDP is used to specify protocols (e.g., codec) in the data transaction. Note that the exchanged multimedia traffic can be supported on either UDP or TCP, whereas the ITU-T H.323 specifies that voice and video use UDP and application data uses the TCP.

The SIP signaling protocol comprises different network component types (see Figure 13.9):

- User agents: consist of the equipments (hardware and software) used in the client/server interaction (for connection admission and control). The client endpoint is called a user agent client (UAC) and can be a personal computer or a PDA running a softphone or a packet-based telephone. The server endpoint

---

* It uses most of the HTTP text-based codes and header fields.
† See Chapter 14 for a detailed description of these cryptographic protocols.
‡ For example, instant messaging, file transfer, application sharing.
§ Both RTP and RTCP were already described for the ITU-T H.323.

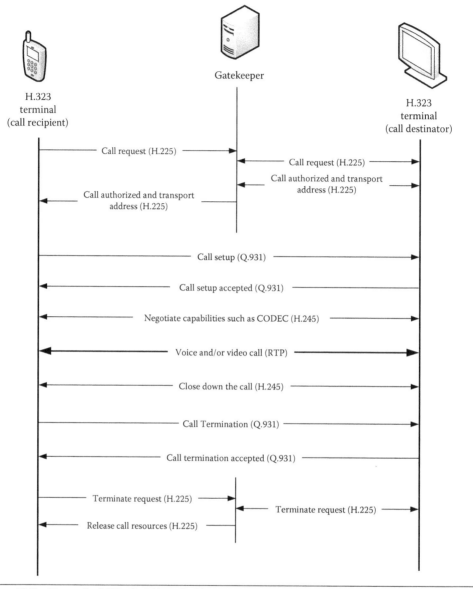

Figure 13.8 Phases of a H.323 voice/video call with corresponding protocols.

is referred to as user agent server (UAS), being used to receive session initiation requests and to return the corresponding responses.

- Registration server (registrar): consists of a server that keeps the location of different SIP users. A certain user with a certain telephone number or username/e-mail address can physically move along an IP-based network. This information is stored in the registrar, as well as the corresponding transport layer address. Similar to the gatekeeper of the ITU-T H.323 protocol, the registrar is responsible for translating telephone numbers or username/e-mail addresses into transport layer addresses.

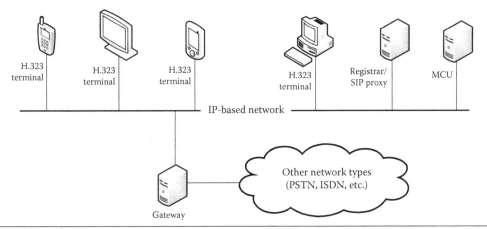

Figure 13.9 Session initiation protocol components.

- Proxy server: it is responsible for forwarding SIP requests and responses to their destination. A LAN implementing a multimedia service using the SIP has its own proxy server.*
- Redirect server: it is responsible for returning the location of another SIP agent, to which the call forwarding must be transfered.
- Gateway: similar to the ITU-T H.323 protocol, it assures interoperability among different network types.
- MCU: similar to the ITU-T H.323 protocol, it allows point-to-multipoint multimedia communications.

Figure 13.10 plots the different phases of a voice and/or video call using the SIP. Comparing the SIP message exchange to those of the ITU-T H.323 protocol (Figure 13.8), it is easy to conclude that the SIP presents a much lower level of complexity.

13.5 Network Management

Several proprietary and standardized protocols were developed to allow the remote control and monitoring of network devices, such as routers, printers, computers, switches, and so on. These protocols were extended to incorporate the remote control and monitoring of other equipments than network devices. Such equipments can be, for example, a satellite transceiver or an electrical diesel generator. The remote control capability may allow powering on a satellite transceiver, changing its transmitting frequency or its transmitting power through a network. The monitoring capability may allow receiving messages in case the satellite antenna is not directed toward the expected position or in case the transmitter fails. For this to be possible, both the controlled equipment (e.g., the satellite transceiver) and the remote control station

---

* It is normally resident in the same server as the registrar.

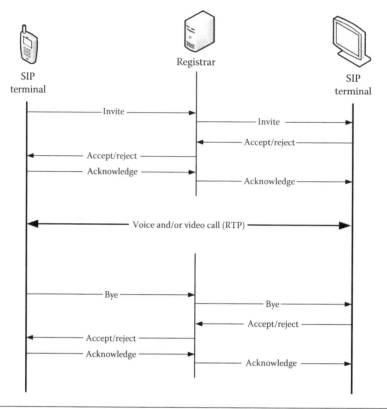

Figure 13.10   Phases of a SIP voice/video call.

need to be interconnected through a network and need to use a network management protocol.

Different network management protocols were standardized. The mostly known is the IP-based simple network management protocol (SNMP) [RFC 1448], which can be found in three different versions (version 1, 2, and 3). The SNMP v3 is commonly referred to as enhanced SNMP due to its improved capabilities. The OSI-RM also standardized a network management protocol entitled common management information protocol (CMIP). Moreover, many proprietary network management protocols have been developed to allow the remote control and monitoring of equipments. In case such equipment is intended to be remotely controlled with a common SNMP, a gateway is required.

13.5.1 Simple Network Management Protocol

The SNMP is an IP-based network management protocol, supported on UDP (see Figure 13.11). The reason for using the connectionless mode is not because data reliability is not a requirement, but because such data reliability can be assured through the repetition of the transmitted data. Since the SNMP comprises the period retransmission of messages, in case a message is received corrupted, the following messages assure that the information is correctly interpreted by the destination.

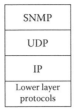

Figure 13.11 Insertion of the SNMP protocol into the TCP/IP stack.

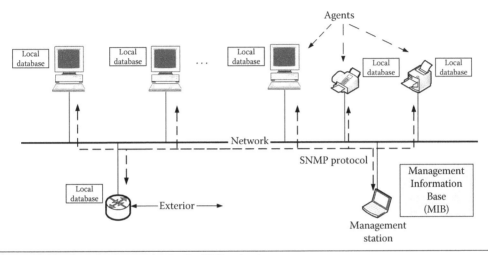

Figure 13.12 Example of a network using the SNMP protocol.

As depicted in Figure 13.12, to allow the remote management of equipments, four different elements need to be implemented*:

- Agent: the device that is intended to be remotely controlled and/or monitored (managed) through the network. For the device to be subject to remote management, it needs to be equipped with a SNMP agent, which comprises a local database with actual and historic variables (object states).
- Management station: the computer or workstation which runs a specific software that allows sending and receiving network management messages to/ from an agent.
- Network management protocol: the communication between an agent and a management station is carried out using the SNMP network management protocol [RFC 1448].
- Management information base (MIB): consists of an hierarchical database, stored in the management station, which compiles the data stored in all different agents (the set of all local databases).

---

* Note that the remotely managed equipments may be located within a local area network (Intranet) or outside it (i.e., in an Extranet).

The three most important SNMP messages exchanged between a network management station and the agents are the following:

- Get: it is used for network monitoring purposes, namely for the management station to request the value of an object from an agent.
- Set: it is used by the management station to send a value of an object to an agent. Consequently, it is used for network control purposes.
- Notify: sent by an agent to the management station to report a failure event (monitoring purposes).

13.6 Names Resolution

Names resolution is the act of translating a numerical address into an alphanumeric name and reciprocally. The most used names resolution protocol is the DNS, as defined in the following.

13.6.1 Domain Name Server

The IP version 4 comprises 32 bit-size addresses used to identify hosts. These addresses are commonly referred to as four groups of decimal numbers. Similarly, the IP version 6 protocol comprises 128 bit-size addresses, which is commonly referred to as eight groups of four hexadecimal numbers. These numbers are difficult to memorize. Consequently, it is easier to memorize the address of a host if a name is used, instead of a sequence of numbers. For the sake of simplicity, the alphanumeric names used to identify the hosts are normally correlated with the organization name and/or department name to which it belongs. These alphanumeric names used in the WWW space are known as domain names.

The DNS [RFC 1034] [RFC 1035] consists of a mapping protocol, which performs the conversion between alphanumeric names into IP addresses. It is an IP-based application layer protocol, which runs over UDP (connectionless).

The conversion of alphanumeric names* (typed by users) into IP addresses (inserted into packet headers, before transmission) is referred to as binding (the most common DNS operation). In some cases, the translation of IP addresses into alphanumeric names is also necessary (operation referred to as reverse mapping).

The DNS mapping protocol can be viewed as the yellow pages of an IP network.

An example of an alphanumeric IP address is www.std.au.edu. The syntax of alphanumeric IP addresses present different hierarchical levels, each one referred to as domains. Higher-level domains are located at the right of alphanumeric addresses. The top-level domain typically identifies a country or an organization type.† This hierarchical organization of alphanumeric names is associated to hierarchical routing‡ and to the corresponding hierarchical organization of IP addresses.

---

* Note that the alphanumeric names used in the World Wide Web are case-sensitive.
† For example, "uk" stands for United Kingdom, "edu" refers to an educational organization, "mil" identifies a military organization, "int" stands for an international organization, and so on.
‡ See Chapter 11.

The top-level domain of the example above is edu (education), which is followed by the second higher domain (au),* and then a third-level domain (std).† The described rules are applicable to both the Internet as well as to private IP-based networks. It is worth noting that the domain names start by the letters www or, alternatively, by the letters http.

The DNS operation requires the existence of a number of key elements, namely the following:

- Domain name space‡: this corresponds to a list of alphanumeric names (which corresponds to a set of IP addresses).
- DNS database: which stores the mapping between alphanumeric names with IP addresses.
- Name server: a server that stores the DNS database and the name resolver, and responds to DNS requests for performing the mapping.
- Name resolver: a program that runs in a server to respond to DNS requests.

The DNS operation can be performed using two different methods:

- Recursive: a host sends a mapping request to a DNS server. If the interrogated server does not have the response, it searches the response in other servers and returns it to the caller.
- Iterative: a host sends a mapping request to a DNS server. If the interrogated server does not have the response, it returns to the host a pointer to the following DNS server to which the request should be sent.

The typical DNS operation comprises a user program that requests the conversion of a domain name into an IP address. Such a request is typically sent to the resolver module resident in the local host or, alternatively, in a local server in the same domain. In case the typed domain name by the user is not in the database (cache), the interrogated server searches it in other servers (recursive method) or returns a pointer to the following DNS server to which the caller should send the request (iterative method). When the IP address is obtained, such information is stored in cache (in the local host of local server) for a certain period of time. Alternatively, if the mapping is not found, the user program receives an error message.

End of Chapter Questions

1. Which methods can be used by the DNS protocol to find the mapping between a domain name and an IP address?
2. What are the key elements of the DNS operation?
3. What is the DNS protocol used for? Why is it used?

---

* Which may refer to a university name.
† Which may refer to a department within the university.
‡ The Internet Assigned Numbers Authority is globally responsible for managing the domain name space.

4. What are the three most important SNMP messages exchanged between the network management station and the agents?
5. What are the four basic elements that need to be in place to allow the remote management of equipments?
6. Why is the SNMP supported on the UDP?
7. What is the SMTP used for?
8. What is the difference between the POP and the SMTP?
9. What is the difference between the POP and the IMAP?
10. What is the MIME protocol used for?
11. What are the differences between the application layer of the TCP/IP model and of the OSI reference model?
12. What are the tasks carried out by the session layer of the OSI reference model?
13. In the scope of the SNMP, what is the MIB used for?
14. Which kind of indirect HTTP connections do you know? Define each one.
15. Which application layer protocols of the TCP/IP model do you know? Define each one.
16. What are the differences between an application layer of a network architecture, an application program, and a data service?
17. What is the application layer of a network architecture used for?
18. What is the presentation layer of the OSI reference model used for?
19. Which are the network components comprised by the ITU-T H.323 standard?
20. Which are the network components comprised by the SIP?
21. Describe the different phases of an ITU-T H.323 call.
22. Describe the different phases of a SIP call.
23. What are the differences between the ITU-T H.323 and the SIP?
24. What are the functionalities carried out by the RTP?
25. What is the RTCP used for?
26. What are the transport layer protocols used by ITU-T H.323 and SIP application layer protocols?
27. What are the different protocols which compose the ITU-T H.323 standard used for?
28. In the scope of the ITU-T H.323 standard, what is the T.120 protocol used for?
29. To which TCP/IP layer does the ITU-T H.323 and the SIP belong?
30. In the scope of the ITU-T H.323 standard, what are the basic differences between the exchange of voice/video traffic and the exchange of data traffic?
31. In the scope of the SIP, what is the registration server (registrar) used for?
32. Which of the two studied IP multimedia protocols present lower level of complexity? Why?
33. Which messages are exchanged by the SIP, since the call establishment up to its termination?

14

NETWORK SECURITY

14.1 Overview of Network Security

Globalization emerged as a result of widespread use of communications and information systems (CIS) by states, organizations, companies, and individuals. The widespread network of networks called the Internet is a great example of globalization and entry into the "Information Society" or "Information Age."

These technological advancements are originating deep changes in organizations emphasizing the human dependence of information (and hence of data) to assist decision making. This dependence on technology has reduced the asymmetries in access to information, enabling criminal and terrorist organizations to gain access to almost as much information as countries. This new paradigm allows that these international players have gained access to almost as much information as states and big organizations, reducing their competitive disadvantages [Arquilla 1997]. In this new context, new players started using the CIS and cyberspace to carry out attacks, obtain information, gain knowledge and neutralize and control systems (e.g., dams, power plants, telecommunication networks, etc.) [Klein 2003]. In addition, criminals who use these new technologies to carry out their activities benefit from a cover-up, because they are difficult to be identified and even, sometimes, the detection of the attacks can be avoided. As an example, in Australia, a hacker managed to shed three million gallons of sewage from a sewage treatment plant [Klein 2003].

In a broad sense, the Information Systems Security (INFOSEC) can be viewed as a set of technical measures and procedures adopted to prevent unauthorized observation, modification or denial of the illegitimate use of knowledge, facts, information, skills, or resources. The information that is intended to be preserved can be stored,* being processed† or in transit‡ [Maiwald 2009] [McClure 2005]. The INFOSEC is decomposed into the following subareas:

- COMPUSEC (computers security)
- COMSEC (communications security)
- NETSEC (network security)
- EMSEC (emanations security)
- Physical security

---

* Paper, disk, tape, diskette, and so on.
† CPU, RAM, virtual memory, and so on.
‡ Using a channel for establishing a communication by electronic, electromagnetic, optical means, and so on.

Table 14.1   Summary of Security Services on Information Systems

ATTACK TYPES AND SERVICES	CONFIDENTIALITY	INTEGRITY	AUTHENTICITY	AVAILABILITY	ACCOUNTABILITY
Access	X				X
Modification		X			X
Denial of Service				X	
Non-Repudiation			X		X

Each one of these subareas needs to be independently protected. It is worth noting that INFOSEC is a process and not a product or technology. It may use various human means, procedures, and/or technologies to enable its implementation, including antivirus, access control, firewalls, authentication mechanisms (passwords, smart cards, tokens, biometrics, etc.), intrusion detection systems, policy management, vulnerability scanning, encryption, physical security, and so on.

The INFOSEC involves the protection of the information systems from attacks against the following attributes (commonly referred to as CIAA) [RFC 2196]:

- Confidentiality (C)
- Integrity (I)
- Availability (A)
- Authenticity (A) and non-repudiation

Table 14.1 summarizes the various security services* and attack types possible to be carried out through the network. These security services and attacks are characterized in the following sections.

14.2 Security Services

14.2.1 Confidentiality

Confidentiality is the protection from unauthorized observation or access of information or data. The access to information or data is limited to those entities (user or process):

- Who have a security clearance equal to higher the security level of the information or data in cause.
- Who need to have access to the information or data in cause (need-to-know principle).

The information, which is the potential target of an attack, can be in transit, stored, or in processing.

The most common type of protection from attacks against the confidentiality is cryptography. Two basic types of cryptography are normally adopted: symmetric and asymmetric. In the symmetric type of cryptography, the same key is used by both the

---

* The accountability is a complementary security service, which adds protection to the following attributes of information: confidentiality, integrity, and authenticity.

source and the destination, whereas the asymmetric scheme comprises two different keys: a public and a private key. A message can be encrypted with the private key and decrypted with the public or the opposite. Moreover, encryption can be applied to the channel (bulk encryption) or to the message (end-to-end encryption). Note that even when encryption is applied, an intruder may extract conclusions from the volume of traffic that is exchanged through a certain channel. The solution to this problem is to use a type of encryption that "fills in" the channel with "empty bits," such that the traffic is kept approximately constant, independent of the volume of messages exchanged.

In addition to encryption, authentication (e.g., use of a password, token, smart cards, biometrics, etc.) and access control mechanisms needs also to be included to assure confidentiality. Access control and authentication methods are part of the accountability, whose purpose is to complement the basic services (confidentiality, integrity, and authenticity), ensuring that INFOSEC is properly established.

The following subsections describe the most common type of attacks against confidentiality.

14.2.1.1 Eavesdropping   The most common type of attack against confidentiality is eavesdropping. It consists of listening to a private communication, which is applicable to phone calls, e-mails, instant messages, and so on. It is an attack against the information in transit, without interfering with it. The eavesdropping is depicted in Figure 14.1.

Since it does not disable the legitimate destination of the message, it is difficult to detect. In case of data communications, it is also referred to as packet sniffing or only sniffing. The execution of such type of attack requires that the network interface card (NIC) allows receiving data (packets) whose destination is a different host. This is possible when a local area nework (LAN) is based on a hub and a host's NIC is configured in promiscuous mode. In case of a hub, all hosts' NIC receive all packets, and the NIC only processes the frames whose destination message authentication code (MAC) address is the own MAC address (in such case the data is transferred to the host). Note that the execution of packet sniffing in a LAN using a switch

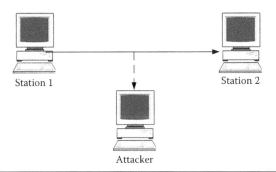

Figure 14.1   Eavesdropping.

becomes more difficult, as this device only forwards frames to the output port where the destination host is located.

Eavesdropping can also be performed by other means. It can be executed by making use of special equipment (e.g., a directional antenna and receiver), which allows receiving the electromagnetic waves generated by local devices (e.g., by a monitor, keyboard, etc., as the cable acts as an antenna that emits electromagnetic waves). Eavesdropping can also be implemented using a laser pointed toward the glass of a window, which allows listening to a conversation taking place in a room. Eavesdropping can also be executed by listening to a WLAN and so on.

14.2.1.2 Snooping   Snooping consists of prying into the private affairs of others, especially by prowling about. The purpose is to read e-mail content, listen to a telephone conversation, or observe a password or a PIN code of others without authorization. It differs from eavesdropping since the information is not in transit.

More advanced snooping capabilities include the remote monitoring of activities of a resource (e.g., a computer) or reading of passwords. This can be performed by using a program installed into the target computer, which automatically sends the captured information through the network to the attacker host (e.g., using an automatic e-mail, etc.). Protective measures such as firewalls, intrusion detection systems, and auditing tools are normally very effective against this type of attack.

14.2.1.3 Interception   Interception consists of the ability to gain access to a private communication, but assuming an active behavior, acting as the source or the destination of a message. The mostly known type of interception is spoofing, where an attacker acts as the source of a data communication (source spoofing) or as the destination of the data communication (destination spoofing). Source and destination spoofings are depicted in Figure 14.2.

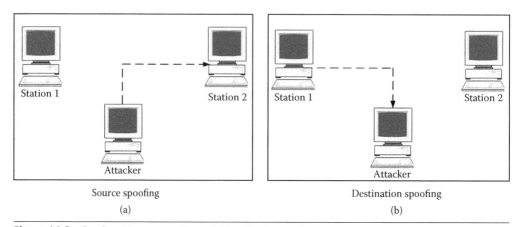

Source spoofing
(a)

Destination spoofing
(b)

Figure 14.2   Spoofing: (a) source spoofing and (b) destination spoofing.

This type of attack requires that the attacker be able to authenticate itself as a third party. The robbery of the authentication (e.g., password) can be initiated by other means such as phishing.* The illegal authentication can also be achieved by performing an attack to a domain name server (modifying the name resolution) and so on. The illegitimate access to a third party bank account using the e-banking can be viewed as a type of source spoofing.

14.2.2 Integrity

Integrity aims to avoid accidental or malicious modification of information or data without knowledge of its rightful owner. This service has two distinct objectives:

- Assuring the owner that the information or data cannot be added, modified, or erased by nonauthorized subjects.
- In case information or data has been added, modified, or erased, the service includes the necessary measures to alert the owner that it is not in the original state.

The key to implementing a protective system from attacks against the modification of information or data relies on adding to the content mechanisms that reveal its modification. This is typically achieved using a hash function. The hash function performs a known processing to the content of a message, and the result of this hash processing (i.e., a hash sum that consists of, e.g., 128 bit string) is added to the message. Once the message is received, the destination applies the same hash function to the message. If the message had been received unchanged, the result of this hash processing is the same as the one received together with the message. Note that the hash function is a one-way processing. This means that it is not mathematically possible to obtain the message from the hash sum.

An example of an open source hash function is the md5sum function. Note that an attacker may be smart enough such that he modifies both the message and the corresponding hash sum transmitted together with a message. To improve the resistance against this advanced integrity attack, either the message or the hash sum should be encrypted. To save processing (and time), the hash sum is normally encrypted (due to its lower length). If the sender encrypts the hash sum with its private key,[†] and if the encrypted hash sum is properly decrypted with the sender's public key the receiver is assured that the message has been sent by its legitimate sender (as the private key is only known by the legitimate sender[‡]). This is the concept of digital signature, which

---

* Phishing consists of inducing a user to contact the wrong entity. This can be performed by sending an e-mail asking for username and passwords.

† Note that an asymmetric encryption comprises a private key and a public key. A message can be encrypted with one of these keys, being decrypted with the other. Nevertheless, for the sake of confidentiality protection, a message is normally encrypted with the destination's public key, being decrypted with the private key. The public key is broadcasted to all potential participants in a communication, whereas the private key is only known by its owner.

‡ There is no possible processing that allows obtaining a private key from the public key.

comprises the use of a hash function and its encryption with the sender's private key. The digital signature provides two levels of security protection as follows:

- Ensures the identification of the producer of the message (authentication)
- Ensures that the signed message was not modified (integrity) [Cross 2003]

The digital signature is the digital equivalent of a handwritten signature. Note that a handwritten signature does not provide any information to the receiver about the integrity of the message, whereas the digital signature does.

Similar to confidentiality, the effectiveness of the protective measures from attacks against the integrity depends on the combination of a hash function with the accountability service (access control, authentication, identification, etc.), namely, to prevent the creation, modification, or deletion of information and data (e.g., permissions to write data).

14.2.2.1 Man-in-the-Middle   The most common attack against the integrity is the man-in-the-middle. It comprises the modification of the content of, for example, a financial transaction. In this case, the attacker captures the bits in transit and modifies, for example, the destination bank account of a financial transaction. This type of attack is depicted in Figure 14.3.

This is of difficult execution, as the attacker has to be in line between the source and destination. Moreover, the attacker must have software that allows him to take charge of the session, so that the legitimate parties do not detect him.

An attack against the integrity may also consist of adding or removing words in a message. Sometimes, adding or removing part of a message may dramatically change the sense of a message.

14.2.3 Availability

Availability aims to ensure authorized access to information or data in a timely manner. The effect of attacks against the availability can be mitigated by using redundancy that is activated in case an attack is detected against the primary system or by using preventive measures (authentication, firewalls, access control mechanisms, packet filtering, etc.). Such attacks are often carried out with the transmission of viruses (spread with human intervention) or worms (which self-propagate through the network). The target of an attack can be an application, a service, a system, a network, bandwidth, and

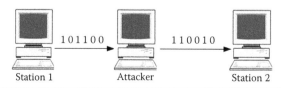

Figure 14.3   Man-in-the-middle.

so on. This type of attack may consist of bandwidth consumption, resources consumption (e.g., server), or even the interruption of support infrastructures.

Depending on the target, there are different protective measures that can be undertaken. In 1999, during the North Atlantic Treaty Organization (NATO) intervention in Serbia (due to the province of Kosovo) both parties used the Internet to carry out computer attacks [Klein 2003]. In this conflict, the Serbs performed a denial of service (DoS) attack against the NATO site (an attack on service availability). This attack stopped the NATO web and mail servers by filling it in with empty messages. The aim of this campaign was to prevent NATO from sending out messages to the world. Meanwhile, the Serbian air defense systems were crippled by a cyber attack. These actions often arise spontaneously by citizens of states or directly executed by the organizations involved in a conflict.

14.2.3.1 Denial of Service The most common type of attack against the availability is the DoS. The purpose is to deny the target the ability to use the applications, services, resources, and so on. The DoS is depicted in Figure 14.4.

A common type of DoS is the SYN Attack. As can be seen from Figure 14.5, it consists of the emission of consecutive requests to establish TCP connections (SYN message), forcing the target to keep responding with SYN ACK messages. In this case, the target is flooded with such handshaking procedure and reaches the maximum number of simultaneous TCP connections possible to be handled by a server.

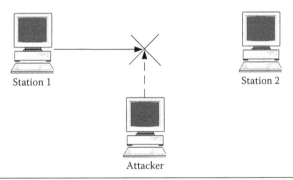

Figure 14.4   Generic diagram of a denial of service attack.

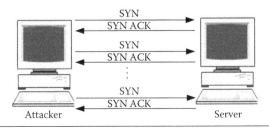

Figure 14.5   SYN attack.

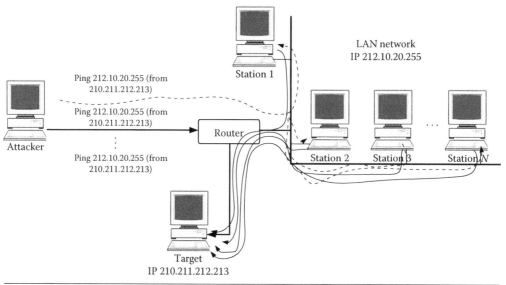

Figure 14.6 Smurf attack.

Another type of DoS is the smurf attack. As can be seen from Figure 14.6, the attacker pings a remote network* composed of a wide range of hosts, while showing the IP address of the target device as a source address in the ping packet (instead of its own IP address). Therefore, it forces each one of the hosts in the pinged network to respond to the ping request with ping response packets to the target device. This target host is flooded and overloaded with packets, being placed out of service.

The DoS attack can also be of distributed mode (distributed DoS, i.e., DDoS). In the DDoS attack, an attacker typically spreads out a worm file along different hosts/ servers. The worm is typically propagated in chain along the network, and the target is typically a server.

The attack is normally triggered (preprogrammed by the worm) to take place at a certain time, being executed against the target server. The attack can be of SYN type, smurf attack, or any other (e.g., buffer overflow). In the DDoS, the attacker is difficult to be identified (see Figure 14.7). Note that, most of the time, these third party hosts do not even have knowledge that they are performing such attack.

A DDoS attack against the White House web server took place on July 31, 2001 (entitled Code Red). The final attack was executed by sending successive ping packets to the White House web server [Klein 2003].

The protection from such type of attacks aims to block the use of information systems and resources to nonlegitimate users (authentication, firewalls, intrusion detection system, access control mechanisms, etc.). A preventive protection also consists of keeping antivirus and other software† up to date.

---

* Using the broadcast (IP) address as the destination IP address of the ping packet.

† Such as operating systems.

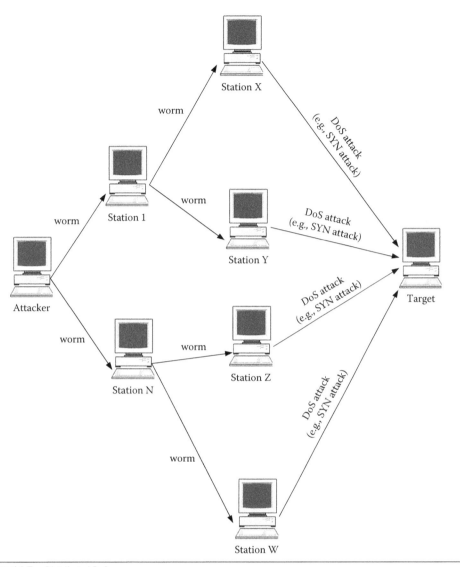

Figure 14.7 Distributed DoS.

14.2.4 Authenticity

The authenticity and non-repudiation aims to ensure that the author of information or data is the declared author (authenticity) or that the author does not come in due course to deny the authorship of the information (non-repudiation). The mechanisms that better ensure the authenticity and non-repudiation consists of using the concept of digital signature. Therefore, the same protection mechanism (digital signature) can be applied for both protection from attacks against the integrity and the authenticity. This is achieved by encrypting the hash sum (normally used for integrity) with the sender's private key. The receiver decodes the hash sum with the sender's public key and compares it against the result of the hash function applied to the received message. If the decoded received hash sum equals the hash sum obtained from the received message, then the receiver may conclude that the message was not modified (integrity) and that the sender is the legitimate (authenticity). If the two hash sums are not equal, then one or both of the attributes are illegal (integrity and authenticity).

Note that there are other mechanisms that can be used to protect from attacks against the authenticity. These mechanisms include the use of authetication mechanisms, such as username and passwords, smart cards, biometrics, access control techniques, and several variations and combinations of these mechanisms. A widely used authentication protocol is the challenge handshaking authentication protocol (CHAP), which is adopted by the PPP as an option, being executed during the connection establishment and periodically (from a server request). Once the authenticator server has received a request to connect from the client, the CHAP is executed as follows (three way handshaking):

- Challenge: the server sends a variable length random stream.*
- Response: the client sends to the server the hash function applied to the digest. The digest corresponds to the concatenation of the challenge (previously received from the server) with the password. This hash sum is sent to the server, together with the username.
- Success or failure: the authenticator server verifies the response from the client against its own calculation. If the response fits with its own calculation, the server sends a success message to the client, otherwise a failure message is sent.

Note that the CHAP requires that both the client and the server know the plaintext password, although it is never explicitly exchanged. The procedure of encrypting the password with a hash function is widely used. An example of its use is the authentication process of computers.

Although not perfect, the CHAP is much better than its predecessor password authentication protocol (PAP), where usernames and passwords are exchanged clearly.

Other authentication procedures exist where the need for the server to have knowledge about the clients' passwords is avoided.† These procedures provide protection from robbery of a passwords file, which is a common type of attack. It is worth noting that most of the attacks come from the interior of the organizations. Therefore, not keeping clear passwords in an authentication server is one important principle of authentication process.‡

Non-repudiation is normally associated with authentication measures combined with registration, such that the user cannot, in the future, deny that a certain action

---

* This is called the challenge, being also generically referred to as token. The challenge provides protection from a replay attack.

† Instead of encrypting the digest with a hash sum (as used by the CHAP), a possibility may rely on encrypting the challenge with the result of the hash function applied to the password. In this case, the authenticator server only needs to store such hash sum (not the plaintext password), and the authentication computation relies on encrypting the challenge with such hash sum, followed by a comparison with the received sequence.

‡ An authentication server should only keep encrypted passwords (with hash function).

was performed by him. An example can be a user who places a bank transfer order. With measures such as authentication combined with registration (accountability), the subject cannot reject the authorship of such bank movement, as the bank keeps a proof (in a record) that such movement was undertaken by the subject.

It is worth noting that preventive measures from attacks against authenticity should also include mechanisms to prevent from using a non-legitimate server, instead of the legitimate one. This attack is normally achieved with a phishing attack (typically an e-mail asking the user to provide username and passwords) or an attack to a DNS server.

14.2.4.1 Replay Attack　In the replay attack, the attacker has access to the bits in transit and records the initial handshaking, which comprises the authentication procedure. Later on, the attacker sends the same sequence of bits to the authenticator machine (e.g., to a server). This can be seen in Figure 14.8.

If the password is sent in clear mode, the replay attack is easy to achieve. On the other hand, since the CHAP sends an encrypted digest (concatenation of the password and the challenge, whereas the challenge consists of a random stream that varies in time), replicating the encrypted digest does not lead to a successful authentication process.

The digital signature by itself does not prevent from a replay attack, as it does not comprise authentication. It only presents authenticity mechanisms, only assuring that a certain message was generated by a subject. In the context of a replay attack, if such

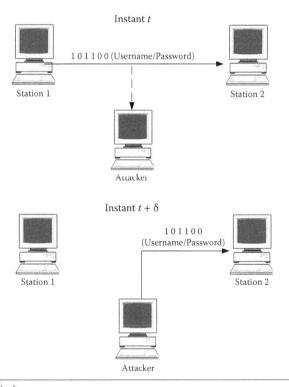

Figure 14.8　Replay attack.

message is resent over the network, the destination party does not have any mechanism to identify this action.

Nevertheless, the authentication protocols of security architecture for IP (IPsec), secure sockets layer (SSL), and transport layer security (TLS) include mechanisms, which enable authentication and protection from a replay attack.

14.3 Accountability

These security services complement the four basic security services commonly referred to as CIAA (see Table 14.1). This service does not bring any direct added value, but its inexistence represents a great reduction in the effectiveness of the confidentiality, integrity, and authenticity services. The main purpose of this service is to assure that the entity (user or process) is legitimate and that the events log reveals the truth of the facts. These complementary services require using additional resources such as processing, memory, bandwidth, and so on. Using these complementary services as protection mechanisms makes an attack more difficult and possible to be registered and a posteriori traced. Therefore, a successful attack against the confidentiality or integrity needs also to include an attack against the authentication process and against the registration (events log). A lack in one of these complementary functions represents a lack of the basic security services.

The accountability can be breakdown as follows:

- I3A: identification, authentication, authorization, and access control
- MRA: monitoring, registering, auditing.

I3A intends to find out who is an entity to prove its authenticity, check its permissions, and allow or deny the access to resources. The MRA has as the main purpose to register events, errors, and accesses; to identify who performed those actions; and to register the moment when those actions were executed. To be effective, it requires an exact log of events, as well as a precise timing reference used in registration.

These individual services are detailed in the following subsections.

14.3.1 Identification

The identification function intends to know who is the entity (user or process) who wants to perform a certain action. The identification is normally associated to the authentication (defined in the following). The combination of these two complementary security services are referred to as I&A.

14.3.2 Authentication

The authentication function intends to prove that the entity (user or process) that wants to perform a certain action is legitimate and effectively corresponds to the claimed identification. Note that authentication differs from authenticity as the former is a process, whereas the latter is a security service. There are different mechanisms used to implement I&A such as the following:

- Something that can be known (password, PIN)
- Something that can be possessed (smart card, token)
- Something that can be seen (biometric, i.e., speech, eye, footprint, etc.)

A more effective I&A process includes a combination of two or more mechanisms.

Authentication may use an encrypted password with a hash function. This procedure prevents the authenticator machine from having to store passwords in clear. This way, an intrusion attack against such machine does not enable him to obtain such password(s). It is also worth noting that it is widely known that most of the attacks come from the interior of the organization. Therefore, preventing employees from having access to passwords is a basic principle. Note that computers normally store passwords encrypted with hash codes. The user inserts the password, which is then encrypted and compared against the encrypted password (which is stored in the computer). In this case, if someone wants to forge the authentication, even though if he can reach the file with encrypted passwords, he cannot succeed from authenticating itself. With such mechanism, an attacker may use one of the following possibilities to forge an authentication using passwords:

- An attack using words from a dictionary
- An attack with variations of the user's name
- A brute force attack

Depending on the system, an attack based on a dictionary may take few minutes. An attack with variations of the user's name may also be quick and easy to achieve. A brute force attack consists of trying all different possible characters, with different lengths. The success of such attack is warranted, as long as there is enough time to perform such attack. Depending on the processing speed, a brute force attack against an authentication system using a password with eight characters may take as much as several months.* Naturally, reducing the password's length leads to a successful brute force attack in a much shorter time frame.

14.3.3 Authorization

It consists of a complementary security service that is used after I&A. It establishes the privileges of each different user. It depends on the organization's plan and policy and defines who may have access to what. As for I&A, compromising this security service represents a compromise of a basic security service.

---

* Note that a brute force attack time increases exponentially with the increase of the password's length.

14.3.4 *Access Control*

It is a complementary security service used after the authorization service. It is used whenever an entity intends to have access to a resource (verifies if the entity is in the access list), allowing or denying it. This service can be implemented in mandatory mode (mandatory access control) or in discretionary mode (discretionary access control).

14.3.5 *Monitoring*

Monitoring consists of an activity that keeps monitoring the actions that are being undertaken, including the identification of who does what and when.

14.3.6 *Registration*

Registration consists of an activity that registers the result of the monitoring activity.

14.3.7 *Auditing*

Auditing consists of an activity that should be implemented periodically to evaluate the effectiveness of the security plans and policies in the network. In addition, an audit may also be performed after the detection of security violations, attacks, intrusion detections, and so on. The purpose of this activity is to identify the authorship of an attack, its consequences, which measures could have been undertaken to prevent such attack, the identification of the procedures that failed, and so on.

14.4 Risk Management

To keep access to own information systems, while protecting them from attacks, there is the need to execute the appropriate INFOSEC measures and to maintain an effective control of CIS systems. Therefore, it is first necessary to identify the goods to protect (software, router, firewall, server, etc.) and their values. Second, it is necessary to determine the threats* likely to occur, namely threats against the confidentiality, integrity, availability, and authenticity of information. Third, determine the vulnerabilities[†] and the risks.

A risk is generally defined as a potential loss of assets and values, which are subject to threats that exploit vulnerabilities in a system, organization, or persons. A risk is often mathematically quantified as

$$Risk = Threat \times Vulnerability \times Value\ of\ Goods \qquad (14.1)$$

---

* A threat consists of a potential action or event that might violate the security of information systems.
[†] A vulnerability is a potential route of exploitation to the threats.

Note that the value of a good is not the monetary value but the impact in the organization, in case such a good fails as a result of an attack.

The risk management should also include the identification of the mitigation measures that can be applied to minimize the risk and the establishment of foundations for a security plan. Each risk is associated with a certain probability and impact, which needs to be noted at this stage. The residual risk corresponds to the initial risk to which the countermeasures are subtracted (patches, firewall, intrusion detection systems, redundancy of systems, cryptography, procedures, policies, etc.). There are measures that aim to minimize the vulnerabilities of information systems (firewall, cryptography, backups, etc.) and measures to protect from threats (patching, hardening, training, etc.). The residual risk should correspond to the acceptable risk, which is a top management decision of the organization.

14.5 Protective Measures

Depending on the information attribute to protect, there are different protective measures that can be implemented. Typically, the confidentiality is protected with cryptography, whereas the digital signature provides protection from attacks against the integrity and against the authenticity. Note that the digital signature also makes use of cryptography. Finally, redundancy is the keyword for allowing availability protection. There are three main types of cryptography:

- Symmetric
- Asymmetric
- Hybrid

The symmetric cryptography uses the same key for encrypting and decrypting the messages, using one or more predefined number of elementary operations based on substitution and transposition. The symmetric cryptography requires the previous distribution of keys, which presents a certain level of vulnerability. Another cryptographic scheme, used very often by terrorist organizations, is called stenography. It consists of a technique that hides messages within images [Klein 2003].

On the other hand, the asymmetric cryptography* makes use of two different keys, namely a private and a public key. Asymmetric cryptography solves the vulnerability problem of key's distribution occurring in symmetric cryptography.

Using an asymmetric encryption scheme, a message is encrypted with a sender's private key, being decrypted with a sender's public key. Alternatively, a message is encrypted with a receiver's public key, while decrypted with a receiver's private key. The private key is kept secret by its owner, whereas the public key is openly distributed. The advantage is that since these keys are different, the distribution of the public key does not involve a high risk. Nevertheless, as a rule of thumb, for a given message

---

* Asymmetric cryptography is also referred to as public key cryptography.

size, the processing time required to encrypt and decrypt a message using asymmetric cryptography is about a thousand times higher (i.e., it is slower) than using symmetric cryptography. As described in Section 14.5.6, hybrid schemes present the advantages of the two previous schemes, while eliminating the corresponding disadvantages.

A hybrid cryptographic scheme requires the processing time of a symmetric scheme, while solving the key's distribution vulnerability. The price to pay is a higher level of complexity, as it makes use of both symmetric and asymmetric schemes simultaneously.

The following sections describe each of these cryptographies.

14.5.1 Symmetric Cryptography

The development of an effective cryptography is a difficult task. Once such a system is implemented, its successive operation relies on the use of difference keys. Consequently, even if an attacker has knowledge about the algorithm being used, as long as the key is unknown (secret), a successful attack against the confidentiality of a message (in transit, processing, stored, etc.) is not possible.

Symmetric cryptographic schemes have been used for centuries. One such example is Caesar's cipher, which relies on a simple substitution of letters. Such technique is known as mono-alphabetic substitution cipher, where the key relies on the substitution of each alphabet letter by another one randomly selected. The effectiveness of such a technique is limited, as long as the text length available to be analyzed by a cryptanalyst[*] is long enough. In fact, for each language, it is known what the most frequent letter is. Analyzing a long cryptogram,[†] it is easy to find the most frequent letter and its correspondence to the plaintext is therefore known. Then, such a process can be repeated for the second most frequent letter, and so on.

To solve the main vulnerability of the mono-alphabetic cipher, the poly-alphabetic substitution cipher, generically referred to as Vigenère cipher, was introduced in the sixteenth century. Such cipher requires a key that is used to convert the plaintext into the ciphertext and reciprocally. As a result of the use of a key, such cipher does not rigidly replace a certain letter by another fixed one. This improves the confusion capabilities of the cipher. Nevertheless, such cipher is also possible to be broken using computational capabilities. This cipher can also be improved by using long length key. In the limit, having a key's length corresponding to the message's length leads us to the "one-time pad," whose resulting cryptogram presents the maximum confusion. Nevertheless, this requires a complex generation and distribution of keys. Since the key is as long as the message, the decoding process is a difficult task.

---

[*] A cryptanalyst is someone who performs cryptanalysis. Cryptanalysis consists of the action of analyzing cryptographic algorithms (using intercepted cryptograms) for the purpose of identifying vulnerabilities and extracting the key.

[†] A cryptogram is the sequence of text that results from the encryption of an original message (clear text). In other words, it consists of a message's ciphertext.

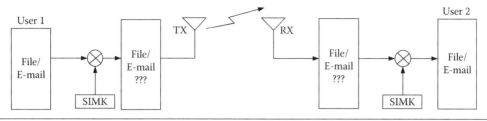

Figure 14.9 Generic block diagram of symmetric cryptography.

Afterwards, symmetric encryption systems started using a combination of two elementary operations as follows:

- Substitution
- Transposition*

The cryptographic Enigma machine was used during World War II by Germany and the Purple machine used by Japan, up to the currently used cryptography machines, all work based on a combination of these two elementary operations. Substitution consists of replacing one letter by another one (currently, this is digitally performed by applying a mathematic function to the bits). Transposition consists of changing the sequence of letters or bits within a message (e.g., write words in columns and read them in line). These operations are performed based on a key, which changes periodically.

Symmetric cryptography presents the advantage that the required processing for encrypting and decrypting messages is much lower than that of asymmetric cryptography. However, since the key used for encryption and decryption is the same, there is the need to distribute the keys along the potential users (senders and receivers of messages). This makes symmetric systems vulnerable.

Figure 14.9 depicts the generic block diagram of a symmetric cryptographic scheme. In Figure 14.9, SIMK stands for symmetric key, user 1 represents the message sender, and user 2 is the message receiver. Moreover, the message is depicted Figure 14.9 as a file/e-mail, but it can be any other media. Note that File/E-mail??? represents the encrypted ciphertext version of the File/E-mail.

14.5.1.1 Symmetric Cryptographic Systems   Currently, most of the existing cryptographic systems are binary based. This means that instead of implementing substitution of one letter by another one using a certain key, the encrypted data is the result of a logic function[†] applied to input data (part of a clear message). Moreover, the transposition operation consists of changing the sequence of bits (this operation is similar to the interleaving operation), whereas the opposite operation is performed at the receiver side. Most of the cryptographic algorithms perform complex iterative sequences of these two binary elementary operations.

---

* Transposition is also known as permutation.

† For example, binary addition (OR), multiplication (AND), bitwise, eXclusive OR (XOR), and so on.

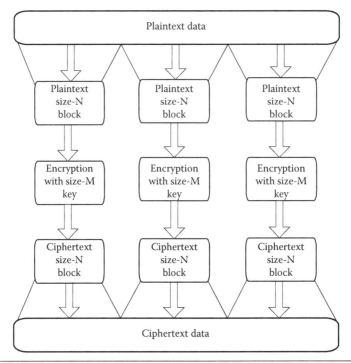

Figure 14.10 Division of plaintext data into blocks, before encryption operation is applied.

As can be seen from Figure 14.10, the basic parameters of cryptographic algorithms are the block size and the key length. In Figure 14.10, plaintext data stands for clear data (before encryption), whereas ciphertext data refers to encrypted data.*

Moreover, the definition of the number and sequence of elementary operations (binary version of substitution and transposition) that composes the encryption operation is also an important issue of cryptographic algorithms.

Among the currently used symmetric cryptographic algorithms, there is the digital encryption standard (DES), the triple data encryption algorithm (TDEA or triple DES), the advanced encryption standard (AES), the international data encryption algorithm (IDEA), the blowfish, and the RC5 algorithm.

The DES algorithm presents a 64-bit block size and a 56-bit key's length. Its cryptographic operations rely on the application of successive transpositions to the "plaintext bits." Due to its reduced key's length, it is considered vulnerable to brute force attacks with existing computation power. Consequently, the TDEA algorithm (commonly referred to as 3DES algorithm) was developed. This comprises the same block size, whereas the key length presents an effective length of 168 bits, corresponding to three times that of the DES. In fact, TDEA consists of the application of the DES cryptographic algorithm three times to each block of data. This results in a much more resistant cryptographic algorithm from brute force attacks, without having to develop a complete new one.

---

* The whole encrypted message is commonly referred to as a cryptogram.

The IDEA algorithm was initially described in 1991 and corresponds to an improvement of the proposed encryption standard (PES). The IDEA algorithm presents a 64-bit block size and a 128-bit key's length and is freely available for noncommercial use. The IDEA is used as a session (symmetric) key by the Pretty Good Privacy (PGP) v.2.0, which is a hybrid encryption protocol.

Another symmetric algorithm is the AES. The AES presents a 128-bit block size, whereas its key's length varies depending on the AES version. Three AES options exist: 128, 192, and 256-bit key's length, which corresponds, respectively, to the AES-128, AES-192, and AES-256. This represents a resistant algorithm against brute force attacks. The AES algorithm was previously published as the Rijndael* algorithm, having been selected by the National Institute of Standards and Technology (NIST), in 2002, by the United States, for governmental use. It is worth noting that the AES was approved by the U.S. National Security Agency (NSA) for encryption of data up to the classification of top secret.

The blowfish algorithm is another symmetric cryptographic algorithm. It presents a 64-bit block size, with a variable key's length up to 448 bits, being easy to implement by software. The blowfish algorithm was developed as an alternative to DES. Nevertheless, its encryption capabilities are much below those of the AES. Finally, the RC5 is a cryptographic algorithm very attractive due to its simplicity. It was initially developed by Ronald Rivest, and the RC5 corresponds to an update of the previous RC4 version. Note that the acronym RC (in RC5) stands for River Cipher. The RC5 has a variable block size of 32, 64, or 128 bits, whereas the key's length is fixed to 255 bits. The RC5 is currently used as session (symmetric) key by hybrid protocols such as the SSL or the TLS.

14.5.2 Asymmetric Cryptography

The concept of asymmetric cryptography was initially proposed by Diffie and Hellman in 1976. Asymmetric cryptography is also referred to as public key cryptography. The main purpose of asymmetric cryptography relies on solving two main issues as follows:

- Mitigate the limitations that results from the distribution of symmetric keys
- Allow the implementation of digital signature, which provides mechanisms to protect from attacks against the integrity and authenticity

Since the public key cryptography makes use of a public key, which is widely distributed, and a private key, which is kept secret, there is no need to implement a complex and vulnerable key's distribution system, as required for symmetric cryptography. Moreover, since the encryption operation can be implemented by any of the

---

* Rijndael stands for the combination of the names of the two Belgian inventors: Joan Daemen and Vincent Rijmen.

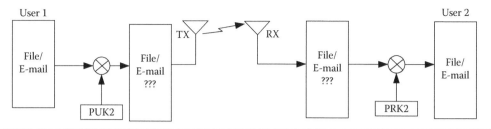

Figure 14.11   Generic block diagram of asymmetric cryptography.

keys* (private of public), this system is well fitted for both data encryption and digital signature.

Public and private keys are mathematically generated, using functions based on the factorization of a number into two prime numbers. The generation of asymmetric keys' pairs, as well as the encryption and decryption processes, are very computationally demanding, which translates in highly time-consuming processes.

There are different types of algorithms used to generate asymmetric keys' pairs. The RSA† is the most common one. RSA was the first publicly described asymmetric algorithm, which allowed the implementation of asymmetric cryptography, as well as digital signatures. The RSA1024 is a frequent version of the RSA algorithm, where 1024 stands for the key's length. Currently, the RSA2048 is already available and recommended for many applications.

There is another public key algorithm that is based on elliptic curves. The elliptic curves algorithm provides similar security protection results as those of the RSA, but with a shorter key length. Consequently, this algorithm tends to be faster than the RSA.

Comparing the asymmetric cryptographic schemes against symmetric in terms of time processing, it is known that the asymmetric schemes are much more demanding than symmetric. As a rule of thumb, for a given message size, the processing time required to encrypt and decrypt a message using an asymmetric cryptographic system is about a thousand times higher than using symmetric cryptography. The solution to this limitation is overcome by hybrid schemes at the cost of a higher system complexity, as described in Section 14.5.6.

Figure 14.11 depicts the generic block diagram of asymmetric cryptography. In Figure 14.11, PUK stands for public key and PRK stands for private key. Furthermore, PUK1 and PUK2 stands for the public key of user 1 (message sender) and of user 2 (message receiver), whereas PRK1 and PRK2 stands for the private key of the user 1 and of user 2.

Although better than in the case of symmetric algorithm, the generation and distribution of public keys still presents some vulnerability, as there is the risk that someone distributes a public key from someone else, claiming he is the owner. The solution

---

* As long as the decryption operation is performed with the other key.

† RSA stands for Rivest, Shamir, and Adleman, the inventors' names of such algorithm.

for this complex problem is provided by the public key infrastructure (PKI) and digital certificates, as described in Sections 14.5.4 and 14.5.5.

14.5.3 Digital Signature

After receiving a ciphered text with symmetric cryptography, the subject cannot be sure that the text was not modified. In case the message is a human readable text, changing the ciphertext while in transit implies that the result of the decryption process does not lead to the original plain text. In fact, this may originate that part or the whole text is not readable. However, in case the message is not text, the detection of the integrity violation becomes more difficult. Such detection is even more difficult in case of machine-to-machine communication. Therefore, another solution for preserving integrity needs to be implemented.

The hash function contributes to protect the message integrity from attacks, while notifying the receiver, in case such attack is executed. The hash function consists of an advanced checksum.* A hash function needs to comply with the following basic properties:

- The hash function is a "one way encryption," that is, it only works in the "encryption" direction, as it is not possible to obtain the clear message from the hash sum. Note that the hash function is not exactly an encryption function, as it is a well-known function that does not make use of a key. However, due to its similarities with cryptography, we may refer to this as an encryption function.
- Taking a message M and its hash sum H(M), it is not computationally possible to find another message that presents the same hash sum.†
- The hash sums of two slightly different messages should be completely different.

Consequently, if one captures a hash sum and intends to find its corresponding "clear text," he needs to compute a brute force attack (try all different possible combination of characters, with different lengths). Since the source text is normally very long, such computation is almost impossible.

The most well-known types of hash functions are the MD5 (message digest) and the SHA (secure hashing algorithm). The MD5 is widely used in the Internet, and it has a length of 128 bits. On the other hand, the SHA has been standardized by the NIST and presents different options depending on the hash sum length: 160, 256, 384, and 512 bits. Hybrid encryption protocols such as SSL and TLS comprises digital signature using a customized hash function and length (negotiated during the initial Handshaking).

---

* A hash function may be implemented, for example, using shift registers.

† A hash sum is a string obtained from applying the hash function to a message. If an hash sum is 128-bit long, the number of possible combinations for the hash sum is $2^{128} \approx 3,40^{+38}$, that is, the possibility of having two messages with the same hash sum is very low ($1/2^{128}$).

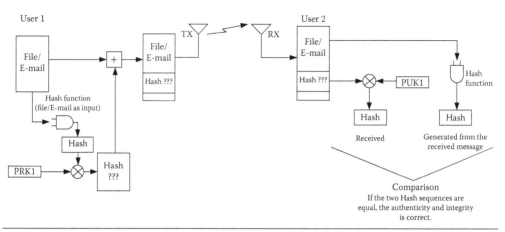

Figure 14.12   Generic block diagram of digital signature.

The hash function is used as a protection mechanism from attacks against the integrity as follows: the message's sender applies a hash function to the original message and sends the message together with its hash sum. The message receiver also applies the hash function to the received message and compares the resulting hash sum against the received hash sum. If they are the same, one may, in principle, conclude that the message was not modified.

Since the hash function is a well-known function, an attacker may capture the message, modify it, compute the hash function of the modified message, and send the new message together with the new hash sum. In this scenario, the receiver is not able to detect such message modification. The solution to this problem relies on encrypting one of the two elements: the message or its hash sum. For the sake of processing saving, it is preferable to encrypt the hash sum, as its length (e.g., 128 bits) is typically lower than the message. Encrypting the hash sum with the asymmetric sender's private key allows it being decrypted at the receiver side with the sender's public key. Once the message and its encrypted hash sum are received, the following operations are performed:

- Decrypt the hash sum with the sender's public key
- Apply the hash function to the received message
- Compare the received decrypted hash sum against the computed hash sum

If the two hash sums are equal, then the receiver may take two conclusions:

- The message was created by its legitimate author.* This corresponds to authenticity.
- The message was not modified. This corresponds to integrity. Note that this mechanism allows the detection, in case the integrity has been violated.

The described sequence of operations comprises the concept of digital signature, being depicted in Figure 14.12.

---

* As only the legitimate author knows the private key used to encrypt the hash sum.

It is worth noting that the digital signature assures authenticity (which is an attribute), while not comprising authentication (which is a process). Note that even though if a third party sends a previously recorded message (replay attack), the receiver is not able to detect it, as the authenticity was not violated (i.e., the claimed author of the message is its legitimate). To avoid such nonlegitimate operation, an authentication process such as the previously described CHAP needs to be implemented. Note that hybrid cryptographic protocols, such as the SSL or the TLS, incorporate their own authentication processes, in addition to the authenticity provided by the digital signature.

Figure 14.13 depicts the simplified block diagram of a system composed of a combination of the digital signature with asymmetric cryptography. Since both schemes make use of public key cryptography (public and private keys), it makes sense to implement these protection mechanisms together. This resulting system incorporates protection mechanisms from attacks against the confidentiality, integrity, and authenticity.

14.5.4 Digital Certificates

Asymmetric cryptography considers the open and wide distribution of public keys, without any level of confidentiality. Nevertheless, it is of high importance to keep the mapping between a public key and its owner. This mapping is achieved with the use of digital certificates.

An attacker that generates a pair of asymmetric keys and advertizes the public key, stating it belongs to a third party, can start sending messages signed by such third party (source spoofing). The destination of the messages thinks such messages were sent by the legitimate third party, while sent by an attacker. Moreover, the intercepted messages sent by the third party to any other parties can easily be decrypted by such attacker. To avoid this type of attacks, the digital certificate is used.

A digital certificate consists of a document emitted by a certification authority (CA), which maps a public key to its owner.* A digital certificate includes a set of information such as website address or e-mail address, the validation date of the certificate, the CA and its digital signature, the owner entity, and its public key.

To assure that the certificate was not modified, the digital certificate is digitally signed by a CA. The application of a client (web browser, e-mail application, etc.) has a list of accepted certification authorities, together with their public keys. When a user wants to initiate an encrypted session with a specific server, he receives the server's digital certificate, and the validation of the certificate is automatically performed by the client's application. This validation is done by applying the stored CA's public key in the client's application to the CA's digital signature located in the digital certificate. If the CA digital signature is validated and the CA belongs to the list of accepted authorities, the encryption handshaking proceeds. Alternatively, if the

---

* A digital certificate works like an identity card, establishing the relationship between an identification and an entity.

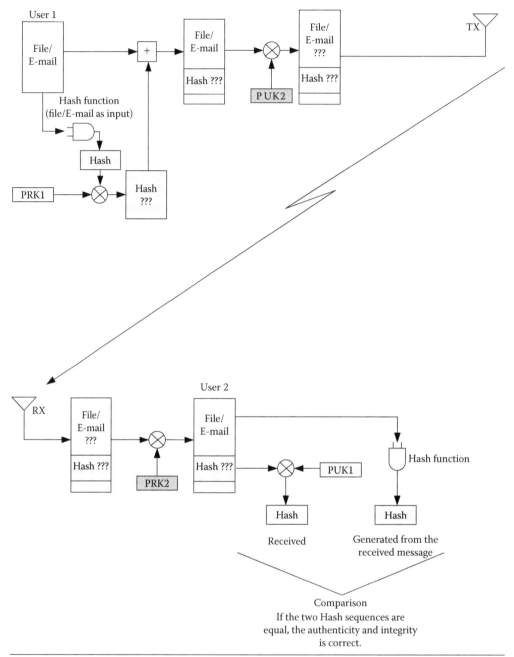

Figure 14.13   Generic block diagram of the digital signature combined with asymmetric cryptography.

digital certificate's CA does not belong to the client's accepted CA list, the client may accept the new authority as an accepted CA. Otherwise, the encryption handshaking ends. It is worth noting that the CA keeps a list of all signed and revoked certificates, possible to be accessed by any entity (client or server).

Finally, let us analyze the situation where an attacker sends a known server's digital certificate, claiming he is the known entity (source spoofing). Since the asymmetric (and hybrid cryptography) includes the data (or session key) encrypted with the destination's (server) public key and since such attacker does not have the corresponding private key, he is not able to decode the sent data (or the session's key). Therefore, even though if an attacker sends a third-party digital certificate claiming he is such entity, he does not succeed to implement an attack.

14.5.5 Public Key Infrastructure

PKI is meant to establish a trust chain between a public key and its owner, as well as to allow a secure public keys distribution system. It consists of an infrastructure using CIS, as well as databases, providing a service to allow the generation of digital certificates, implementing the corresponding keys management and distribution system.

As previously described, the trust chain is achieved with the use of digital certificates, together with certification authorities. Moreover, the PKI also includes registration authorities (RA), validation authorities (VA), certificate revocation lists (CRL), valid certificate lists, and so on.

Each CA stores its signed digital certificates in a public repository (database), being accessible for the public in general. Similarly, when the certificate ends the validity or if an error is detected in any distributed certificate, the CA revokes the digital certificate and advertizes it using a CRL. These databases are accessible by entities using the online certificate status protocol.

The PKI follows a hierarchical structure, with several levels of certificate authorities. Upper layer CAs are normally used to validate the work developed by lower layer CA, as well as to validate the certificates generated or signed by lower level CA. Consequently, these entities are also referred to as VA. On the other hand, the lower level CA, responsible for directly interacting and identifying end entities, are referred to as RA. In this scenario, only the lower level RA are responsible for sending digital certificates directly to the entities, on a daily basis, while such certificates have to be sent from a RA to an upper layer CA, for validation.

Before an entity starts using a digital certificate (e.g., the owner of a web server), he first needs to have the digital certificate properly certified and signed by a CA. This process can be carried out using one out of two procedures:

- Procedure 1: The CA produces and distributes private and public keys. In this case, the entity sends a certificate request, together with its personal data. The CA sends the signed certificate and the private and public keys to the entity. Moreover, the CA stores the public key in a CA public repository for public consultation. The drawback of this procedure relies on how to be sure that the CA does not keep a copy of the generated private key, using it illegitimately.
- Procedure 2: The entity produces a private and public key. In this scenario, the entity signs his public key and his identification information with the corresponding hash function and his private key. Then, the entity sends its clear

identification, his clear public key, and the corresponding digital signature to the CA, which verifies the identification information with the received clear public key. Once it has been verified, the CA creates a certificate and signs it with the CA private key and returns the new certificate to the entity, posting it in a public repository. The drawback of this procedure relies on how to be sure that the entity is exactly who he is claiming to be. This procedure can easily be improved by using one or more authentication mechanisms such as using personal presentation of Id Card, or other identification documents. Note that procedure 2 has overcome the drawback of procedure 1, as the CA does not have knowledge about the entities' private keys.

The most used digital certificate standard, also adopted by the Internet Engineering Task Force [IETF], is the ITU-T X.509 standard. This standard defines a format for the content of the several certificate's fields, including version, serial number, owner identification, CA digital signature, validity data, and the entity's public key.

A PKI can be implemented within an organization (for internal use) or as an open infrastructure for generation, distribution, and storage of digital certificates over the Internet.

14.5.6 Hybrid Cryptography

Due to the highly demanding mathematical computation, the processing time required to encrypt and decrypt a message using asymmetric cryptography is approximately one-thousand times higher than using symmetric cryptography. On the other hand, symmetric cryptography presents a vulnerability associated to key's distribution. Hybrid schemes aim to explore the advantages of symmetric and asymmetric cryptography, while overcoming their disadvantages. This is achieved by using both symmetric and asymmetric keys in an efficient manner. As can be seen from Figure 14.14, the hybrid cryptography works as follows:

- When a message* needs to be sent, the sender generates a symmetric (session) key and encrypts the message with such key.
- Then, the symmetric key is encrypted with the receiver's public key and the result is added to the previously encrypted message. These two components are sent together to the receiver.
- The receiver takes the encrypted symmetric (session) key and decrypts it with its private key.
- Then the "plaintext" symmetric key is finally used to decrypt the ciphertext message.

Similar to asymmetric cryptography, the hybrid cryptography requires the previous distribution of public keys.

---

* In fact, sending a message should be understood as the establishment of a session.

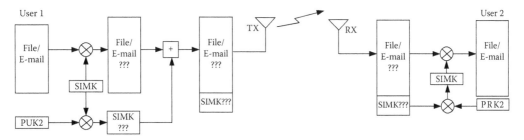

Figure 14.14 Generic block diagram of hybrid cryptography.

Note that all described operations are automatically performed just by pushing the "encrypt" button in the browser or in the e-mail application.

Similar to the diagrams of the symmetric and asymmetric cryptography, in Figure 14.14 SIMK stands for symmetric key, PUK stands for public key, and PRK stands for private key. Furthermore, PUK1 and PUK2 stands for the public key of user 1 (message sender) and user 2 (message receiver), while PRK1 and PRK2 stands for the private key of the user 1 and user 2.

With hybrid cryptography, since the message is encrypted with a symmetric key, the required processing time is approximately one-thousand times lower than that of asymmetric cryptography. Moreover, encrypting the symmetric key with the receiver's public key is not time demanding, as the key's length is much lower than the message's length. Although hybrid schemes use symmetric keys for encrypting messages, the previous distribution of such keys is avoided. This translates in a measure that mitigates the vulnerabilities associated to the distribution of symmetric keys. Consequently, the hybrid cryptography takes advantage of the reduced encryption and decryption time provided by the symmetric cryptography, while taking advantage of the easy and secure way public keys are distributed. Moreover, since each symmetric key is only used in a single session, this type of cryptography is very reliable.

Figure 14.15 depicts the generic block diagram of a system that combines the hybrid cryptography with the digital signature. Examples of hybrid cryptographic protocols include the SSL, TLS, and the IPsec,* which are defined in the following subsections.

14.5.6.1 Secure Sockets Layer and Transport Layer Security   SSL and TLS are examples of two important hybrid cryptographic protocols widely used in the Internet. These protocols make use of different symmetric and asymmetric algorithms, as well as hash functions, in a customized manner. The combination of the three selected algorithms is called a cipher suit.

The initial handshaking between a client and a server includes the negotiation of the protocols possible to be used by both parties. In case one of the parties does not accept an initially proposed protocol, either they have to re-negotiate the protocols to use or the secure connection cannot be established. Note that the server generally

---

* The pretty good privacy (PGP) is another hybrid protocol.

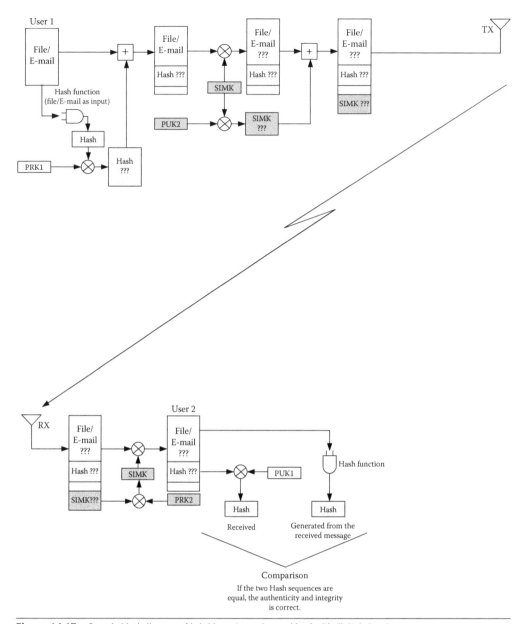

Figure 14.15 Generic block diagram of hybrid cryptography combined with digital signature.

chooses the strongest cipher suite possible to be used by both parties (by the client and the server). As an example, a SSL or TLS session may include the AES128 symmetric algorithm (as a session key), together with the RSA1024 asymmetric algorithm, along with the 128-bit SHA1 as an hash function. Note that in online secure connections, the keys are negotiated at the start of each session.

As can be seen from Figure 14.16, the SSL/TLS protocols are used to implement an end-to-end encryption, interacting directly with the application layer. Some authors refer to the SSL/TLS protocols as transport protocols, while others refer to these protocols as application layer protocols. This latter approach is justified by the fact that the

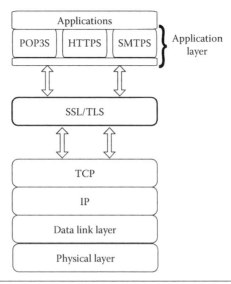

Figure 14.16 Location of SSL/TLS protocols into TCP/IP protocol stack.

SSL/TLS protocols implement HTTPS, POP3S, and SMTPS secure protocols* in the application layer (which include cryptography, digital signature, etc.). The implementation of these protocols by the application layer facilitates the implementation of high-level security functions such as user authentication.

As can be seen from Figure 14.16, the application layer protocols implemented in association with the SSL/TLS protocols are similar to those of unencrypted, but with a suffix S added to its designation (e.g., POP3 becomes POP3S, HTTP becomes HTTPS, SMTP becomes SMTPS). Moreover, the well-known ports assigned to the secure layer 4 protocols (in the server side) are different from those assigned to nonsecure protocols (e.g., HTTP uses, by default, the port 80, whereas the HTTPS uses the port 443).

The implementation of the HTTPS secure protocol is achieved following a client/server architecture, using a browser in the client side and a web server in the server side. A similar process is implemented for secure e-mail. Current browsers and e-mail applications support, by default, the ability to establish secure sessions using SSL/TLS.

The SSL protocol was created in 1994 by Netscape Communications Corporation to provide some security to financial transactions through the Internet. To improve security functions, the Netscape Communications Corporation deployed, in 1996, the SSL v3.0. During that year, the IETF used the SSL v3.0 as the basis for the development of the TLS v1.0,† which is the most used cryptographic protocol in the Internet. Although both TLS v1.0 and SSL v3.0 is supported by most web browsers, the TLS v1.0 is normally the default secure protocol. Note that the protocol used in a secure session is negotiated by both parties.

---

* These protocols aim to provide secure services like web browsing and e-mail.
† TLS v1.0 is also known as SSL v3.1.

The following paragraphs provide a generic description of the SSL v3.0 protocol (generically referred to as SSL protocol). Since the TLS v1.0 is very similar, this latter protocol is not described here.

Figure 14.17 depicts the SSL handshaking that is implemented before the exchange of encrypted application data takes place. The description of each step of the handshaking is described as follows (the reader should refer to the numbers in Figure 14.17):

1. The client application* that aims to establish a secure session with a server application initializes the handshaking by sending a "Hello" message followed by a 28-byte random number R_{client},† as well as a list of cryptographic algorithms supported (asymmetric, symmetric and hash functions).

2. The server‡ also sends a "hello" message followed by another 28-byte random number R_{server}, its list of cryptographic algorithms supported, as well as a session identification (session ID). The cipher suite selected for the ciphered session is the one that provides the higher level of security, while possible to be implemented by both parties. In addition, the server also sends its public key along with its ITU-T X.509 certificate (digital certificate). Note that the validity of the digital certificate needs to be confirmed by the client.

3. If the server requires a digital certificate§ from the client, for allowing the establishment of the secure session, a "digital certificate request" message is sent to the client along with the list of acceptable certificate authorities.

4. If the server requires a digital certificate, the client sends it, which includes the client's public key. If such digital certificate is not available at the client side, a "no digital certificate" message is sent, which may result in ending the session.

5. After all the previously described operations, the server sends a "hello done" message.

6. The client verifies the validity of the server's digital certificate and verifies all parameters proposed by the server. Afterwards, the client sends to the server the "client key exchange" message, which contains the 48-byte pre-master secret key¶ S, a 46-byte random number, and the MAC** key, encrypted with the server's public key. Note that the master secret key K is generated by both

---

* For example, the e-mail or web browsing application.
† This random number together with the server's random number and with the pre-master secret key S are used to generate the master secret key. The master secret key is then used to generate the session's key (symmetric). Note that these random sequence numbers do not consist of any authentication process.
‡ For example, the e-mail or web server.
§ Currently, most of the bank transactions do not require a digital certificate from the client. Normally, only companies have digital certificates.
¶ This key is computed by the client and sent encrypted to the server, using the server's public key.
** The MAC keys are used to assure the authenticity and integrity of the exchanged messages. It consists of a digital signature, using hash functions.

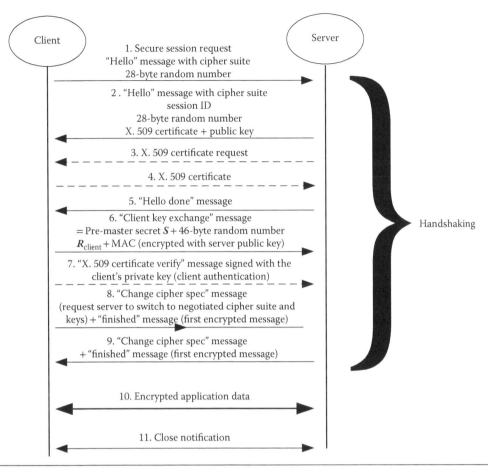

Figure 14.17 SSL handshaking, secure data exchange, and session closeout.

the client and the server independently from random numbers and pre-master secret key S (which is generated by the client and sent to the server encrypted). Moreover, the master secret key K is used by both parties to generate their own session keys (symmetric keys). Note that master secret keys K are never explicitly exchanged. The session keys and the MAC keys are different in the two directions.

7. In case the server requires a digital certificate from the client, the client sends a "digital certificate verify" message signed with the client's private key. This corresponds to the authentication process of the client, assuring that the client's digital certificate belongs and is sent by the legitimate client. In case such digital certificate is not required by the server, the authentication of the client is only performed after the establishment of the secure session (and therefore, once the session is already encrypted) using a protocol such as the CHAP. Note that the server does not need to send the digital certificate to the client properly signed with the server's private key. This server's authentication process is not required because in case the server does not have a private key corresponding to the previously sent pre-master secret S, it cannot

decrypt it* and the server cannot create the master secret key.† In such case, the handshake fails.

8. Once the master secret has been generated, the client sends the "change cipher spec" message to switch to the accepted cipher suite. Then, the client sends the "finished" message, which is the first encrypted message and which finalizes the SSL handshaking in this direction.

9. The server also sends the "change cipher spec" message along with the encrypted "finished" message. This ends the SSL handshaking in both directions.

10. The encrypted session is finally established and both parties can exchange encrypted data.

11. When one of the parties wants to finalize the ciphered session, it sends a close notification message to the other parties.

14.5.6.2 Security Architecture for IP   As previously described for the SSL and TLS, since these protocols are implemented between the session and the application layer,‡ they are utilized to provide end-to-end security (authentication, integrity, and confidentiality). Conversely, if a point-to-point encryption in a certain path of a connection (e.g., between two different LAN or between a host and a LAN) is necessary, the IPsec is the appropriate protocol (in tunnel mode). Although the IPsec may also be used to provide end-to-end (host-to-host) encryption (in transport mode), since the original IP headers are exchanged in clear (in this mode), the level of protection is reduced.

The IPsec was developed to provide security to the IPv6, while kept as an option for IPv4 [RFC 4301] [RFC 4309]. As can be seen from Figure 14.18, the IPsec protocol is implemented between layer 3 and layer 4, being transparent to the application layer.§ Consequently, user interaction is not included, which may also present some advantages. In fact, the encryption can even be implemented without the knowledge of the end users. The IPsec is directly implemented in the operating system. This way, the security service is provided to all applications that interact with the transport layer (TCP or UDP). Therefore, IPsec is well fitted for uses such as remote access of an Intranet from the Internet or for the interconnection of office delegations with the headquarters (HQ) through a virtual private network (VPN). These IPsec utilizations

---

* This is encrypted with the server's public key.

† This rule is applicable to all transactions where a digital certificate is used. Even though if an attacker sends a known server's digital certificate, claiming he is the known entity, since the asymmetric and hybrid cryptography includes the data or session's key encrypted with the destination's (server) public key, and since such attacker does not have the corresponding private key, he is not able to decode the transmitted data or the session's key.

‡ This facilitates, for example, the user authentication process.

§ Note that, in the case of the IPsec protocol, the application layer protocols are not added with the S suffix (contrary to the SSL/TLS protocols).

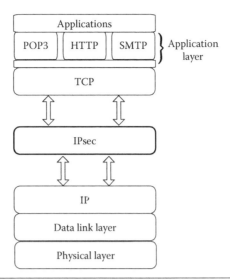

Figure 14.18 Location of IPsec protocol into TCP/IP stack.

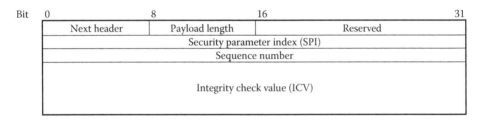

Figure 14.19 Header of IPsec packet using the authentication header mode.

can be seen from Figures 14.27 and 14.28. The reader should refer to Section 14.7 for a detailed description of VPN and to figures therein.

The IPsec security services are provided through the utilization of either one or both of the following protocols:

- Authentication header (AH): it provides authenticity and integrity to the exchanged packets, protecting the point-to point connections from attacks such as replay attack, spoofing, and connection hijacking. This is achieved through the use of the integrity check value (ICV), together with the identification of packets with a sequence number (see Figure 14.19).
- Encapsulation security payload (ESP): it provides authenticity, integrity, and confidentiality to the exchanged packets, protecting the point-to-point connections from attacks such as replay attack, packets partitioning, and packets sniffing. This is achieved through the use of authentication and integrity protection mechanisms (provided by the ICV field, as well as with the identification of packets with a sequence number), together with hybrid cryptography*

---

* Asymmetric protocol combined with DES or 3DES symmetric (session) protocol.

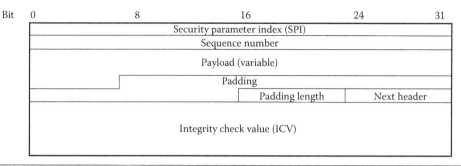

Figure 14.20   Header of IPsec packet using the encapsulation security payload mode.

(see Figure 14.20). An option includes the combination of the ESP and AH protocols. In such option, the ESP is encapsulated into the AH protocol (externally), allowing the destination station to verify the authenticity of the packet, before it is deciphered.

The IPsec can be implemented to protect one or more connections between a pair of hosts, between two security gateways* or between a host and a gateway. The protection is achieved through the creation of a security association (SA), which is identified by three different parameters:

- Security parameter index (SPI): a unique number defined during the previous handshaking of a connection that identifies the SA. The parties involved in the encrypted connection need to use the SPI during the data exchange.
- Destination IP address: although the connection can be unicast, multicast, or broadcast type, the IPsec utilizes an unicast address, assuming the other cases as an extension of the unicast.
- Protocol identification (AH or ESP): includes the identifier 50 for the ESP and 51 for the AH protocol.

In case both the AH and ESP protocols are used together, two SA must be defined.
In addition to the AH and ESP protocols, the Internet key exchange (IKE) protocol [RFC 2409] is used to create, manage, and exchange different keys between hosts or security gateways.

The IPsec can be implemented either in transport or tunnel mode:

- Transport mode: only the packet payload is protected. This mode establishes a SA† between two hosts (end-to-end), inserting the IPsec header between the IP header and the higher level protocol header (TCP or UDP). Finally, the IP payload is encapsulated into the IPsec payload;

---

* A firewall or a router running the IPsec protocol.
† In fact, two SA can be established, in case the AH and ESP are used together.

- Tunnel mode: both the payload and the header of the packet are protected. In this mode, a SA is established between two security gateways* or between a host and a security gateway located at the entrance of a network. In the tunnel mode, a new IPsec header is externally added. The original IP packet (including its header) is encapsulated into the IPsec payload. Then, the whole IPsec packet (including its header) is encapsulated into another IP packet. The source and destination IP addresses of this external IP packet correspond to the source and destination IP addresses of the security gateways, whereas the source and destination IP addresses included in the internal IP header specifies the real source and destination of the IP packet.

14.6 Network Architectures

Depending on the services and security levels, there are different network architectures that can be implemented within an organization such as an enterprise. Two different network types may exist within an organization:

- Intranet: only used by the organizational employees
- Extranet: used by both customers and business partners

These two network types are typically separated by a firewall, consisting of a network device that implements a group of rules to verify the conformity of the network with the established security rules, procedures, and policies. The firewall can be either implemented above a traditional operating system in a regular computer or implemented as a dedicated and stand-alone system. To reduce the level of vulnerabilities, the firewall implementation using a stand-alone system tends to be more resilient. This results from the fact that increasing the amount and software complexity results in higher vulnerabilities.[†] Note that a typical network attack consists of exploring the vulnerabilities of operating systems. To solve a problem, the network administrator normally switches off applications, leaving room for an attacker to penetrate in the target network and to implement the desired actions. Depending on the depth of conformity verification of the rules implemented with the established policy, there are different types of firewalls:

- Packet filtering—stateless: it inspects layer 3 and 4 headers (source and destination IP and port addresses) and sequence numbers. Note that it does not keep the state of the connections (stateless), that is, it does not take into account the established connections to allow or deny new connections. Moreover, it does not interrupt connections (in opposition to the proxy firewall).

---

* A security gateway can be either a firewall or a router that runs the IPsec protocol. It is typically located at the entrance of a LAN, such that the interconnection between two LANs (through unsecure WAN or MAN) or between a remote host and a LAN.

† In average, a software program presents a certain number of bugs per million of instructions. Therefore, reducing the software complexity results in a more resilient system.

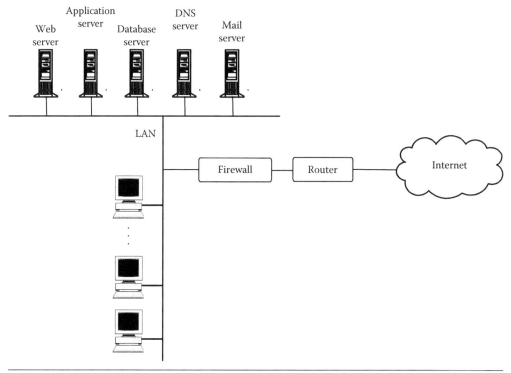

Figure 14.21   Generic corporate architecture composed of a single network.

- Packet filtering—stateful: similar to the stateless firewall type, it inspects layer 3 and 4 headers. However, contrarily to the previous firewall type, the stateful firewall keeps the state of the established connections, taking this information into account in the decision to allow or deny new connections. It does not interrupt connections.
- Application layer gateway—stateful inspection: since this firewall interrupts the connections, it is also referred to as "proxy firewall" (similar to the proxy server). Moreover, it works on the application layer, performing a deep inspection of the packet's contents, while taking information about the state of the established connections to allow or deny new connections.

Figure 14.21 depicts a generic network architecture normally implemented in small companies. In this case, a single network is used as both Intranet and Extranet, and the firewall protection is normally placed at the output of the network. In fact, most common architectures such as small business networks include a common device that acts as both router and firewall. This architecture is very vulnerable, as the access to the servers is achievable from the Internet (to be used by the customers and partners). Since the servers used by both customers (Extranet) and employees (Intranet) are the same, the level of vulnerability of this network architecture is very high.

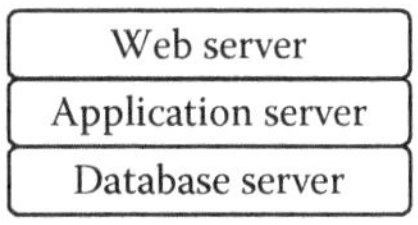

| Web server |
| Application server |
| Database server |

Figure 14.22 Hierarchical structure of the web service.

Note that a web service is provided following a hierarchical structure, as depicted in Figure 14.22. In fact the web server is responsible for establishing HTTP connections with remote client hosts. Any specific service required by a client is indirectly provided by the application server. In fact, the client only connects to the web server, but this server makes use of the application server to provide the desired service to the client (e.g., a bank transfer). In addition, the application server processes data that is stored in a database server, and a service provided to a client may result in a change in the database. Such example is a bank transfer, where a HTTP connection is established with the client using the web server, the transaction itself is implemented using the application server, and the application server requests the database server to check whether or not the client has enough balance in his current account. In case the response is positive, the transaction is carried out by the application server, and the database server reduces the current account balance by an amount corresponding to the amount transferred. During this transaction, the database server provided a service to the application server, and the application server provided a service to the web server. The client only establishes a connection with the web server, and all other functions are performed in a transparent manner to the client using the hierarchical structure of the web service. These three servers are included in the several network architectures presented in this chapter. Nevertheless, in many cases, the different server functionalities are implemented by a single or two servers.*

To improve the security level, a common procedure consists of separating the Intranet from the Extranet using a firewall. This can be seen from Figure 14.23. This architecture includes the duplication of servers, some of them dedicated to the Intranet and others dedicated to the Extranet. In the depicted architecture, the database server exists only in the Intranet. In this case, the external application server connects to it every time it is needed. Such permission needs to be properly included in the list of firewall allowed connections. Moreover, the two e-mail servers are normally synchronized, which needs also to be allowed by the firewall. The architecture of Figure 14.23 already includes a proxy (placed in the Extranet), which avoids the direct connection from the Intranet to the Internet. A host in the Intranet, which aims to establish a connection with a web server placed in the Internet, initiates a connection with the proxy and, in order to provide the desired web service

---

* For example, by the Internet Information System (IIS) server or by the Apache application server, which include both the web and the application server functionalities.

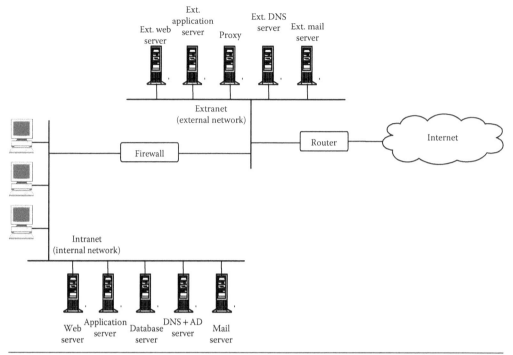

Figure 14.23   Generic corporate architecture composed of an Intranet and an Extranet.

to the host, the proxy establishes a second connection with the destination server. Therefore, the proxy establishes two different HTTP connections. Then, the proxy "downloads" the desired HTML file from the web server and, in a second step, uses the HTTP connection established with the host to transfer such file. Note that the proxy stores HTML files during a certain period, which can be accessed by internal hosts, without having to get access to the destination server.

From Figure 14.23, the existence of the active directory (AD) server is also seen in the Intranet. This is responsible for allowing or denying requests (permissions) from hosts connected in the Intranet, in accordance with the established organizational security policy.

To further improve the security level, the external access to the client servers needs to be protected by a firewall. This is achieved by creating a demilitarized zone (DMZ), as depicted in Figure 14.24. A DMZ consists of a network that accommodates the external servers, and whose access needs to be properly authorized by the firewall. Contrarily, in the previous architecture, all external users could get access to all external servers. Therefore, using a DMZ architecture, the Extranet is left empty. It is worth noting that a basic principle of this architecture is that direct connection from/to the Intranet into/from Internet is not allowed.

This architecture could even be improved by using two firewalls, instead of a single one. As depicted in Figure 14.25, one firewall is placed between the Intranet and the DMZ, whereas the second is placed between the DMZ and the empty external network (which then gives access to the Internet).

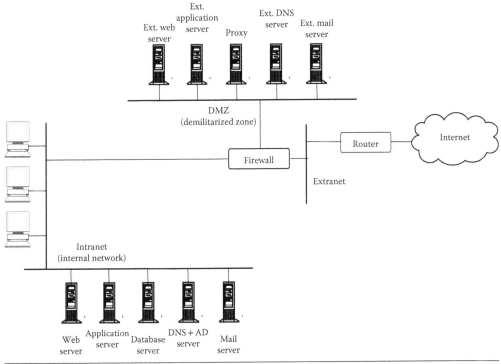

Figure 14.24   Generic corporate architecture with demilitarized zone.

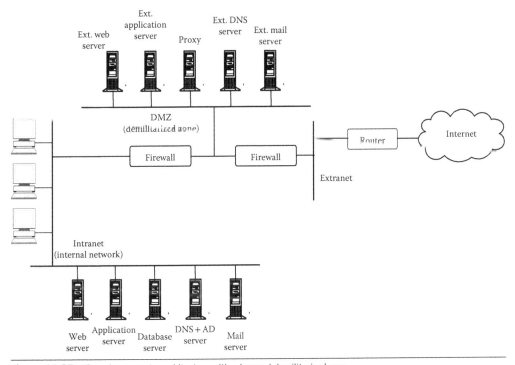

Figure 14.25   Generic corporate architecture with advanced demilitarized zone.

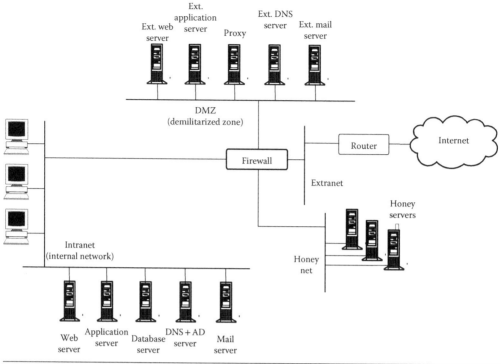

Figure 14.26 Generic corporate architecture with demilitarized zone and Honey Net.

Another possible architecture implemented by organizations that want to provide higher level of security makes use of a honey net, which acts as a decoy for attackers (see Figure 14.26). The firewall permissions to the honey servers are made easier, such that an attacker thinks he is hacking the organizational servers, while only getting access to decoy servers. In addition, the network may use other mechanisms such as an intrusion detection system (IDS) to detect such intrusion, identify the author, register his actions, and track him.

14.7 Virtual Private Networks

A VPN is used by both employees and partners to allow access from a public communications network (typically, from the Internet) to the Intranet or to a partners' DMZ. A partner's DMZ is commonly referred to as partners' network. As can be seen from Figure 14.27, this consists of an additional network configured in the firewall, in parallel with the regular DMZ. The partners' DMZ is the location where the servers used by partners are installed. The information stored in these servers is part of the information stored in internal servers (Intranet). This is achieved by synchronization of the desired servers through the firewall.

The remote access VPN can be secure or unsecure. The ability for the employees to get access to Intranet and for the partners to get access to partners' DMZ from the Internet, while preventing Internet attackers from getting access to the exchanged

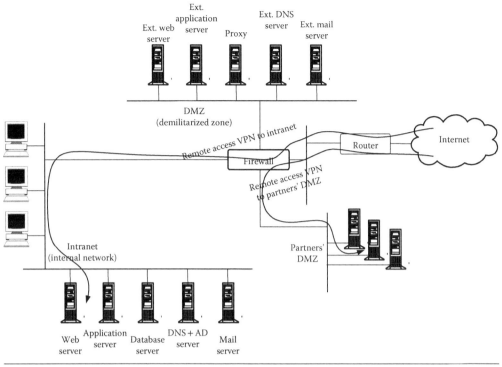

Figure 14.27   Remote access VPN to Intranet or to partners' demilitarized zone from the Internet.

data, is achieved by using a cryptographic tunnel configured in firewalls or routers that implement the IPsec cryptographic protocol. These mechanisms assure the confidentiality, authenticity, and integrity of exchanged data through the Internet.

Currently, organizations such as banks have a HQ where all important servers are located and a high number of remote delegations spread out over the territory (close to the customers). The remote delegations are typically connected to the HQ's Intranet through the Internet using a secure VPN. The IPsec in tunneling mode (using the ESP protocol) is the most used cryptographic protocol adopted to allow such connectivity. In this case, a secure IPsec tunnel is configured at both the HQs' and remote delegations' firewall. The firewalls act as security gateways, encapsulating the original IP packet (including the original IP header) into an IPsec packet. The IPsec packet is then encapsulated into the payload of another IP packet, whereas its source and destination IP address refers to the IP address of the security gateways (firewalls) of the remote delegation and HQ. This is depicted in Figure 14.28.

In case a user is located anywhere in a public communications network (Internet),* instead of in a remote delegation, the same IPsec VPN in tunneling mode can be implemented, but the tunnel is established directly between the remote host and the HQs' security gateway.

---

* Physically, a remote user can be anywhere (e.g., at home, in a hotel).

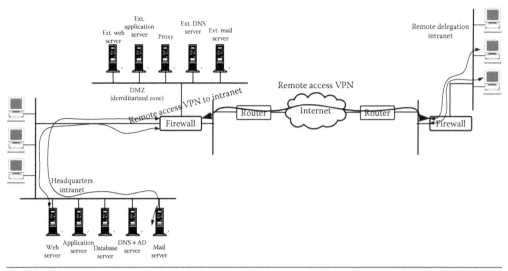

Figure 14.28   Remote access VPN configured between a remote delegation's and headquarters' firewall.

The exchange of data (connectivity) can be assured by any transmission means depending on the service requirements, such as the required data rates, accepted delays, and so on. A remote user from home may access the Intranet using the regular home network (e.g., a wireless LAN connected to the Internet through xDSL modem). A very small remote delegation may opt for a low cost xDSL or cable modem connection, whereas larger remote delegations may use MPLS connectivity to get access to the headquarters' router with high-speed connectivity.

End of Chapter Questions

1. How is the authentication process implemented in the SSL protocol?
2. Enumerate the types of attacks against the confidentiality.
3. Which are the security areas included in INFOSEC?
4. Enumerate the types of attacks against the integrity.
5. What is the difference between authentication and authenticity?
6. Describe the CHAP protocol.
7. Describe the way the accountability protects attack against the confidentiality, integrity, and authenticity.
8. What is a distributed denial of service (DDoS) attack?
9. What are the mechanisms that can be used to protect the information from attacks against the confidentiality?
10. Define hybrid cryptography.
11. What are the mechanisms that can be used to protect the information from attacks against the integrity?
12. What is the difference between a digital signature and a digital certificate?

13. What are the requirements of the accountability of information?
14. What is the difference between an attack against to confidentiality and against the integrity? Enumerate types of attacks against these two information attributes.
15. Describe the man-in-the-middle attack.
16. Which procedures can be used to generate digital certificates? What are their advantages and disadvantages?
17. Which types of authentication mechanisms do you know?
18. What is the difference between authenticity and non-repudiation?
19. What is the residual risk? How can it be quantified?
20. What are the mechanisms that can be used to protect the information from attacks against the authenticity?
21. What does the public key cryptography (PKI) stands for? What is its purpose?
22. Which elementary cryptographic operations do you know?
23. To provide authenticity, why is it important to encrypt the hash sum? How is it normally encrypted?
24. Why is it important to employ long passwords?
25. What is the hash function used for?
26. Which hybrid protocols do you know?
27. What is a brute force attack?
28. What is a digital signature? What does it protects from?
29. What is the public key cryptography?
30. In the scope of the risk analysis, what is the difference between vulnerability and threat? Give examples of both.
31. What are the advantages of symmetric cryptography, relating to asymmetric?
32. What does a VPN stand for?
33. What are the advantages and disadvantages of asymmetric cryptography?
34. What is a DMZ?
35. How can a VPN be established between a remote delegation and a headquarters?
36. What is the public key infrastructure used for?
37. What is the advantage of using the SSL/TLS relating to the IPsec?
38. Which type of firewalls do you know?
39. What is the different between an Intranet and an Extranet?
40. From which type of attacks do hybrid cryptographic schemes give protection?

Annex A

Fourier Transforms

The Fourier transform of the time domain variable $x(t)$ is mathematically defined by

$$X(f) = \mathcal{F}[x(t)]$$

$$= \int_{-\infty}^{+\infty} x(t) \cdot e^{-j2\pi ft} \, dt \tag{A.1}$$

Similarly, the inverse Fourier transform of the frequency domain variable $X(f)$ is mathematically defined by

$$x(t) = \mathcal{F}^{-1}[X(f)]$$

$$= \int_{-\infty}^{+\infty} X(f) \cdot e^{j2\pi ft} \, df \tag{A.2}$$

Note that lower and upper case signal variables correspond to time and frequency domain variables, respectively. The mapping between one and the other is achieved through Fourier transforms ($\mathcal{F}[x]$ denotes "Fourier transform" of x) and inverse Fourier transform operations ($\mathcal{F}^{-1}[x]$ denotes "inverse Fourier transform" of x), that is, $X = \mathcal{F}[x]$ and $x = \mathcal{F}^{-1}[X]$.

A list of Fourier transforms of common functions is listed in Annex A. As an example, the Fourier transform of the sinc function corresponds to the rectangular pulse, that is, $\mathcal{F}[\mathrm{sinc}(2Wt)] = \dfrac{1}{2W} \Pi\left(\dfrac{f}{2W}\right)$ (see Table A.1). This means that, with the sinc function in the time domain, its frequency spectrum is the rectangular pulse. The Fourier transform of a signal gives us information about which frequency components are present in the signal, as well as their relative strength.

Table A.1   Common Fourier Transforms

FUNCTION	$v(t)$	$V(f)$
Rectangular	$\Pi\left(\dfrac{t}{\tau}\right)$	$\tau\,\mathrm{sinc}(f\tau)$
Triangular	$\Lambda\left(\dfrac{t}{\tau}\right)$	$\tau\,\mathrm{sinc}^2(f\tau)$
Gaussian	$e^{-\pi(bt)^2}$	$(1/b)e^{-\pi(f/b)^2}$
Causal exponential	$e^{-bt}u(t)$	$\dfrac{1}{b+j2\pi f}$
Symmetric exponential	$e^{-b\lvert t\rvert}$	$\dfrac{2b}{b^2+(2\pi f)^2}$
Sinc	$\mathrm{sinc}(2Wt)$	$\dfrac{1}{2W}\Pi\left(\dfrac{f}{2W}\right)$
Sinc squared	$\mathrm{sinc}^2(2Wt)$	$\dfrac{1}{2W}\Lambda\left(\dfrac{f}{2W}\right)$
Constant	1	$\delta(f)$
Phasor	$e^{j(w_c t+\phi)}$	$e^{j\phi}\delta(f-f_c)$
Sinusoid	$\cos(w_c t+\phi)$	$\dfrac{1}{2}\left[e^{j\phi}\delta(f-f_c)+e^{-j\phi}\delta(f+f_c)\right]$
Impulse	$\delta(t-t_d)$	e^{-jwt_d}
Sampling	$\displaystyle\sum_{k=-\infty}^{+\infty}\delta(t-kT_s)$	$\displaystyle f_s\sum_{n=-\infty}^{+\infty}\delta(f-nf_s)$
Signum	$\mathrm{sgn}(t)$	$1/(j\pi f)$
Step	$u(t)$	$\dfrac{1}{j2\pi f}+\dfrac{1}{2}\delta(f)$

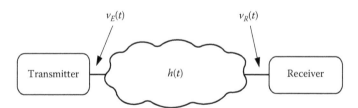

Figure A.1   Generic communication system with the signals depicted in the time domain.

As defined by Equation A.1, the channel's frequency response $H(f)$ is related to the channel's impulse response $h(t)$ (plotted in Figure A.1) through the Fourier transform defined by $H(f)=\displaystyle\int_{-\infty}^{+\infty}h(t)e^{-j2\pi ft}\,dt$.

Table A.2 Theorems of Fourier Transforms

OPERATION	FUNCTION	TRANSFORM
Superposition	$a_1 v_1(t) + a_2 v_2(t)$	$a_1 V_1(f) + a_2 V_2(f)$
Time delay	$v(t - t_d)$	$V(f)e^{-jwt_d}$
Scale change	$v(\alpha t)$	$\dfrac{1}{|\alpha|}V\left(\dfrac{f}{\alpha}\right)$
Conjugation	$v^*(t)$	$V^*(-f)$
Duality	$V(t)$	$v(-f)$
Frequency translation	$v(t)e^{jw_c t}$	$V(f - f_c)$
Modulation	$v(t)\cos(w_c t + \phi)$	$\dfrac{1}{2}\left[V(f - f_c)e^{j\phi} + V(f + f_c)e^{-j\phi}\right]$
Differentiation	$\dfrac{d^n v(t)}{dt^n}$	$(j2\pi f)^n V(f)$
Integration	$\displaystyle\int_{-\infty}^{t} v(\lambda)d\lambda$	$\dfrac{1}{j2\pi f}V(f) + \dfrac{1}{2}V(0)\delta(f)$
Convolution	$v * w(t)$	$V(f)W(f)$
Multiplication	$v(t)w(t)$	$V * W(f)$
Multiplication by t^n	$t^n v(t)$	$(-j2\pi)^{-n}\dfrac{d^n V(f)}{df^n}$

The channel's impulse response $h(t)$ is extracted at the output of the channel when a Dirac function is injected at the channel's input.

The resulting frequency domain signal at the channel's output is

$$V_R(f) = V_E(f) \cdot H(f) \tag{A.3}$$

Using Equation A.2, the time domain signal at the channel's output can be obtained performing the inverse Fourier transform to the frequency domain signal as

$$v_R(t) = \int_{-\infty}^{+\infty} V_R(f)e^{j2\pi ft}\,df \tag{A.4}$$

Instead of using Equation A.3 followed by Equation A.4 to compute the signal at the channel's output, this can also be obtained mathematically in the time domain by performing the convolution operation of the input signal $v_E(t)$ with the channel's impulse response $h(t)$ defined by

$$v_R(t) = \int_{-\infty}^{+\infty} v_E(t)h(t - \tau)d\tau \tag{A.5}$$

The computation of $v_R(t)$ may be simpler to perform in the frequency domain using Equation A.3, as it consists of a multiplication, while the convolution operation defined by Equation A.5 may be very demanding for complex channel impulse responses.

Table A.1 lists the most common Fourier transforms, while Table A.2 lists the important Fourier transforms theorems.

Annex B

Common Functions Used in Telecommunications

Table B.1 Common Functions Used in Telecommunications

Gaussian probability	$Q(k) = \dfrac{1}{\sqrt{2\pi}} \displaystyle\int_{k}^{\infty} e^{-\lambda^2/2} \, d\lambda$
Exponential	$\exp(t) = e^t$
Sinc	$\mathrm{sinc}(t) = \dfrac{\sin(\pi t)}{\pi t}$
Sign	$\mathrm{sgn}(t) = \begin{cases} 1 & t \geq 0 \\ -1 & t < 0 \end{cases}$
Step	$u(t) = \begin{cases} 1 & t \geq 0 \\ 0 & t < 0 \end{cases}$
Rectangle	$\Pi\left(\dfrac{t}{\tau}\right) = \begin{cases} 1 & |t| < \tau/2 \\ 0 & |t| > \tau/2 \end{cases}$
Triangle	$\Lambda\left(\dfrac{t}{\tau}\right) = \begin{cases} 1 - \dfrac{|t|}{\tau} & |t| < \tau \\ 0 & |t| > \tau \end{cases}$

Annex C

Example of Configuration of Routers and Switches Using Cisco IOS

Annex C lists the configuration commands in different routers and switches for the network depicted in Figure C.1. As can be seen, the network comprises three routers and four switches. Contrarily to the case described in Chapter 11, where two virtual local area networks (VLAN) are configured in the same switch, the network of Figure C.1 considers that hosts belonging to the same VLAN are connected to different switches. The configuration of the switches is similar with the difference that the interfaces used to interconnect the two switches need to be configured in trunk mode. Moreover, a crossover* fast Ethernet cable is used to interconnect the two switches. The connectivity between the switch 1 and router 1 needs also to be configured in trunk mode (in the switch side) and with two sub-interfaces (in the router side). It is assumed that the lines are already configured and connected to the devices.

In the router1:

```
Router>enable
Router#conf t
Router(config)#hostname router1
router1(config)#int se0/0
router1(config-if)#ip address 196.136.235.1 255.255.255.0
```

---

* As described in Chapter 4, a crossover Ethernet cable is used to interconnect devices of the same type (two computers, two switches, two routers, etc.).

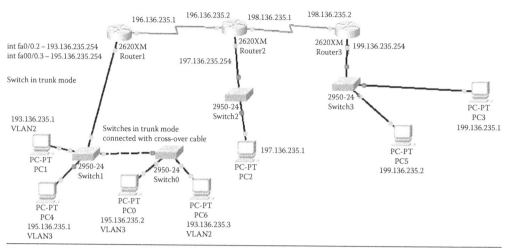

Figure C.1 Example of a network to be configured.

router1(config-if)#clock rate 64000 (the clock rate needs to be configured in the DCE terminal of the serial cable)

router1(config-if)#no shutdown

router1(config-if)#^Z

router1#conf t

router1(config)#int fa0/0

router1(config-if)#no ip address (the interface has no IP address)

router1(config-if)#no shutdown

router1(config-if)#exit

router1(config)#int fa0/0.2 (creation of a sub-interface)

router1(config-subif)#encapsulation dot1q 2 (the sub-interface is connected to VLAN 2, using the IEEE 802.1q protocol)

router1(config-subif)#ip address 193.136.235.254 255.255.255.0 (assignment of an IP address to the sub-interface)

router1(config-subif)#^Z

router1#conf t

router1(config)#int fa0/0.3

router1(config-subif)#encapsulation dot1q 3

router1(config-subif)#ip address 195.136.235.254 255.255.255.0

router1(config-subif)#^Z

router1#conf t

router1(config)#ip route 197.136.235.0 255.255.255.0 196.136.235.2 (static route configuration)

router1(config)#ip route 198.136.235.0 255.255.255.0 196.136.235.2

router1(config)#ip route 199.136.235.0 255.255.255.0 196.136.235.2

router1(config)#^Z

router1#

In the router2:

Router>enable
Router#config t
Router(config)#hostname router2
Router2(config)#int fa0/0
Router2(config-if)#ip address 197.136.235.254 255.255.255.0
Router2(config-if)#no shutdown
Router2(config-if)#^Z
Router2#config t
Router2(config)#int se0/0
Router2(config-if)#ip address 196.136.235.2 255.255.255.0
Router2(config-if)#no shutdown
Router2(config-if)#^Z
Router2#config t
Router2(config)#int se0/1
Router2(config-if)#ip address 198.136.235.1 255.255.255.0
Router2(config-if)#clock rate 64000
Router2(config-if)#no shutdown
Router2(config-if)#^Z
Router2#config t
Router2(config)#ip route 193.136.235.0 255.255.255.0 196.136.235.1
Router2(config)#ip route 195.136.235.0 255.255.255.0 196.136.235.1
Router2(config)#^Z

In the router3:

Router>enable
Router#conf t
Router(config)#hostname router3
router3(config)#int fa0/0
router3(config-if)#ip address 199.136.235.254 255.255.255.0
router3(config-if)#no shutdown
router3(config-if)#int se0/0
router3(config-if)#ip address 198.136.235.2 255.255.255.0
router3(config-if)#no shutdown
router3(config-if)#^Z
router3#config t
router3(config)#ip route 0.0.0.0 0.0.0.0 198.136.235.1 (default static route
    configuration)
router3(config)#^Z
router3#

In the switch0:

Switch>conf t
Switch>enable
Switch#conf t
Switch(config)#hostname switch0
switch0(config)#^Z
switch0#vlan database
switch0(vlan)#vlan 2 name finances
VLAN 2 added:
  Name: finances
switch0(vlan)#vlan 3 name human_resources
VLAN 3 added:
  Name: human_resources
switch0(vlan)#exit (save and exit when the switch is in the privileged mode)
switch0#int fa0/2
switch0#switchport access vlan 2
switch0#conf t
switch0(config)#int fa0/2
switch0(config-if)#switchport access vlan 2
switch0(config-if)#int fa0/3
switch0(config-if)#switchport access vlan 3
switch0(config-if)#int fa0/1
switch0(config-if)#switchport mode trunk
switch0(config-if)#int fa0/4
switch0(config-if)#switchport mode trunk
switch0(config-if)#^Z (save and exit when the switch is in the configuration mode)
switch0#

In the switch1:

Switch>enable
Switch#conf t
Switch(config)#hostname switch1
switch1(config)#exit
switch1#vlan database
switch1(vlan)#vlan 2 name finances
VLAN 2 added:
  Name: finances
switch1(vlan)#vlan 3 name human_resources
VLAN 3 added:
  Name: human_resources

```
switch1(vlan)#exit
switch1#conf t
switch1(config)#int fa0/1
switch1(config-if)#switchport access vlan 2
switch1(config-if)#int fa0/2
switch1(config-if)#switchport access vlan 3
switch1(config-if)#int fa0/3
switch1(config-if)#switchport mode trunk
switch1(config-if)#^Z
switch1#
```

References

Chapter 1

[Marques da Silva et al. 2010] Marques da Silva, M., A. Correia, R. Dinis, N. Souto, and J. C. Silva. 2010. *Transmission Techniques for Emergent Multicast and Broadcast Systems*. 1st ed. Boca Raton, FL: CRC Press Auerbach Publications.

[Khanvilkar 2005] Khanvilkar, S., F. Bashir, D. Schonfeld, and A. Khokar. 2005. Multimedia networks and communication. In *The Electrical Engineering Handbook*, ed. W-K. Chen, 401–26. Burlington, MA: Elsevier Academic Press.

[Monica 1998] Monica, P. 1998. *Comunicação de Dados e Redes de Computadores*. Lisboa, Portugal: ISTEC.

[Raj 2010] Raj, M., A. Narayan, S. Datta, and S. Das. 2010. Fixed mobile convergence: Challenges and solutions. *IEEE Commun Mag* 48(12):26–34.

[Stallings 2010] Stallings, W. 2010. *Data and Computer Communications*. 8th ed. Upper Saddle River, NJ: Prentice Hall.

[Vitthaladevuni 2001] Vitthaladevuni, P. K., and M. S. Alouini. 2001. BER computation of 4/M-QAM hierarchical constellations. *IEEE Trans Broadcast* 47(3):228–39.

[Vitthaladevuni 2003] Vitthaladevuni, P. K., and M. S. Alouini. 2003. A recursive algorithm for the exact BER computation of generalized hierarchical QAM constellations. *IEEE Trans Inform Theory* 49:297–307.

Chapter 2

[RFC 2460] Deering, S., and R. Hinden. 1998. *RFC 2460—Internet Protocol Version 6 (IPv6) Specification*. Network Working Group. http://www.ietf.org/rfc/rfc2460.txt

[Forouzan 2007] Forouzan, A. F. 2007. *Data Communications and Networking*. 4th ed. New York: McGraw Hill Higher Education.

[Marques da Silva et al. 2010] Marques da Silva, M., A. Correia, R. Dinis, N. Souto, and J. C. Silva. 2010. *Transmission Techniques for Emergent Multicast and Broadcast Systems*. 1st ed. Boca Raton, FL: CRC Press Auerbach Publications. ISBN: 9781439815939.

[Monica 1998] Monica, P. 1998. *Comunicação de Dados e Redes de Computadores*. Lisboa, Portugal: ISTEC.

[RFC 826] Plummer, D. C. 1982. *RFC 826—An Ethernet Address Resolution Protocol*. Network Working Group.

[Shannon 1948] Shannon, C. E. 1948. A mathematical theory of communication. *Bell Syst Tech J* 27:379–423, 623–56.

[Stallings 2010] Stallings, W. 2010. *Data and Computer Communications*. 8th ed. Upper Saddle River, NJ: Prentice Hall.

Chapter 3

[Benedetto 1987] Benedetto, S., E. Biglieri, and V. Castellani. 1987. *Digital Transmission Theory*. New Jersey: Prentice-Hall Inc.

[Burrows 1949] Burrows, C. R. 1949. *Radio Wave Propagation*. New York: Academic Press.

[Carlson 1986] Carlson, A. B. 1986. *Communication Systems*. 3rd ed. New York: McGraw Hill.

[CCIR 322] CCIR Report 322. 1995. Noise Variation Parameters. Singapore.

[Foschini 1998] Foschini, G. J., and M. J. Gans. 1998. On limits of wireless communications in fading environments when using multiple antennas. *Wireless Pers Commun* 6:315–35.

[Marques da Silva 2010] Marques da Silva, M., A. Correia, R. Dinis, N. Souto, and J. C. Silva. 2010. *Transmission Techniques for Emergent Multicast and Broadcast Systems*. 1st ed. Boca Raton, FL: CRC Press Auerbach Publications. ISBN: 9781439815939.

[Proakis 1995] Proakis, J. G. 1995. *Digital Communications*. 3rd ed. New York: McGraw Hill.

[Shannon 1948] Shannon, C. E. 1948. A mathematical theory of communication. *Bell Syst Tech J* 27:379–423, 623–56.

[Rappaport 1996] Theodore, S. 1996. *Rappaport—Wireless Communications*. Upper Saddle River, NJ: Prentice Hall.

Chapter 4

[ANSI/EIA/TIA-568] ANSI/EIA/TIA-568. 1991. *Commercial Building Telecommunications Standard*.

[ANSI/TIA/EIA-568-A] ANSI/TIE/EIA-568-A. 1995. *Commercial Building Telecommunications* Standard.

[ANSI/TIA/EIA-568-B] ANSI/TIE/EIA-568-B. 2001. *Commercial Building Telecommunications Standard*.

[ISO/IEC 11801] ISO/IEC 11801. 2002. *Information Standard–Information Technology–Generic Cabling for Customer Premises*. 2nd ed.

[ITU-T G.621] ITU-T Recommendation G.621. 1993. *Transmission media characteristics of 0.7/2.9 mm coaxial cable* pairs.

[ITU-T G.622] ITU-T Recommendation G.622. 1993. *Transmission media characteristics of 1.2/4.4 mm coaxial cable* pairs.

[ITU-T G.623] ITU-T Recommendation G.623. 1993. *Transmission media characteristics of 2.6/9.5 mm coaxial cable* pairs.

[Keiser 1991] Keiser, G. 1991. *Optical Fiber Communications*. 2nd ed. New York: McGraw-Hill.

[Stallings 2010] Stallings, W. 2010. *Data and Computer Communications*. 8th ed. Upper Saddle River, NJ: Prentice Hall.

[Winzer 2010] Winzer, P. 2010. Bell labs, alcatel-lucent, beyond 100G ethernet. *IEEE Commun Mag* 48(7):26–30.

Chapter 5

[Burrows 1949] Burrows, C. R. 1949. *Radio Wave Propagation*. Academic Press.

[Carlson 1986] Carlson, A. B. 1986. *Communication Systems*. 3rd ed. Singapore: McGraw Hill.

[Fernandes 1996] Fernandes, C. A. 1996. *Aspectos de Propagação na Atmosfera—Anexo*. Secção de Propagação e Radiação IST.

[Ha 1990] Ha, T. T. 1990. *Digital Satellite Communications*. 2nd ed. Singapore: McGraw Hill Communications Series.

[Kadish 2000] Kadish, J. E., and T. W. R. East. 2000. *Satellite Communications Fundamentals*. London: Artech House.

[ITU-R P.368-7] ITU-R Recommendation P. 368–7. 1992. *Ground-Wave Propagation Curves for Frequencies between 10 kHz and 30 MHz*.

[Marques da Silva 2010] Marques da Silva, M., et al. 2010. *Transmission Techniques for Emergent Multicast and Broadcast Systems*. 1st ed. Boca Raton: CRC Press Auerbach Publications. ISBN: 9781439815939.

[Parsons 2000] Parsons, J. D. 2000. *The Mobile Radio Propagation Channel*. 2nd ed. New York: Wiley.

[Proakis 1995] Proakis, J. G. 1995. *Digital Communications*. 3rd ed. Singapore: McGraw Hill.

[NBS 1967] Rice, P. L., et al. 1967. Transmission loss predictions for tropospheric communication circuits. *Technical note 101*, Vol. II. U.S. Departament of Commerce, National Bureau Standards (NBS).

Chapter 6

[Carlson 1986] Carlson, A. B. 1986. *Communication Systems*. 3rd ed. Singapore: McGraw Hill.

[MIL-STD-3005 1999] MIL-STD-3005 Department of Defense—Telecommunications Systems Standard. 1999. *Analog-to-Digital Conversion of Voice by 2,400 bit/second Mixed Excitation Linear Prediction (MELP)*.

[Goldsmith 1997] Goldsmith, A., and S. G. Chua. 1997. Variable-rate variable power M-QAM for fading channels. *IEEE Trans Commun* 45(10):1218–30.

[ITU-T G.711] ITU-T Recommendation G.711. 1972. *Pulse Code Modulation (PCM) of Voice Frequencies*.

[ITU-T G.726] ITU-T Recommendation G.726. 1990. *40, 32, 24, 16 kbit/s Adaptive Differential Pulse Code Modulation (ADPCM)*.

[Chamberlain 2001] Chamberlain, M. W. 2001. A 600 bps MELP vocoder for use on HF channels. *Mil Commun Conf* 1:447–53.

[Marques da Silva 2009] Marques da Silva, M., A. Correia, and R. Dinis. 2009. On transmission techniques for multi-antenna W-CDMA systems. *Eur Trans Telecomm* 20(1):107–21.

[Proakis 1995] Proakis, J. G. 1995. *Digital Communications*. 3rd ed. New York: McGraw Hill.

[Wang 2000] Wang, T., K. Koishida, V. Cuperman, A. Gersho, and J. S. Collura. 2000. A 1200 bps speech coder based on MELP *Acoust Speech Signal Process* 3:1376.

[Watkinson 2001] Watkinson, J. 2001. *The MPEG Handbook: MPEG-I, MPEG-II, MPEG-IV*. Oxford, UK: Focal Press.

[Webb 1994] Webb, W. T., and L. Hanzo. 1994. *Modern Quadrature Amplitude Modulation: Principles and Applications for Fixed and Wireless Channels*. New York: IEEE Press.

[Webb 1995] Webb, W. T., and R. Steele. 1995. Variable rate QAM for mobile radio. *IEEE Trans Commun* 43(7): 2223–30.

Chapter 7

[3GPP 25.211-v5.2.0] 3GPP TR 25.211-v5.2.0. 2003. Physical Channels and Mapping of Transport Channels onto Physical Channels (FDD).

[3GPP 25.214-v5.5.0] 3GPP, TR 25.214-v5.5.0. 2003. *Physical Layer Procedures (FDD)*.

[3GPP TR 25.803-v6.0.0] 3GPP TR 25.803-v6.0.0. 2005. *S-CCPCH Performance for MBMS*.

[3GPP 25.101-v6.6.0] 3GPP, TS 25.101-v6.6.0. 2004. *User Equipment (UE) Radio Transmission and Reception (FDD)*.

[3GPP 25.213-v6.1.0] 3GPP, TS 25.213-v6.1.0. 2004. *Spreading and Modulation (FDD)*.

[Alamouti 1998] Alamouti, S. M. 1998. A simple transmitter diversity scheme for wireless communications. *IEEE JSAC* 16(8):1451–58.

[Benvenuto 2002] Benvenuto, N., and S. Tomasin. 2002. Block iterative DFE for single carrier modulation. *IEE Electronic Lett* 39(19):1144–45.

[Cimini 1985] Cimini. L. 1985. Analysis and simulation of a digital mobile channel using orthogonal frequency division multiplexing. *IEEE Trans Comm* 33.

[Cover 1972] Cover, T. 1972. Broadcast channels. *IEEE Trans Inform Theory* IT-18:2–14.

[Dinis 2003] Dinis, R., A. Gusmão, and N. Esteves. 2003. On broadband block transmission over strongly frequency-selective fading channels. In *Wireless 2003*. Calgary, Canada.

[Dogan 2004] Dogan, S., et al. 2004. Video content adaptation using transcoding for enabling UMA over UMTS. In *Proc of Wiamis 2004*. Lisbon, Portugal.

[Engels 1998] Engels, V., and H. Rohling. 1998. Multi-Resolution 64-DAPSK Modulation in a Hierarchical COFDM Transmission System. *IEEE Trans Broadcast* 44(1):139–49.

[ETSI 2904] ETSI. 2004. *ETS 300 744: Digital Video Broadcasting (DVB) Framing Structure, Channel Coding and Modulation for Digital Terrestrial Television (DVB-T) V1.5.1*. European Telecommunication Standard.

[Falconer 2002] Falconer, D., S. Ariyavisitakul, A. Benyamin-Seeyar, and B. Eidson. 2002. Frequency domain equalization for single-carrier broadband wireless systems. *IEEE Commun Mag* 4(4):58–66.

[Foschini 1996] Foschini, G. J. 1996. Layered space-time architecture for wireless communication in a fading environment when using multiple antennas. *Bell Labs Tech J* 1(2):41–59.

[Foschini 1998] Foschini, G. J., and M. J. Gans. 1998. On limits of wireless communications in fading environments when using multiple antennas. *Wirel Pers Commun* 6:315–35.

[Glisic 1997] Glisic, S., and B. Vucetic. 1997. *Spread Spectrum CDMA Systems for Wireless Communications*. Norwood, MA: Artech House Publishers.

[Goldsmith 1997] Goldsmith, A., and S. G. Chua. 1997. Variable-rate variable power M-QAM for fading channels. *IEEE Trans Commun* 45(10):1218–30.

[Holma 2000] Holma, H., and A. Toskala. 2000. *WCDMA for UMTS*. New York: John Wiley & Sons.

[Holma 2007] Holma, H., and A. Toskala. 2007. *WCDMA for UMTS: HSPA Evolution and LTE*. 4th ed. New York: John Wiley & Sons.

[Hottinen 2003] Hottinen, A., O. Tirkkonen, and R. Wichman. 2003. *Multi-antenna Transceiver Techniques for 3G and Beyond*. Chichester, UK: John Wiley & Sons.

[Johansson 1999] Johansson, A. L., and A. Svensson. 1999. On multi-rate DS/CDMA schemes with interference cancellation. *J Wirel Pers Commun Kluwer* 9(1):1–29.

[Kudoh 2003] Kudoh, E., and F. Adachi. 2003. Transmit power efficiency of a multi-hop virtual cellular system. *Proc IEEE Vehicular Technol Conf 2003 (VTC2003-Fall)* 5:2910–14.

[Li 2001] Li, W. 2001. Overview of fine granularity scalability in MPEG-4 video standard. *IEEE Trans CSVT* 11(3):301–17.

[Liang 2003] Liang, X., and X. Xia. 2003. On the nonexistence of rate-one generalized complex orthogonal designs. *IEEE Trans Inform Theory* 49(11):2984–88.

[Liu 2005] Liu, H., and G. Li. 2005. *OFDM-Based Broadband Wireless Networks*. New Jersey: John Wiley & Sons.

[Liu 2003] Liu, J., B. Li and Y.-Q. Zhang. 2003. Adaptive video multicast over the Internet. *IEEE Multimed* 10(1):22–33.

[Marques da Silva 2005c] Marques da Silva, M. 2005c. *Emergent Radio Systems*. PhD Thesis. Lisbon, Portugal: Instituto Superior Tecnico.

[Marques da Silva 2000] Marques da Silva, M., and A. Correia. 2000. *Parallel Interference Cancellation with Commutation Signaling.* New Orleans: IEEE—International Conference on Communications—ICC2000 Spring.

[Marques da Silva 2001] Marques da Silva, M., and A. Correia. 2001. *Space Time Diversity for the Downlink of WCDMA.* Aalborg-Denmark: IEEE—Wireless Personal and Mobile Communications—WPMC'01.

[Marques da Silva 2002a] Marques da Silva, M., and A. Correia. 2002a. *Space Time Block Coding for 4 Antennas with Coding Rate 1.* Prague–Check Republic: IEEE—International Symposium on Spread Spectrum Techniques and Application—ISSSTA.

[Marques da Silva 2002b] Marques da Silva, M., and A. Correia. 2002b. *Space Time Coding Schemes for 4 or More Antennas.* Lisbon–Portugal: IEEE—International Symposium on Personal Indoor and Mobile Radio Communications—PIMRC'02.

[Marques da Silva 2003a] Marques da Silva, M., and A. Correia. 2003a. Joint multi-user detection and intersymbol interference cancellation for WCDMA satellite UMTS. *Int J Satellite Commun Netw (Spec Issue on Interference Cancellation–Wiley)* 21(1):93–117.

[Marques da Silva 2003b] Marques da Silva, M., and A. Correia. 2003b. Combined transmit diversity and beamforming for WCDMA. In *IEEE EPMCC'03.* Glasgow, Scotland.

[Marques da Silva 2005a] Marques da Silva, M., A. Correia, and R. Dinis. 2005a. A Decorrelating MUD approach for the downlink of UMTS considering a RAKE in the receiver. In *Proceeding of the 16th IEEE Personal Indoor and Mobile Radio Communications 2005 (PIMRC'05),* 11–4. Berlin, Germany.

[Marques da Silva 2009] Marques da Silva, M., A. Correia, and R. Dinis. 2009 On transmission techniques for multi-antenna W-CDMA systems. *Eur Trans Telecomm* 20(1):107–21.

[Marques da Silva 2004] Marques da Silva, M., A. Correia, J. C. Silva, and N. Souto. 2004. Joint MIMO and parallel interference cancellation for the HSDPA. In *Proceedings of IEEE International Symposium on Spread Spectrum Techniques and Applications 2004 (ISSSTA'04).* Sydney, Australia.

[Marques da Silva 2010] Marques da Silva, M., A. Correia, R. Dinis, N. Souto, and J. Silva. 2010. *Transmission Techniques for Emergent Multicast and Broadcast Systems,* 1st ed., ISBN: 9781439815939. New York: CRC Press Auerbach Publications.

[Marques da Silva 2005b] Marques da Silva, M., R. Dinis, and A. Correia. 2005b. A V-BLAST detector approach for W-CDMA signals with frequency-selective fading. In *Proceeding of the 16th IEEE Personal Indoor and Mobile Radio Communications 2005 (PIMRC'05),* 11–4. Berlin, Germany.

[Nam 2002] Nam, S. H., and K. B. Lee. 2002 Transmit power allocation for an extended V-BLAST system. In *PIMRC'2002.* Lisbon, Portugal.

[Ojanpera 1998] Ojanperä, T., and R. Prasad. 1998. *Wideband CDMA for Third Generation Mobile Communications.* Norwood, MA: Artech House Publishers.

[Patel 1994] Patel, P., and J. Holtzman. 1994. Analysis of simple successive interference cancellation scheme in DS/CDMA. *IEEE J Sel Areas Commun* 12(5):796–807.

[Price 1958] Price, R., and P. E. Green. 1958. A communication technique for multipath channels. *Proc IRE* 46(3):555–70.

[Proakis 2001] Proakis, J. 2001. *Digital Communications.* 4th ed. New York: McGraw-Hill.

[Pursley 1999] Pursley, M. B., and J. M. Shea. 1999. Non-uniform phase-shift-key modulation for multimedia multicast transmission in mobile wireless networks. *IEEE J Sel Areas Commun* 17(5):774–83.

[Ramchandran 1993] Ramchandran, K., A. Ortega, K. M. Uz, and M. Vetterli. 1993. Multiresolution broadcast for digital HDTV using joint source/channel coding. *IEEE J Sel Areas Commun* 11(1):6–23.

[Rooyen 2000] Rooyen, P. V., M. Lötter, and D. Wyk. 2000. *Space-Time Processing for CDMA Mobile Communications.* Boston, MA: Kluwer Academic Publishers.

[Sari 1994] Sari, H., et al. 1994. *An Analysis of Orthogonal Frequency-division Multiplexing for Mobile Radio Applications*, IEEE VTC'94. Stockholm, Sweden.

[Schacht 2003] Schacht, M., A. Dekorsy, and P. Jung. 2003. *Downlink Beamforming Concepts in UTRA FDD*. Kleinheubacher Tagung 2002. Kleinheubacher Berichte.

[Silva 2003] Silva, J. C., N. Souto, A. Rodrigues, F. Cercas, and A. Correia. 2003. Conversion of reference tapped delay line channel models to discrete time channel models. In *Proc IEEE Vehicular Technology Conference-VTC2003 Fall*. Orlando.

[Souto 2007] Souto, N., R. Dinis, and J. C. Silva. 2007. Iterative decoding and channel estimation of MIMO-OFDM transmissions with hierarchical constellations and implicit pilots. In *International Conference on Signal Processing and Communications (ICSPC'07)*. Dubai (United Arab Emirates).

[Souto 2005a] Souto, N., et al. 2005a. Iterative Turbo Multipath Interference Cancellation for WCDMA Systems with Non-Uniform Modulations. In *Proc. IEEE Vehicular Technology Conf-VTC2005-Spring*. Stockholm, Sweden.

[Souto 2005b] Souto, N., et al. 2005b. Non-Uniform Constellations for Broadcasting and Multicasting Services in WCDMA Systems. In *Proc. IEEE IST Mobile & Wireless Communications Summit*. Dresden, Germany.

[Sydir 2009] Sydir, J., and R. Taori. 2009. An evolved cellular system architecture incorporating relay stations. *IEEE Commun Mag* 47(6):150–5.

[Tarokh 1999] Tarokh, V., et al. 1999. Space-time block codes from orthogonal designs. *IEEE Trans Inform Theory* 45(5):1456–7.

[Telatar 1995] Telatar, I. E. 1995. Capacity of multiantenna Gaussian channels. *AT&T Bell Lab Tech.* Memo.

[Telatar 1999] Telatar, I. E. 1999. Capacity of multiantenna Gaussian channels. *Eur Trans Commun* 10(6):585–95.

[Tuchler 2002] Tuchler, M., R. Koetter, and A. Singer. 2002. Turbo equalization: Principles and new results. *IEEE Trans Comm* 50:754–67.

[Varanasi 1990] Varanasi, M. K., and B. Aazhang. 1990. Multistage detection in asynchronous CDMA communications. *IEEE Trans Commun* 38(4):509–19.

[Vetro 2003] Vetro, A., C. Christopoulos, and H. Sun. 2003. Video transcoding architectures and techniques: An overview. *IEEE Sig Proc Mag* 20(2):18–29.

[Viterbi 1990] Viterbi, J. 1990. Very low rate convolutional codes for maximum theoretical performance of spread-spectrum multiple-access channels. *IEEE J Sel Areas Commun* 8(4):641–9.

[Webb 1994] Webb, W. T., and L. Hanzo. 1994. *Modern Quadrature Amplitude Modulation: Principles and Applications for Fixed and Wireless Channels*. New York: IEEE Press.

[Webb 1994] Webb, W. T., and R. Webb. 1995. Variable rate QAM for mobile radio. *IEEE Trans Commun* 43(7):2223–30.

[Wei 1993] Wei, L.-F. 1993. Coded modulation with unequal error protection. *IEEE Trans Commun* 41(10):1439–49.

Chapter 8

[Marques da Silva 2010] Marques da Silva, M., A. Correia, R. Dinis, N. Souto, and J. C. Silva. 2010. *Transmission Techniques for Emergent Multicast and Broadcast Systems*. 1st ed. Boca Raton, FL: CRC Press Auerbach Publications.

[3GPP 25.814-v1.2.2] 3GPP TR 25.814-v1.2.2. 2006. *3rd Generation Partnership Project, Technical Specification Group Radio Access Network; Physical Layer Aspects for Evolved UTRA (Release 7)*.

[3GPP 25.913-v7.0.0] 3GPP TR 25.913-v7.0.0. 2005. *Requirements for Evolved UTRA (E-UTRA) and Evolved UTRAN (E-UTRAN)*. http://www.3gpp.org (accessed May 21, 2011).

[3GPP 36.913-v8.0.0] 3GPP TR 36.913-v8.0.0. 2008. *Requirements for Further Advancement for E-UTRA (LTE-Advanced).*

[Andrews 2007] Andrews, J. G., A. Gosh, and R. Muhamed. 2007. *Fundamentals of WiMAX: Understanding Broadband Wireless Networking.* Prentice-Hall: New Jersey.

[Ansari 2009] Ansari, A., S. Dutta, and M. Tseytlin. 2009. S-WiMAX: Adaptation of IEEE 802.16e for mobile satellite services. *IEEE Commun Mag* 47(6):150–5.

[Astely 2006] Astély, D., et al. 2006. A future radio-access framework. *IEEE JSAC* 24(3): 693–706.

[Astely 2009] Astély, D., et al. 2009. LTE: The evolution of mobile broadband. *IEEE Commun Mag* 7(4):44–51.

[Bogineni 2009] Bogineni, K., et al. 2009. LTE Part II: Radio access. *IEEE Commun Mag* 47(4):40–42.

[Boudreau 2009] Boudreau, G., et al. 2009. Interference coordination and cancellation for 4G networks. *IEEE Commun Mag* 47(4):74–81.

[Dahlman 2008] Dahlman, E., et al. 2008. *3G Evolution: HSPA and LTE for Mobile Broadband.* 2nd ed. Oxford, UK: Academic Press.

[Eklund 2002] Eklund, C., et al. 2002. IEEE Standard 802.16: A technical overview of the wirelessMAN air interface for broadband wireless access. *IEEE Commun Mag* 40(6):98–107.

[Holma 2007] Holma, H., and A. Toskala. 2007. *WCDMA for UMTS: HSPA Evolution and LTE.* 4th ed. New York: John Wiley & Sons.

[IEEE 802.16] IEEE 802.16 Relay Task Group. 2008. http://www.ieee802.org/16/relay (accessed May 18, 2011).

[IEEE 802.16-2004] IEEE Standard 802.16-2004. 2004. *Air Interface for Fixed Broadband Wireless Access Systems.*

[IEEE 802.16e] IEEE Standard 802.16e-2005. 2005. *Amendment to Air Interface for Fixed and Mobile Broadband Wireless Access Systems–Physical and Medium Access Control Layers for Combined Fixed and Mobile Operations in Licensed Bands.*

[IEEE 802.16m] IEEE Standard 802.16m-07/002r8. 2009. *IEEE 802.16m System Requirements.* http://ieee802.org/16/tgm/index.html (accessed February 5, 2011).

[ITU-R M.2133] ITU-R Recommendation M.2134. 2008. *Requirements, Evaluation Criteria, and Submissions Templates for the Development of IMT-Advanced.*

[Liu 2005] Liu, H., and G. Li. 2005. *OFDM-Based Broadband Wireless Networks.* New Jersey: John Wiley & Sons.

[Marques da Silva 2000a] Marques da Silva, M., and A. Correia. 2000a. *MAI Cancellation with Commutation Signaling.* Tokyo-Japan: IEEE–Vehicular Technology Conference–VTC2000 Spring.

[Marques da Silva 2000b] Marques da Silva, M., and A. Correia. 2000b. *Parallel Interference Cancellation with Commutation Signaling.* New Orleans: IEEE–International Conference on Communications–ICC2000 Spring.

[Marques da Silva 2003a] Marques da Silva, M., and A. Correia. 2003a. Interference Cancellation using joint Pre-distortion filtering and Selective Transmit Diversity. In *Proceedings of 4th Conference on Telecommunications.* Aveiro, Portugal.

[Marques da Silva 2003b] Marques da Silva, M., and A. Correia. 2003b. Interference Cancellation with combined Pre-distortion filtering and Transmit Diversity. In *Proceedings of 14th IEEE International Symposium on Personal Indoor and Mobile Radio Communications.* Beijing, China.

[Marques da Silva 2004a] Marques da Silva, M., and A. Correia. 2004a. Joint MIMO and MAI Suppression for the HSDPA. In *International Conference on Telecommunications 2004 (ICT'04).* Fortaleza, Brasil.

[Marques da Silva 2004b] Marques da Silva, M., A. Correia, N. Souto, J. C. Silva, F. Cercas, and A. Rodrigues. 2004b. Interference Suppression consisting of Pre-Distortion Filtering and Beamforming with joint Transmit Diversity. In *SEACORN Workshop on Simulation of Enhanced UMTS Access and Core Networks*. Cambridge.

[NGMN 2006] NGMN. 2006 *Next Generation Mobile Networks Beyond HSPA and EVDO*. V. 1.3. http://www.ngmn.org (accessed December 25, 2010).

[NGMN 2007] NGMN. 2007. *NGMN Radio Access Performance Evaluation Methodology*. V. 1.2. http://www.ngmn.org (accessed December 12, 2010).

[Ohrtman 2008] Ohrtman, F. 2008. *WiMAX Handbook*. New York: McGraw-Hill Communications.

[Osseiran 2009] Osseiran, A., et al. 2009. The road to IMT-advanced communication systems: State-of-the-art and innovation areas addressed by the WINNER+ Project. *IEEE Commun Mag* 47(6):38–47.

[Oyman 2007] Oyman, O., J. N. Laneman, and S. Sandhu. 2007. Multihop relaying for broadband wireless mesh networks: From theory to practice. *IEEE Commun Mag* 45(11): 116–22.

[Peters 2009] Peters, S. W., and R. W. Heath. 2009. The future of WiMAX: Multihop relaying with IEEE 802.16j. *IEEE Commun Mag* 47(1):104–11.

[Sydir 2009] Sydir, J., and R. Taori. 2009. An evolved cellular system architecture incorporating relay stations. *IEEE Commun Mag* 47(6):150–5.

[Yarali 2008] Yarali, A., and S. Rahman. 2008. WiMAX broadband wireless access technology: services, Architecture and Deployment Models. In *IEEE CCECE2008*, 77–82. Niagara Falls, ON.

[Zhang, 2010] Zhang, J., and G. de la Roche. 2010. *Femtocells-Technologies and Deployment*. 1st ed. Chichester, UK: John Wiley & Sons.

Chapter 9

[ANSI T1.105] ANSI Standard T1.105.07. 1996. Synchronous Optical Network (SONET)— Sub-STS-1 Interface Rates and Formats Specifications, Revised 2005.

[RFC 2702] Awduche, D., et al. 1999. *RFC 2702—Requirements for Traffic Engineering over MPLS*. Network Working Group.

[Costa 1997] Costa, C. 1997. Apontamentos de Redes de Telecomunicações—ATM, IST— Departamento de Engenharia Electrotécnica e de Computadores. Lisbon, Portugal.

[Handel 1991] Handel, R., and M. Hubber. 1991. *Integrated Broadband Networks. An Introduction to ATM-Based Networks.* Munich: Addison-Wesley.

[ITU-T G.711] ITU-T Recommendation G.711. 1993. General aspects of digital transmission systems—Terminal equipments—Pulse code modulation (PCM) of voice frequencies.

[ITU-T G.732] ITU-T Recommendation G.732. 1988. Characteristics of primary PCM multiplex equipment operating at 2048 kbit/s.

[ITU-T G.733] ITU-T Recommendation G.733. 1988. Characteristics of primary PCM multiplex equipment operating at 1544 kbit/s.

[ITU-T G.707] ITU-T Recommendation G.707. 2007. Network node interface for the synchronous digital hierarchy (SDH).

[ITU-T T.121] ITU-T Recommendation I.121. 1991. Integrated services digital network (ISDN) general structure and service capabilities.

[ITU-T T.113] ITU-T Recommendation I.113. 1997. Vocabulary of terms for broadband aspects of ISDN.

[RFC 3270] Le Faucheur, F., et al. 2002. *RFC3270—MPLS Support of Differentiated Services*. Network Working Group.

[Prycker 1991] Prycker, M. 1991. *Asynchronous Transfer Mode*. New York: Ellis Horwood.

[RFC 3031] Rosen, E., A. Viswanathan, and R. Callon. 2001. *RFC3031—Multi-protocol Label Switching Architecture*. Network Working Group.

[Sexton 1992] Sexton, M., and A. Reid. 1992. *Transmission Networking: Sonet and the Synchronous Digital Hierarchy*. Norwood, MA: Artech House.

Chapter 10

[IEEE 802.11] ANSI/IEEE Std 802.11. 1999. *Wireless LAN Medium Access Control (MAC) and Physical Layer (PHY) Specifications*. (E) Part 11, ISO/IEC 8802-11.

[Benedetto 1997] Benedetto, S., E. Biglieri, and V. Castellani. 1997. *Digital Transmission Theory*. New Jersey: Prentice Hall.

[Ferro 2005] Ferro, E., and F. Potorti. 2005. Bluetooth and Wi-Fi wireless protocols: A survey and a comparison. *IEEE Wirel Commun* 12(1):12–26.

[Geier 2002] Geier, J. 2002. *Wireless LANs*. Indiana: SAMs Publishing.

[IEEE 802.5] IEEE Standard for Information Technology. 1998a. *Specific Requirements: Token Ring Access Method and Physical Layer Specifications*.

[IEEE 802.2] IEEE Standard for Information Technology. 1998b. *Telecommunications and Information Exchange between Systems*. Local and Metropolitan Area Networks; Specific Requirements for Logical Link Control, Revision 2003.

[IEEE 802.3] IEEE Standard for Information Technology. 2008. *Specific Requirements: Carrier Sense Multiple Access with Collision Detection (CSMA/CD) Access Method and Physical Layer Specifications*. revision 2005.

[ANSI X3.139] Information Systems. 1997. *Fiber Distributed Data Interface (FDDI)*. Token Ring Media Access Control (MAC). Replaces ANSI X3.139-1987 R.

[ANSI X3.148] Information Systems. 1999. *Fiber Distributed Data Interface (FDDI)*. Token Ring Physical Layer Protocol (PHY). Replaces ANSI X3.148-1988 R.

[ISO 9314-1] International Standard. 1989a. *Information Processing Systems—Fibre Distributed Data Interface (FDDI)*. Part 1: Token Ring Physical Layer Protocol (PHY).

[ISO 9314-2] International Standard. 1989b. *Information Processing Systems—Fibre Distributed Data Interface (FDDI)*. Part 2: Token Ring Media Access Control (MAC).

[ISO 3309] ISO International Standard (IS) 3309. 1979. *Data Communications—High-Level Data Link Control Procedures—Frame Structure*.

[ISO 4335] ISO International Standard (IS) 4335. 1979. *Data Communications—High-Level Data Link Control Procedures*. Specification for Elements of Procedures.

[ISO/IEC 8802-5:1998] ISO/IEC Standard for Information Technology. 1998a. *Specific Requirements: Token Ring Access Method and Physical Layer Specifications*.

[ISO/IEC 8802-2:1998] ISO/IEC Standard for Information Technology. 1998b. *Telecommunications and Information Exchange between Systems*. Local and metropolitan area networks; Specific requirements for Logical Link Control.

[Ohrtman 2003] Ohrtman, F., and K. Roeder. 2003. *Wi-Fi Handbook: Building 802.11b Networks*. New York: McGraw Hill.

[RFC 1663] Rand, D. 1994. *RFC 1663—PPP Reliable Transmission*. Network Working Group.

[Watkinson 2001] Watkinson, J. 2001. *The MPEG Handbook: MPEG-I, MPEG-II, MPEG-IV*. Waltham, MA: Focal Press.

Chapter 11

[Blanchet 2006] Blanchet, M. 2006. *Migrating to IPv6: A Practical Guide to Implementing IPv6 in Mobile and Fixed Networks*. England: Wiley.

[RFC 1752] Bradner, S., and A. Mankin. 1995. *RFC 1752—The Recommendation for the IP Next Generation Protocol*. Network Working Group.

[RFC 3927] Cheshire, S., B. Aboba, and E. Guttman. 2005. *RFC 3927—Dynamic Configuration of IPv4 Link-Local Addresses*. Network Working Group.

[RFC 2740] Coltun, R., D. Ferguson, and J. Moy. 1999. *RFC 2740—OSPF for IPv6*. Network Working Group.

[Davies 2008] Davies, J. 2008. *Understanding IPv6*. 2nd ed. Washington, DC: Microsoft press.

[RFC 2460] Deering, S., and R. Hinden. 1998. *RFC 2460—Internet Protocol, Version 6 (IPv6) Specification*. Network Working Group.

[Dijkstra 1959] Dijkstra, E. W. 1959. A note on two problems in connexion with graphs. *Numerische Math* 1:269–71.

[RFC 1058] Hedrick, C. 1988. *RFC1058—Routing Information Protocol*. Internet Engineering Task Force.

[RFC 2373] Hinden, R., and S. Deering. 1998. *RFC 2373—IP Version 6 Addressing Architecture*. Network Working Group.

[RFC 2374] Hinden, R., O. O'Dell, and S. Deering. 1998. *RFC 2374—An IPv6 Aggregatable Global Unicast Address Format*. Network Working Group.

[RFC 1723] Malkin, G. 1994. *RFC1723—RIP Version 2 Carrying Additional Information*. Internet Engineering Task Force.

[RFC 4260] McCann, P. 2005. *RFC 4260—Mobile IPv6 Fast Handovers for 802.11 Networks*. Network Working Group.

[RFC 792] Postel, J. 1981 *RFC 792—Internet Control Message Protocol*. Network Working Group.

Chapter 12

[RFC 2213] Baker, F., J. Krawczyk, and A. Sastry. 1997. *RFC 2213—Integrated Services Management Information Base using SMIv2*. Network Working Group.

[RFC 2475] Blake, S., et al. 1998. *RFC 2475—An Architecture for Differentiated Services*. Network Working Group.

[RFC 2205] Braden, R., S. Berson, and S. Herzog. 1997. *RFC 2205—Resource Reservation Protocol (RSVP)*. Network Working Group.

[RFC 1633] Braden, R., D. Clark, and S. Shenker. 1994. *RFC 1633—Integrated Services in the Internet Architecture: An Overview*. Network Working Group.

[RFC 2215] Krawczyk, J. 1997. *RFC 2215—General Characterization Parameters for Integrates Services Network Elements*. Network Working Group.

[RFC 2212] Shenker, S., C. Partridge, and R. Guerin. 1997. *RFC 2212—Specification of Guaranteed Quality of Service*. Network Working Group.

[RFC 2211] Wroclawski, J. 1997. *RFC 2211—Specification of the Controlled-Load Network*. Network Working Group.

Chapter 13

[RFC 114] Bhushan, A. 1971. *RFC 114—A File Transfer Protocol*. Network Working Group.

[RFC 1448] Case, J., et al. 1993. *RFC 1448—Protocol Operations for Version 2 of the Simple Network Management Protocol (SNMPv2)*. Network Working Group.

[RFC 2616] Fielding, R., et al. 1999. *RFC 2616—Hypertex Transfer Protocol—HTTP/1.1*. Network Working Group.

[RFC 2045] Freed, N., and N. Borenstein. 1996. *RFC 2045—Multipurpose Internet Mail Extensions (MIME), Part One: Format of Internet Message Bodies*. Network Working Group.

[RFC 2543] Handley, M., et al. 1999. *RFC 2543—SIP: Session Initiation Protocol*. Network Working Group.

[ITU-T H.310] ITU-T Recommendation H.310. 1995. *Broadband Audiovisual Communication Systems and* Terminals.

[ITU-T H.320] ITU-T Recommendation H.320. 2004. *Narrow-Band Visual Telephone Systems and Terminal Equipment*.

[ITU-T H.321] ITU-T Recommendation H.321. 1998. *Adaptation of H.320 Visual Telephone Terminals to B-ISDN Environments*.

[ITU-T H.323] ITU-T Recommendation H.323. 2009. *Packet-Based Multimedia Communications Systems*. Version 7.

[ITU-T H.324] ITU-T Recommendation H.324. 2009. *Terminal for Low Bit-Rate Multimedia Communication*.

[Liu 2000] Liu, H., and P. Mouchtaris. 2000. Voice over IP signaling: H.323 and beyond. *IEEE Commun Mag* 38(10)142–8.

[Minoli 2002] Minoli, D., and E. Minoli. 2002. *Delivering Voice over IP Networks*. 2nd ed. New York: Wiley.

[RFC 1034] Mockapetris, P. 1987a. *RFC 1034—Domain Names—Concepts and Facilities*. Network Working Group.

[RFC 1035] Mockapetris, P. 1987b. *RFC 1035—Domain Names—Implementation and Specification*. Network Working Group.

[RFC 821] Postel, J. B. 1982. *RFC 821—Simple Mail Transfer Protocol*. Network Working Group.

[RFC 959] Postel, J., and J. Reynolds. 1985. *RFC 959—FTP*. Network Working Group.

[RFC 1225] Rose, M. 1991. *RFC 1225—Posts Office Protocol—Version 3*. Network Working Group.

[RFC 3261] Rosenberg, J., et al. 2002. *RFC 3261—SIP: Session Initiation Protocol*. Network Working Group.

[Schulzrinne 2000] Schulzrinne, H., and J. Rosenberg. 2000. The session initiation protocol: Internet-centric signaling. *IEEE Commun Mag* 38(10):134–41.

[Thom 1996] Thom, G. A. 1996. H.323: The multimedia communications standard for local area networks. *IEEE Commun Mag* 34(12):52–6.

[RFC 326] Westheimer, E. 1972. *RFC 326—Network Host Status*. Network Working Group.

Chapter 14

[Arquilla 1997] Arquilla, J. 1997. In *Athena's Camp: Preparing for Conflict in the Information Age, RAND*. Santa Monica, CA: National Defense Research Institute.

[Cross 2003] Cross, M., et al. 2003. *Security +*. Rockland, MA: Syngress Publishing, Inc.

[RFC 2196] Fraser, B. 1997. *RFC 2196—Site Security Handbook*, Network Working Group.

[RFC 2409] Harkins, D., and D. Karrel. 1998. *RFC 2409—The Internet Key Exchange (IKE)*. Network Working Group.

[RFC 4301] Kent, S., and K. Seo. 2005. *RFC 4301—Security Architecture for the Internet Protocol*. Network Working Group.

[Klein 2003] Klein, J. 2003. *CyberTerrorism*. Movie from http://www.history.com/ (accessed May 23, 2010).

[Maiwald 2009] Maiwald, E. 2003. *Network Security—A Beginner's Guide*. 2nd ed. Osborne, New York: McGraw-Hill.

[McClure 2005] McClure, S., and J. Scambray. 2005. *Hacking Exposed: Network Security, Secrets & Solutions*. Osborne, New York: McGraw-Hill.

[RFC 4309] Wousley, R. 2005. *RFC 4309—Using Advanced Encryption Standard (AES) CCM Mode with IPsec Encapsulating Security Payload (ESP)*. Network Working Group.

Index